Contents

Merging Design Cultures

Sustainable Development

Design Education Philosophy and Practice

International Comparisons

Authors' Index

Organizing Committee

E K Brodhurst
Institution of Engineering Designers

P R N Childs
Sharing Experience in Engineering Design and University of Sussex

M Evatt
Institution of Engineering Designers and Coventry University

P Hamilton
Sharing Experience in Engineering Design

C McMahon
Sharing Experience in Engineering Design and University of Bristol

J Poole
Institution of Engineering Designers

National Advisory Group

E Billett
Brunel University

R V Chaplin
Lancaster University

P R N Childs
University of Sussex

C Dowlen
South Bank University

M Evatt
Coventry University

P Hamilton
Sharing Experience in Engineering Design

B Hollins
Institution of Engineering Designers

B Ion
University of Strathclyde

C Ledsome
Imperial College

C McMahon
University of Bristol

J Poole
Institution of Engineering Designers

S Thorpe
Coventry University

J Vogwell
University of Bath

Preface

The Engineering and Product Design Education Conference 2000 is an important event for educators and students of design from industry, higher education, further education, schools, and training establishments. The conference serves to bring together an international assembly of experts to share their experience and develop new strategies. The Engineering and Product Design Education Conference 2000 incorporates the 22nd annual conference on Engineering Design Education under the auspices of SEED (Sharing Experience in Engineering Design) and the 7th National Conference on Product Design Education under the auspices of the Institution of Engineering Designers (IED). The conference was hosted by the School of Engineering and Information Technology at the University of Sussex in the attractive seaside town of Brighton, UK on 6–7 of September 2000.

The subject of this year's conference, *Integrating Design Education Beyond 2000*, has resulted in the collection of articles, which are presented here. Some striking themes have emerged:

- The benefit of reflection on past experience to identify the way forward in the use of new tools and techniques to facilitate and augment the design process;
- The development of successful strategies for integrating industrial experiences in the education of designers;
- The revolutionary impact of information technology on the design process, which while promising the potential for productivity improvement, requires educators to consider and redefine modes of teaching;
- The necessity for direct exposure to design and rapid prototyping technologies to provide a hands-on and thereby inspirational learning experience;
- The ever increasing demand for designers to consider the place of products in both global and regional markets.

The programme addresses the full spectrum of design education from primary, through secondary and tertiary sectors, to continual professional development. The conference aims to provide a forum for the sharing of experiences and will make an important contribution to the body of knowledge in design education.

About the Editors

Peter Childs is a Reader in the School of Engineering and Information Technology, University of Sussex and is Joint Assistant Director of the Rolls-Royce University Technology Centre for Aero-Thermal Systems. He lectures mechanical design, fluid mechanics, and heat transfer. He has general interests in gas turbine and spark ignition engines, medical ventilators, and temperature instrumentation. He completed his doctorate in 1991 and has written text books on mechanical design and practical temperature measurement. He is actively involved in research and development contracts for ABB Alstom, Ford, Rolls-Royce plc, Johnson Matthey, and Ricardo Consulting Engineers Limited. He is a Fellow of the Institution of Mechanical Engineers and in 1999 he was the winner of the American Society of Mechanical Engineers – International Gas Turbine Institute John P Davis award for exceptional contribution to the literature of gas turbine technology.

Elizabeth Brodhurst joined the Institution of Engineering Designers in 1997. She is responsible for education and training, continuing professional development, course accreditation, university liaison, and conference organization.

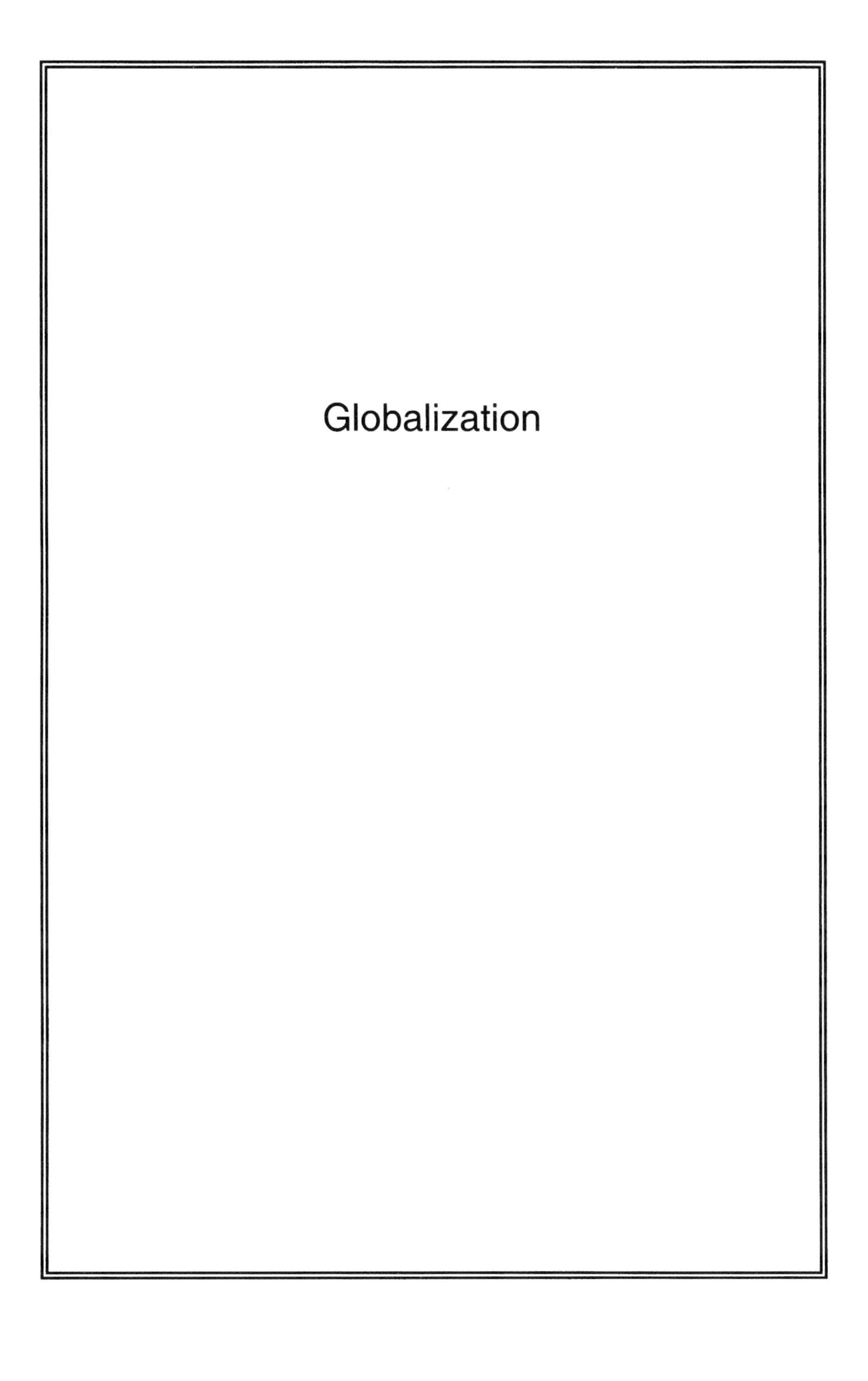
Globalization

Design and globalization – the role of the designer in relation to questions of cultural diversity

M M ONO
Department of Industrial Design and Concurrent Engineering R&D Laboratory, Centro Federal de Educação Techológica do Paraná, Curitiba, Brazil

SYNOPSIS

This paper discusses the role of the designer in relation to questions of cultural diversity. It analyses the symbolical, practical and technical function of objects, the paradox of cost in the diversification of product design, and the influences of imported products and conceptual design in the creation of wishes. Moreover, it points out tendencies to individualization, global standardization and hybridism in industrial design. Taking into account the capitalistic system and the globalisation process, it considers the designer as (1) an important mediator between the material constructions, the practical and the symbolical universe of individuals and societies, and (2) as co-responsible for the preservation of identities and cultural heritages of mankind.

1. INTRODUCTION

Facing the globalisation process, that has been reinforced from the 90's, an important question has been revealed to the designers, specially in the companies that had been incorporated or associated to transnational groups: Which is the role of designer in relation to questions of cultural diversity? Should he conceive products following a global standardisation, a local particularisation, or some other approach?

This question brings about direct implications to people who consume, use, or are affected, in any way, by the insertion of objects in the markets, considering that the practice of design becomes decisive in the translation and materialisation of its symbolical, practical and technical functions, in the core of the productive process and facing the strategies of development that have been adopted by the companies.

The adoption of a "global" design demands objects to express a universal language, capable to penetrate all social, economical and cultural barriers of the various markets. In this case, we would have standardised products world-wide, disrespecting the local particularities. Consequently, we would have a reorganisation of cultures of individuals and societies, subordinated to the dominant groups, in a unified system of symbolical production, within a transnational organisation of the culture linked to the multinational capital. Or, in a less extreme situation, design could be relegated to a mere handling of the appearance of the objects, for a superficial tuning with the various segments of marketing.

However, it has also been verified, since the 80's, the existence of a process of stagnation of the mass production, and, moreover, consumers have become more demanding in relation to the quality of services and products. In this context, the resistance to "the global" standardisation of products is reinforced, and divergent approaches are placed in the scope of culture and design. One of them, based on a nationalistic ethnocentric vision, defends the development of a typical "national" design to be projected as a world-wide reference (an "Italian", "German", or a "Brazilian" design, for instance). This considers the development of local products as an autonomous expression of its culture as possible. Another one, of a relativist nature, considers that each person must develop his own approach, in accordance with his particular values, necessities and yearnings, without any interference, or external judgements. And a third approach, which we could consider a moderate relativism, admits the cultural multiplicity, the existence of inter-relationships and the influences between societies, as well as, simultaneously, defending the preservation of the contexts and local initiatives (in this case, the use of "local" must be understood not only as the concept of "state-nation", but also considering the diverse social groups and individuals). These approaches are, in general, contrary to the total homogenisation of values and attitudes.

The national companies have been face to face with the process of globalisation, and, in the last years, several of them have been incorporated, when not swept away, by large multinational companies, as a result of the social, cultural, technological, political and economical context of the various countries within the world. In this scenario, design has come across with the question of "global design" versus "local design", specially in national companies which have been incorporated by multinationals, taking into account that the introduction of structures and organisational objectives, distinct and conflicting in relation to the local values, can be reflected in the creation of devices and practices, and even of distinct meanings of the reality, which consists the existential basis and philosophy of living for individuals and societies.

Analysing examples of a study of case (1) conducted at Electrolux of Brazil S.A. and Multibrás S.A. - originally Brazilian companies, which have become, respectively, subsidiary of the multinationals corporations Electrolux[1], operating in the South America, and Whirlpool[2], operating in Latin America - this paper discusses the role of the designer in relation to the symbolical, practical and technical functions of products. Moreover, it brings out a reflection concerning to the influences of the imported products and conceptual design in the creation of wishes, as well as about the paradox of the factor "cost" in the diversification of product design. Moreover, it points out tendencies to individualisation, global standardisation and hybridism in design.

---

[1] Electrolux Group is the world's largest manufacturer of large appliances such as washing machines, refrigerators, and freezers. Its brand names include AEG, Eureka, Frigidaire, Zanussi, Arthur Martin, Rex, Tappan, Kelvinator, Husqvarna, Electrolux, among others. In Brazil, the company is the second largest white-goods producer.

[2] Whirlpool Corporation is the world's second producer of major home appliances, after Sweden's Electrolux. In Brazil, its subsidiary Multibrás is the largest producer of white-goods.

2. DESIGN AND CULTURAL DIVERSITY: THE TRANSLATION OF THE SYMBOLICAL, PRACTICAL AND TECHNICAL FUNCTION OF THE OBJECTS

Ever since human beings started to create the first devices, searching to facilitate their daily activities, the world has become more and more crowded with objects, which have been suitable for a multiplicity of functions, and have also brought a series of implications to the cultural dimension, helping to reproduce and to create values, necessities and actions, in individuals and societies.

The understanding of the functions assumed by the objects within people's lives, and considering the social and cultural contexts where they are inserted, is one of the basic factors for the definition of the role of the designer, within the process of development of concepts and material supports.

The basic objective of an object is to assist to one or more necessities of its users, and, therefore, it must assist to the required functions. From the producer's or user's point of view, the quality of an object consists in a condition that responds to their necessities, be it a large-scale product, or a unique (exclusive) product.

Considering the semantic point of view, objects can be understood as shapes impregnated by symbolic qualities, within a social, cultural and cognitive context. When perceiving an object, a person may consider it attractive, little attractive, or repulsive. Such judgement results from the experienced context and from the memory of emotions, feelings and experiences related to other perceived objects during this person's existence; the symbolic meaning of the object. In this way, the process of attractiveness of the objects results not only from the incorporation of aspects of the visual perception, but also from the social and cultural characteristics of the individuals. In this sense, it can be said that the symbolic function of objects surpasses the notion of style or trend, consisting of the meaning of the devices for the observers and users, manifesting their thoughts, rites, gestures, artistic expressions and languages, amongst other phenomenon's.

From the sociological point of view, as Bourdieu (2) understands it, the aesthetic preferences, as well as life styles, vary according to the "relations of power" and the social stratification that are established between the individuals and societies. The "aesthetic disposal", the "taste", the preferences and options are distinguished between the different social classes, and are strictly linked to the "cultural capital"[3], reproducing relations of power that are established in the social sphere. In this sense, the style of living and the consuming of products are invested of a symbolic character of distinction in the social structure, representing the social status of the individuals, that is determined by the possessed "economic and cultural capital". Thus, through "distinction marks", the social citizens constitute and express, for themselves and for other people, their positions in the social structure (3).

Therefore, a simple object can represent the identity of an individual, group, entity, organisation, etc, based on its aesthetic, organisational characteristic and value of marketing. By the same way, it is possible to establish not only a differentiated process of legitimisation

[3] The cultural capital qualify the individuals or specific social groups to understand, classify and to use adequately the products (Pierre Bourdieu, *Distinction: A Social Critique of the Judgement of Taste* (London: Routledge & Kegan Paul, 1984).

of symbolical goods, but also a hierarchical stratification in the distribution of them, among the different social classes.

Therefore, developing products with "a universal" approach would be to ignore the differences of perception of meanings, that can be even antagonistic, among the various individuals and social groups. Thus, for example, what is "nice" for someone, can be considered "ugly" by someone else; what is "sophisticated" for someone, can be considered "ordinary" by someone else. From this point of view, reinforcing Cuche's vision (4), amongst others, it can be said that "mass production" and "mass consumption", promoted by the capitalist system, do not correspond, necessarily, to a "mass aesthetics".

Case studies, such as the one conducted within the designers of Electrolux of Brazil and Multibrás, make evident the necessity of differentiation of product design, considering the distinct markets, although there is, in general, a vision that delimits the diversity to the concept of "nations", mentioning, for example, the "Italian" or "German" design, amongst others. This causes that, in many occasions, one conforms to stereotypes, by the generic nature of the affirmations. Actually, diversity is founded in the various instances of the local societies, and it is even manifested among individuals of a same family.

The Electrolux and Whirlpool corporations have adopted a strategy of differentiation of products through brand names and lines of products. It remains, however, that there are various social groups, with specific necessities which have not yet been well assisted. In Brazil, for instance, Electrolux and Multibrás have developed products with the same design for the whole country, despite the notable social, cultural, economical and environmental differences between the several regions and localities. There are, for example, different preferences of colours between the Northeast and South regions of Brazil. We can also verify cases of adaptations in the practical function of products, made by their own users, such as in the case of the settlement of a tap at the horizontal freezer, in the Northeast region[4], for the extraction of "coconut water"[5].

Multibrás - Whirlpool's subsidiary active in Brazil - acts with two brand names, Consul and Brastemp, that reach distinct social classes, differentiating the translation of the symbolical, practical and technical functions of their products. Those from Consul, for example, present rounded shapes, with graphical allegories (variation of colours, pictograms), searching to convey a "snugger" design, that appeal more to their target users, the middle class housewives, who work directly with the home appliances. On the other hand, Brastemp's products search to pass an image of advanced technology and present a more discrete graphic design (more textual, with neutral colours), using ampler arcs. It is destined to vanguard women, who belong to the upper classes, working outside their homes and not having a close contact with the home appliances. In terms of technical functions, these two brand names have differentiated quality of the materials between each other. Brastemp uses nobler materials than Consul, resulting in higher costs in its products.

The washing machines illustrate the diversity in the acceptance of types of products of a same general function, among distinct markets. In Brazil, for example - which suffered a great North American influence in relation to washing machines, because the first ones were

[4] It belongs to a hot weather region.

[5] Juice obtained from the coconut, generally sold at the beaches of the Northeastern region of Brazil.

imported from the United States – a notable preference for the "Top-load" type exists, while in Europe and in Argentina (that is a Brazil neighbouring country) the "Front-load" type, economically and technically superior, is preferred. It attends to a preference, in the case of Brazilian market, basically linked to the symbolic function.

In relation to the graphic design of the products, it is also verified the presence of the diversity in the use function. In the case of the command panels of microwave-ovens, for instance, the graphic design of the products destined to the Brazilian market needs to be more textual than pictographic, because people do not like and do not use to read the manuals of using instruction. Therefore, models of microwave-ovens, that use too many pictograms in the command panel, are not well accepted in Brazil.

In terms of shapes, we can verify that the German product design, as the one from AEG, for example, use to be traditional, robust, simplistic, rational, as a result of the profile of the German consumer, who demands ergonomic and elegant products, as well as good quality. On the other hand, French products, such as those from Arthur Martin, the Italian ones, from Zanussi, or the Brazilian ones, from Electrolux and Consul, usually explore more details and flexible lines. [6]

The North American products, in the opinion of Brazilian designers from Electrolux and Multibrás, are usually "immense" and possess a more "recherché" finishing, generally with golden and chromed friezes, with textured external surfaces, differently from the Brazilian products. Some of them have been imported by Electrolux of Brazil, and by Multibrás, besides other companies. It is the case of the Side-by-Side refrigerator, whose adaptation to the local contexts has not been quite easy. Some problems have occurred, partly because of its sizing - that makes it difficult to pass through the doors of elevators and residences - and partly due to the necessity of an outlet of water near the refrigerator, which normally is not foreseen in the Brazilian constructions.

In what concerns to the symbolic function, differences in the formal composition between the imported and national products can also be perceived. In Brazil, for example, several imported products have a predominance of straight lines and simplistic design, while the national ones, in general, are less rigid and present more symbolic appeals. Brazilian market demands, for example, lids on the cookers, something that is not used in European countries. The lid on the cookers basically fulfils a symbolic function for the Brazilian consumers, not presenting a reasonable practical functionality. For the fact that these consumers are generally more "impulsive" in the purchase act, the companies have inserted elements sometimes superfluous in the products, such as the arched component of the lid. [7]

Divergences between the design of national and imported products, as the ones mentioned, illustrate some of the implications of the questions of cultural diversity in the interpretation of the functions of the objects. It is undeniable, however - although the existence of considerable barriers against the process of massive introduction of imported products in the countries - the social influence of these products, as well as that of conceptual design products, in the

---

[6] AEG, Arthur Martin and Zanussi are companies and brand names that belong to Electrolux Corporation.

[7] Such characteristic of the Brazilian market causes that various cookers with a simplistic design, like the German ones as example, are rejected in marketing researches.

construction of desires and values in individuals and social groups, as the elite groups, that by acquiring them, influence, in certain aspects, the formation of the opinion of other groups.

We may verify, in companies in general, that the biggest obstacle to the diversification of products has been the factor "cost". This has, actually, assumed a paradoxical position in this question. As a matter of fact, it has promoted the standardisation of products and components, facing the strategies of rationalisation and of profits magnifying of the capitalist companies. On the other hand, it has led to the diversification of products, due to the economic inequalities and unbalanced access to the goods between the various social groups. In one country alone , such as Brazil, for example, there are differentiated products, because of the great economical contrasts existing between the social groups.

In the world-wide scenario, where competition between companies has become more and more exasperated, we can identify a trend toward the coexistence of individualised, hybrid and "global" products, in the core of the dynamics that moves the cultural process of the human societies, where both the movements in direction to the diversity, as well as those leading to standardisation, are present.

At Electrolux, for example, there have been differences in the design of the products developed for the European, Asian, North American and Latin American markets. However, the diversity of products developed for determined markets, by virtue of the cultural specificity's, coexists with the interchange of components, adaptations and products directed to the global market (as it is in the case of the vacuum cleaners Mondo and Clario, for example), that constitute part of the strategies of rationalisation of the production and of the increase in profits.

A strong trend toward the development of hybrid design is revealed, resulting of the influences brought by the new horizon of information and inter-organisational relations, joint work between centres of design, partnerships with companies of other countries and corporate strategies, that aim to reach the global market, including interchange of components and exploitation of tools already developed.

An example of hybrid design, resulting of a partnership between Electrolux of Brazil and an Asian company, is a line of microwave-ovens, launched in 1998 in Brazil, whose cabinet and technology have been preserved, including, however the alteration of the graphical design and the adaptation of the program (due to the alimentary diversity) for the Brazilian market.

Another marketing trend can also be observed , that emphasises the "sensorial experience", through the creation of an emotional bond between the consumer and the product, which demands great attention to the cultural diversity. Facing this perspective, technology is one strong lever for the viability of customised products, allowing the increase of the flexibility and quickness of the productive organisation.

One can therefore verify the coexistence of forces both in the direction of the diversification, and standardisation in design, ratifying the agreement of Lévi-Strauss (5), amongst others, concerning the existence of mankind as an essentially contradictory process.

The culture has shown itself as an essentially dynamic phenomenon - intrinsically tied to the process of development of the economical and social aspects of societies (6) - being capable of representing, reproducing and transforming the tangible and intangible elements that conform the social system and life, influencing and being influenced by the economical practices and by the symbolic relations (7). In this way, we must be attentive to the local particularities and to the questions of cultural diversity, respecting other cultures, and having, at the same time, a critical vision and a moral commitment to the societies. In the transcultural relations, individuals and societies must search for the understanding of the experiences of other cultures, assimilating or reinterpreting those that bring benefits to them, in terms of quality of life, preserving, however, their cultural inheritance and their identity.

3. FINAL CONSIDERATIONS

Design, as a cultural practice, as well as its implications in the life of individuals and societies, has been sub-researched. We have until now lived involved by devices that are available and help us to model our lives, and, in this context, their design consists of symbolic signs, that brings meanings to the social relationships and referential relations to the identity of the people. Design participates in the construction of values, yearnings, thoughts and actions, contributing to the satisfaction of the necessities of individuals, influencing and being influenced by the social relations.

Within the economic sphere, the design participation in the strategies of competitiveness of the companies distinguishes it as an instrument of accumulation of capital, being co-responsible for the development of a wide range of products that have been introduced in societies along history.

In the globalisation context, that is characterised essentially as a paradoxical process, where both forces toward homogenisation and to standardisation coexist, the action of the designer becomes decisive in the interpretation of the functions of products, that must reflect the characteristics and necessities of individuals and societies. It therefore becomes necessary to have a continuous and thorough research on users and local contexts, as well as on the cultural dimension of design.

It is necessary to have a clear view of the role of the designer, that it should not be limited to a terminal, peripheral or superficial action. It is fundamental to have the concept of the designer as an important mediator between the products, individuals and societies, and as co-responsible for the preservation of the identities and cultural inheritances of mankind. In this sense, a stronger emphasis to the social and cultural dimension of objects must be given, and, moreover, achieve a more complete integration between the various areas of knowledge involved in the development of products, in such a way that it reduces the existing gap between the cultural and the technical dimension.

ACKNOWLEDGEMENTS

I would like to thank Prof. Dr. Marilia Gomes de Carvalho - who oriented my studies on Globalization and Cultural Diversity - as well as Mr. Roberto Azevedo Smolka and Mr. James Alexandre Baraniuk – who contributed to review the English version of this paper.

REFERENCES

(1) Ono, Maristela M. *Industrial Design and Cultural Diversity: a Case Study at Electrolux of Brazil S.A. and Multibrás S.A.* (Dissertation developed at Technology Post-graduation Program from Centro Federal de Educação Tecnológica do Paraná), Curitiba, 1999.

(2) Bourdieu, Pierre. *Sociologia*. São Paulo: Ática, 1983.

(3) *A economia das trocas simbólicas*. São Paulo: Perspectiva, 1974.

(4) Cuche, Denis. *La noton de culture dans les sciences sociales*. Paris: Éditions La Découverte, 1996.

(5) Levi-Strauss, Claude *et al.* Raça e História. In: Raça e Ciência. São Paulo: Perspectiva, 1970, p. 170-231.

(6) Canclini, Néstor García. *As culturas populares no capitalismo*. São Paulo: Editora Brasiliense, 1983.

(7) Geertz, Clifford. *A interpretação das culturas*. Rio de Janeiro : LTC – Livros Técnicos e Científicos S.A., 1989.

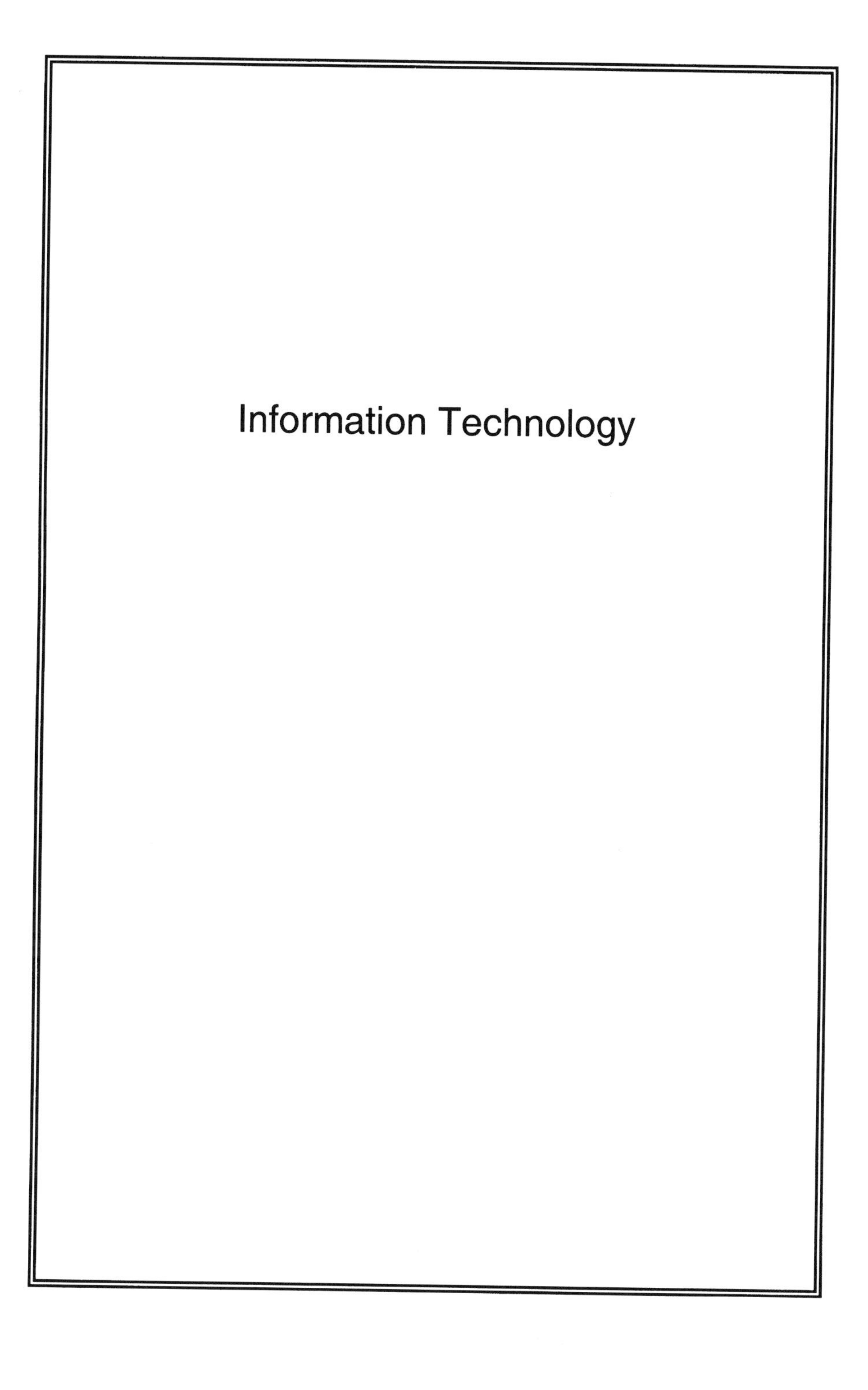

Information Technology

The impact of information and communication technologies on design teaching

C McMAHON
Department of Mechanical Engineering, University of Bristol, UK

SYNOPSIS

We are currently experiencing a revolution in information and communication technologies (ICT) at a time of pressure and change in education. This paper argues that ICT may be an avenue for productivity improvement, or it may add to the burdens on educators, depending on whether or not academic institutions change their *modus operandi* to take advantages of the economies of scale that ICT allows. The paper first reviews the methods for productivity improvement in commerce and industry, and explores the difficulties in trying to translate these methods into education. It then gives an overview of the current capabilities of ICT in educational applications, and discusses implications for teaching practice of these methods. Finally, a speculative scenario for a future pattern of engineering education is presented.

1 INTRODUCTION

We are in the middle of a revolution in information and communication technologies (ICT). The Internet and the World Wide Web allow us to access information from around the world at our desktops. Video conferencing capabilities now allow people to view lectures and participate in tutorials remotely. School pupils now use computers and computer-aided learning (CAL) material from the age of 5 onwards, and many young people have access to computers at home. The continuing improvement in the performance of computing equipment suggests that the penetration of computing into our lives will continue for some time to come.

The dramatic changes in ICT come at a time of general pressure and change in education. In higher education in particular the number of students has generally been increasing for some years, but without a commensurate increase in funding levels. There has been change in the style and manner of teaching, with more emphasis on continuous assessment and project work, the introduction of formal quality assurance and assessment procedures, and with a move away from "chalk and talk" to increased use of handouts and audio-visual aids. Changes in teaching have combined with an increased effort required on research – through increased reporting requirements, higher paper production rates, lower proportions of approved projects, a greater need to disseminate and publicise and, of course, performance assessment exercises.

The various pressures in education lead in the author's view to a very genuine threat to the ability of educational institutions to provide a high quality educational system in the medium and long term. Already we are seeing concerns expressed by many academics about the long-term viability of our present *modus operandi* [1]. Without significant change in the way that educational institutions operate, it is also the author's view that information and computing technologies may simply add to the pressures on education. Staff effort will be dispersed in learning new languages and computer packages and in maintaining existing computer-based educational material. Institution budgets will be eaten up with the cost of computing and communications equipment, presentation equipment and so on. Yet this need not be the case. With a significant change in pedagogical practice, information and computing technologies may be the way for the productivity issue in education to be addressed. This paper will argue that information and communication technologies can allow for economies of scale in education, and may offer a real improvement in the efficiency and effectiveness of some types of teaching, but with the corollary that the preparation and delivery of the material will have to be undertaken by larger groups with less variety than present academic units and institutions.

The arguments in the paper will be developed by first reviewing the mechanisms used in industry and commerce for productivity improvement, and commenting on the ease or difficulty of their application in educational institutions. Information and communication technologies that may be applied in teaching will be briefly reviewed, and teaching practices will then be discussed to identify what aspects of present practice should be retained and developed, and what aspects may be replaced or supplemented by new modes of delivery. The paper will conclude with a speculative scenario for a future pattern of engineering education.

2. MECHANISMS FOR PRODUCTIVITY IMPROVEMENT

Mechanisms for productivity improvement in industry and commerce include investment in equipment to reduce labour costs; modification to the product or product range to simplify its production; modification to the process of production and economies of scale. In almost every case, techniques that may be applied commercially are difficult to apply in education.

The classic method for productivity improvement in manufacturing industry is investment in machinery that has higher output rates and/or lower labour requirements. The possibility for productivity gain in teaching has, however, so far been rather limited. Word processors have assisted in the "production" of academic papers, and software has for example been used for processing of examination marks, but many impacts of new technology have in fact been negative on teaching load. My father lectured with "chalk and talk". I now have computer-produced overheads and handouts, that increase the volume of material I can deliver but also require significant effort in their production and maintenance. The difficulty is that new equipment has not yet begun to replace teacher contact, and until this happens the opportunities for productivity improvement will be small[1].

The second approach to productivity improvement is modification to the product. For example, in manufacturing the improvement in productivity from use of machines has meant that the labour-intensive part of the process has become more important, and hence the great emphasis in recent years on design for assembly – often the most labour intensive aspect of manufacture. In academia

[1] Overhead projectors allow lecturing to large classes, and this has a significant impact where buildings allow.

the scope for taking such an approach has been much more limited, and therefore because teaching is so labour intensive pressure on costs has meant increased workloads and pressure on real wages and on participation in such activities as professional societies. In some cases of course changes have been made: numbers of courses have been reduced, lecture programmes combined, laboratory write-ups eliminated and so on, but again there has been a very limited effect. Often, for the student, the result has been to limit choice or reduce opportunity for them to develop skills through practice and repetition and for this reason many institutions have resisted pressure to change and have retained significant option programmes and high project and laboratory class loads.

Another manufacturing approach to reduce assembly cost has been to decrease labour costs by process changes such as carrying out assembly overseas, eliminating inspection and so on. Such an approach is again difficult to implement in academia. Where process changes have been made they have often had negative effects – inspection, assessment and documentation have been increased, and committee structures have proliferated, while changes made to balance these have often actually had a negative effect on the real quality of education by, for example reducing the numbers of markers for project work, replacing academics by postgraduate assistants or increasing student:staff ratios in laboratory and project classes.

The final driver to be considered is economies of scale, and these have been very much in the news recently. Motor manufacturers now need to be building millions of cars a year to achieve the economies of scale necessary to be profitable, and amalgamation and take-overs are being seen in insurance, in banking, in supermarkets and so on. Again, there has been limited scope for the same thing in academic life because of the independent status of academic institutions. Instead, we have a proliferation of separately developed courses and curricula, of university administrations, of computing services and so on. Perhaps these are rather like the proliferation of craft-based manufacturers of motor vehicles at the start of this century: will they be swept away by "mass-producers" to be replaced by educational corporations?

3. THE IMPACT OF COMPUTING AND COMMUNICATION TECHNOLOGIES

Throughout section 2 it has been suggested that there have been few opportunities for productivity change and improvement in academic life, but I believe that computing and information technologies offer the possibility to change this. A generation of young people is growing up that is familiar and comfortable with computers, and that may be prepared to use computer–based techniques for learning. The technologies are getting to the point that we may now be able to routinely:

- Lecture on-line to large audiences using video conferencing and the Internet [2].
- Use shared computing workspaces combined with video telephony for on-line tutoring [3]. Together with video conferencing of lectures this would largely free us from geographical constraints on the location of students.
- Use much enhanced example material using virtual reality and on-line videos
- Use computer simulation technology to supplement and to train for laboratory activities
- Mark and assess by computer – e.g. random choice multiple assessment is a current capability [4], and text analysis allows checking of the style of written work, automatic checking for plagiarism and so on [5].
- Use newsgroups and mailbases for the sharing of problems.

However, although we can do all of the above, it may only be feasible if we have a significant change in our pedagogical approach. Small academic units may find it prohibitively expensive to set up video conferencing facilities with good enough production values and equipment for it to have a significant impact (as in manufacturing where the most sophisticated testing and analysis techniques demand large teams and resources). Computer-based assessment and teaching support must be supported by programming and support teams that can produce software of the required reliability and quality and can make the required maintenance effort over a sustained period. Piecemeal effort by small groups will not work: it is useful to establish implementation ideas and to experiment, just as in new product areas "new lines of development each start with a period of experiment and groping, during which a wide range of types is evolved" [6]. Widespread application of computer-based technologies will require concentration by the academic community on small ranges of software, and it also implies a more common curriculum, in early years of study at least, to justify the economies of scale.

There is of course an implication in this for the independence of academic institutions. In order to gain economies of scale in the production of teaching material, institutions will have to forego some of their independence. Once they have made this move the next step to amalgamation and collaboration to gain economies of scale in administrative and marketing effort is perhaps only a small one.

4. IMPLICATIONS FOR TEACHING PRACTICE

Although we may be able to carry out all of the ICT based activities listed in section 3, which might we actually want to use? For many the prospect of teaching using the technologies described above is an appalling prospect, but perhaps not using them may be equally unpalatable. If labour-intensive teaching is not funded properly then the prospect is for increasingly large class sizes, gradual erosion of small-group, laboratory and project work and a substitution of postgraduates and teaching assistants for academic staff in many classes. The academic community needs to identify what in a student's experience of university is important from a pedagogical point of view and as a formative experience, and it is suggested that this should be the subject of research effort in the near future. The views presented here are personal opinions combined with the results of informal discussions with students and colleagues on the topics presented in this paper.

The first reaction of students to a suggestion that teaching might move towards computer-based delivery was not an objection to that method as such, but a stress that the social aspects of the university experience are very important – probably more so than the pedagogical experience for many. The social experience provides a largely benign transition from adolescence to adulthood, and the experience of studying away from home is very important in the UK. There are, however, a number of facets of this social experience that are debatable: for example, does it require three or four years away from home, and should it continue to be funded from the public purse if there are lower cost ways of meeting the pedagogical objectives?

The second reaction of students is that a structured programme is very important to academic progress: the pressures of a timetabled activities and regular deadlines encourage methodical working patterns and maintain motivation. We must be careful that the same opportunities for structured progress are maintained in any new system. The community of students working together on the same subjects allows sharing of experience between students that is also valuable. Conversely, the availability of teaching material on-line would allow students to proceed at a pace that is more closely matched to their ability, and this opportunity could be exploited also. For example, students willing to work through the calendar year might reduce the duration of their studies, while others might choose to study part-time while working and thus take the course over an extended period. This latter point would be of particular value in extending access, for example to mothers wishing to prepare to return to the workplace who could fit their study around their other commitments, or families that could not afford a loss of income over three or four years.

A third reaction of students is that there are many different lecturing styles, and that different subjects require quite different approaches. Some styles and subjects will adapt quite easily to new media, but for others the transition will be difficult. For some lecturers also it is the asides and personal examples that they give – some of which they might not want to be widely broadcast, that make their lectures interesting and useful. Students also want to be able to give feedback to the lecturer – to be able to pose questions, to indicate that pace is too slow or too fast and so on. Research will be needed into ways of making the transition to new media, and especially on how personal touches can be incorporated and the issue of feedback addressed.

The principal reaction of some academic colleagues was that "lectures are not that important" and that it is the pupil-teacher contact that is vital – there is "no substitute for the one-one contact in examples classes and tutorials". The opportunity for the pupil to experience the way of thinking and working of the teacher is seen as most important in the educational experience, as is the opportunity of contact for the teacher to assess the progress and particular circumstances of the pupil. If this view is correct then it is a matter of concern that this style of teaching is what is threatened by pressures for productivity improvement in education, and that this aspect of teaching is not well measured in quality assessment procedures.

A reaction shared by academics and students is that contact with physical artefacts, laboratory activities, and project and group work are important elements of high quality engineering education, and that meeting these elements in an educational context is important – because of the need for "freedom to make mistakes" that might not be present in an industrial context.

The final point here is a personal one. In my view, academics are paid by society to be experts in their fields, and yet society appears to make relatively poor use of their expertise. For the most part the expertise is exposed to a small group of peers in the academic community through conferences and the academic literature, to undergraduate and postgraduate students with whom the academics have direct contact, and perhaps to industrial research and consulting collaborators – in all cases quite small numbers in absolute terms. Historically, the best exposure to a wider community was perhaps by writing a book, but there is relatively little kudos and credit in this in the present academic community (at least in engineering in the UK). The specialist expertise of academics is often presented in courses taken in the later years of undergraduate programs and in postgraduate taught courses. Perhaps new technologies will allow these to reach a much larger audience by allowing people from the wider community to join in lecture courses and demonstrations through video links. These same technologies may also facilitate a greater involvement in university work by the wider community: for an engineer to join in a discussion with a group of students about a design project by video conferencing would be much less disruptive than having to travel to the university.

5. CONCLUDING REMARKS

To conclude, let us speculate on what the future pattern of typical courses in engineering tertiary education might be:

(1) Routine lecture teaching will be increasingly carried out with the support of video conferencing and programmed electronically delivered teaching material, and aspects of this work will be examined electronically also. This content will be created and delivered by specialist units that can afford the economies of scale that will allow development of high quality material. Within academic institutions, personalised "insight" type lectures and direct pupil-teacher tutorial contact will support this material. Mailbases and video telephony will also support question/answer sessions.

(2) In their lecture-based teaching, educational institutions will concentrate on their specialist courses, generally delivered in the later years of an undergraduate course and in post-graduate taught courses, which they will also offer to a wider audience via the Internet through video links and through well-prepared web content.

(3) Educational institutions will also concentrate on the hands-on design, laboratory and project elements of their courses, and it is these aspects that will distinguish between courses and will give the high value course features that students will seek out and be prepared to pay for. It is the design thinking, and the group and individual design and laboratory work that will be the stimulating conventional experience for many students. For some students, these elements of the course may be delivered in concentrated blocks of work, while they study other material remotely over the Internet.

(4) Design and laboratory work will be increasingly supported by virtual reality and computer-based team support software, for example that would facilitate students working in distributed groups and with industrial collaborators

DISCLAIMER

The views expressed in this paper are of the author only, and do not represent those of the Department of Mechanical Engineering of the University of Bristol.

REFERENCES

(1) Carvel, J, *Fees get big thumbs down*, Guardian Education, 13 March 2000

(2) Burns, J, Lander, R, Ryan, S and Wragg, R, *Practical guidelines for teaching with video conferencing*, www.jtap.ac.uk/reports/htm/jtap-037.html

(3) Lee, M and Thompson, R, 1999, *Teaching and learning at a distance: building a virtual learning environment*, www.jtap.ac.uk/reports/htm/jtap-033.html

(4) Sims Williams, J H, 1999, *The Teach and Learn project*, www.dig.bris.ac.uk/tal/tal.htm

(5) www.plagiarism.org

(6) Ricardo H R and Hempson J G G, 1968, *The high-speed internal combustion engine*, 5th Edition, London, Blackie

Using the Internet to manage an undergraduate interdisciplinary design course

R CLOUTMAN and **M YANG**
Faculty of Engineering and Information Sciences, University of Hertfordshire, Hatfield, UK

SYNOPSIS

The management and delivery of an interdisciplinary design course for engineering undergraduates has been significantly enhanced via a dedicated website. The site supports an interactive database for course management, reporting mechanisms for monitoring student progress, communication facilities, an area for support documentation and a delivery point for course assessments. The website was successfully launched in September 1999.

An overview of the site structure and content is presented. An evaluation of the effectiveness of the website is made based upon feedback received. Subsequent site developments and enhancements are described in preparation for its re-launch in September 2000.

1. INTRODUCTION

The Interdisciplinary Engineering Design Project, or IEDP, is a one-semester course which forms an integral part of all second-year BEng Undergraduate programmes at the University of Hertfordshire. The course has been developed extensively since its original introduction in 1989 as an intensive, one-week project (1, 2). The format of the present course requires student teams to develop design proposals for a specific engineering product or system that they have selected, on-line, from a list published at the start of the semester (3).

It has been estimated that the recruitment needs of major industrial companies will be for some 75 - 80% of engineers to be broad-based rather than specialists. The requirement in all cases will be for the graduate to possess multidisciplinary skills along with the ability to function as a team member (4). The IEDP has been developed to reflect this need by providing students with an excellent learning experience whilst building their team working, presentation and interpersonal skills.

The IEDP website provides every team with their own secure area for project management, message passing and the submission of progress reports and coursework to Supervisors. Supervisory staff may also access a dedicated area which hosts a number of on-line tools for monitoring team progress, communicating with their teams and managing assessments. Access to all course support information and interactions by both staff and students with the course database are performed via a standard web browser. The IEDP Website is mounted on a departmental server which runs Linux and the NCSA Web server software.

2. WEBSITE STRUCTURE AND CONTENT

The course website has been designed to meet two fundamental aims;

- to provide a focal point for all text-based course materials along with links to additional support materials,
- to streamline the management and execution of the course, both in terms of the performance of student teams and the ability of staff to monitor their progress.

The objective of the site therefore has been to provide an environment that is quick and easy to navigate. At the heart of the system is a central database which may be interrogated using forms imbedded in HTML pages and a standard web browser. Course administration is also integrated into the system via a password-protected area of the site (5). The main architecture of the system is illustrated in Figure 1.

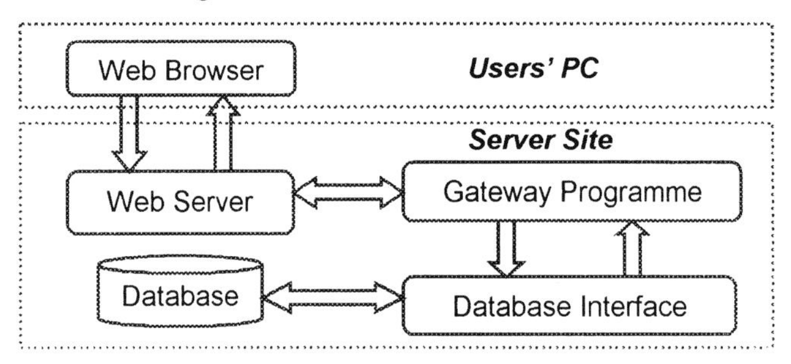

Fig. 1 Website System Architecture

The website provides four areas;

- Common Area: unrestricted access to all course support material
- Teams' Area: password access for each student project team
- Supervisors' Area: a restricted area for team management by staff
- Site Manager's Area the 'top-level' restricted area for overall course management.

The website has been tailored to reflect the existing IEDP course format and to replace paper-based procedures. The course is led by a co-ordinator who works closely with 9 supervisory staff, three from each of the three engineering departments within the Faculty. Each Supervisor is responsible for overseeing a 'Section' of 6 multidisciplinary student teams, as shown in figure 2, hence there are fifty-four teams in total with 6 to 8 students in each team. The total cohort size is therefore approximately 400 students.

2.1 Design Activity

Students use the website initially to register on-line for a project title that must be chosen from a list published on the site. Before registering, students check an on-line distribution table showing the number of places available, for their specific course-code, against each project title. When the site server receives a valid registration form submission, the distribution table is decremented automatically to reflect the current availability of titles.

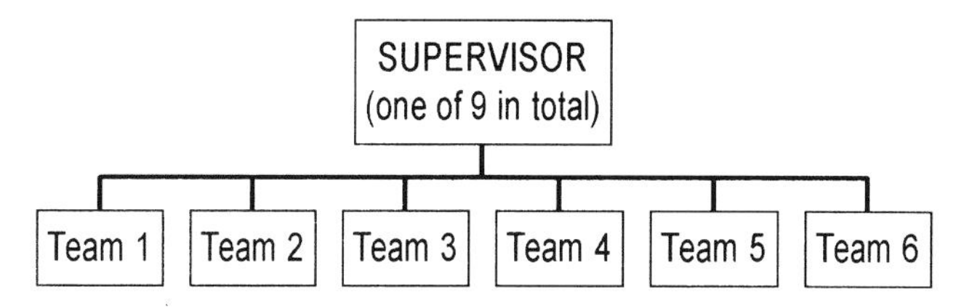

Fig. 2 Structure of each Project Section

Project teams undertake four design phases, based upon the SEED design activity model (6), which span a 12-week period and consist of:

- **Information Gathering:** to identify and obtain all possible information, opinion or examples which may be relevant to the project topic.
- **Specification:** where the information gathered is transformed into a statement of goals and constraints relevant to the design to form the Product Design Specification (PDS).
- **Concept Generation and Evaluation:** when ideas are generated, evaluated and developed with reference to the specification phase. This third phase must clearly demonstrate development towards an optimum concept.
- **Presentation:** the chosen concept must be presented in a readily understood way as a design proposal and accompanied by evidence of the way in which decisions were reached.

A key innovative feature of the IEDP course is the *Design Week* near the end of the semester during which students work full-time on their final design proposals, no other courses are scheduled during this week.

2.2 Progress Reporting

Students must ensure that good inter-team communication exists and that their Supervisor is kept fully informed of team progress. A paper-based system for generating weekly team progress reports has been refined over a number of years and has now been implemented as a site-based HTML form template. This form requires responses to specific questions to be entered into text boxes. Every team member is automatically e-mailed a copy of the completed form when it is submitted to the database. The respective team Supervisor also receives an e-mailed notification that a report has been submitted. The contents of each report provide the starting point for weekly discussions with Supervisors. An acknowledgement is e-mailed back to the team once their Supervisor has logged an assessment of their progress in the database.

3. EVALUATION

In January 2000 students were invited to respond to an on-line questionnaire about the IEDP course and its website. The on-line form linked back to the course database to provide a cumulative total for each response to the 25 questions posed (5). The questionnaire was answered by 62% of those registered for the course. Comments were also sought from supervisory staff. An analysis of this feedback flagged the following points:

- Over 70% of students felt the website helped them find relevant course information,

- About 50% felt the site helped them communicate better with other team members and their Supervisor,
- Approximately 65% of students felt that team members did not participate equally,
- Approximately 60 e-mails were received which commented on the tutorial exercises,
- Almost 60% of students found the weekly tutorial meetings with their Supervisors to provide sufficient guidance for the course.

An analysis of the feedback indicates that the textual content of the site is satisfactory. It was felt that further site development should concentrate on refining the methods used to monitor individual student performance and the provision of on-line tutorial exercises.

4. CURRENT DEVELOPMENTS

The site layout has redesigned to enhance access from the home page and provide an improved course notice board (7). The team weekly report form has also been improved to include a peer review section where students must collectively assess the performance of individual team members in meeting targets set on their team plan. The team plan, or Gantt chart, is created via the website. Supervisors will also be required to provide a more detailed assessment of team progress by generating a weekly mark which will be automatically merged into an accumulative total and e-mailed to every team member. Tutorial exercises will be placed on-line to encourage greater site usage and will replicate a range of paper-based ones already developed. Supervisors will also monitor student access and usage of the site.

5. SUMMARY

The Internet has been successfully used to provide support for a large-scale interdisciplinary undergraduate engineering design project during the 1999/2000 academic session. Analysis of course feedback received has indicated which areas of the site should be improved. Current site development is concentrating on the provision of on-line tutorial support and enhanced monitoring of team performance. The revised site will be demonstrated at the Conference.

REFERENCES

(1) Gregory, R.D., Hamilton, P.H., 'Interdisciplinary design project', *Innovative Teaching in Engineering*, Ed: Smith, R.A., Ch 67, pp. 409-414, Prentice Hall, 1991.

(2) Cloutman, R.W.G., 'An Interdisciplinary Engineering Design Project for second year undergraduates', *4th World Conference on Engineering Education*, Vol 2, pp 85-88, 1995.

(3) Cloutman, R.W.G., Yang, M.; Bullen, P., 'A large-scale interdisciplinary design project', *21st Annual SEED conference*, pp 223-226, 1999.

(4) John, V., 'A future path for Engineering Education - educating engineers for Europe', *IEE Science and Education Journal*, pp 99-103, June 1995.

(5) Yang, M., Cloutman, R.W.G., 'Web-based Interactive System to support an Interdisciplinary Engineering Design Project', *Engineering Education for the 21st Century*, pp 271-275, 2000.

(6) Curriculum for Design - Engineering Undergraduate Courses, SEED 1985.

(7) IEDP web-site address for September 2000, *http://dragon.herts.ac.uk/iedp*

Collaborative engineering experiences

P A M VAN KOLLENBURG and **D VAN SCHENK BRILL**
IPA Research Centre, Fontys University of Professional Education, The Netherlands
G SCHOUTEN and **P MULDERS**
Philips CFT, The Netherlands
J B OCHS
Integrated Product, Process and Project Development Program (IP^{3}D), Lehigh University, USA
M ZIRKEL, I KIMURA, and **G KLETTE**
Otto-von-Guericke-Universitaet, Germany

SUMMARY

In the fall of 1999, an international integrated product development pilot project based on collaborative engineering was started with team members in two international teams from the United States, The Netherlands and Germany. Team members interacted using various Internet capabilities, including, but not limited to, ICQ (means: I SEEK YOU, an internet feature which immediately detects when somebody comes "on line"), web phones, file servers, chat rooms and Email along with video conferencing. For this study a control group with all members located in the USA only also worked on the same project.

Lessons learned as well as areas in need of improvement include 1) planning, 2) communications, 3) project definition, 4) project leadership and 5) teamwork. In general team cohesion stayed low (because students did not meet in real life). Chatting and Email were in this project by far the most important communication media. Despite many and varied set backs, the experience was valuable for all involved and will be repeated in the fall of 2000.

1. INTRODUCTION

Technical education is commonly based on specialised courses on different topics. In such a course the lecturer may operate independently from colleagues or even from other courses. As a result students are trained to solve technical problems only in relation to a given course. In the 1980's and into the 1990's only rarely students were educated in solving a problem in a broader and multi-disciplinary context. In concurrent engineering and integrated product development lecturers must cooperate and communicate with lecturers of other disciplines; this yields more work (meetings), more planning and raises more problems with the assessment of the students' and team member's progress. So it does not surprise anyone that multi-disciplinary teamwork is not common in education yet [1,2].

On the other hand there is a great demand for well-trained people especially in concurrent and collaborative engineering. Océ-van der Grinten (development and production of copiers and printers) states, "People applying for a job at our company are in a better position to get that job when they have been trained in concurrent engineering!" Industrial participants in the USA have stated that students with the IPD experience and education can be effective and productive team leaders in one half the time of those who do not, saving the companies millions of dollars each year [3]

Engineers face the era of globalisation where projects are planned, implemented and have impacts across national and cultural boundaries. With the new communication possibilities (Internet and videoconferencing), Integrated Product Development (IPD) teams are comprised of participants who are located in geographically dispersed areas of the globe. With this IPD - Collaborative Engineering (IPD-CE) project we balance the need to create an environment that encourages student innovation, creativity and entrepreneurial spirit with the realities of today's global engineering, business and design.

We started in the international project in the fall of 1999. Two teams were formed with mechanical engineering students from Lehigh University in the US, electrical engineering and business engineering students from the Fontys University in Eindhoven, the Netherlands and mechanical engineering students from the Otto-von-Guericke University in Magdeburg, Germany. A third team at Lehigh University consisted of only Lehigh engineers. At each location we had 3- 4 students per team. In this pilot the international student teams focused on the development of a lighting system to improve the quality of watching TV. The topic of the project was provided by Philips Electronics NV, in the Netherlands. Funding for the project and course expenses for Lehigh's participation in the pilot was provided by NASA/CAPE and Lehigh's Ventures initiative [4].

In this paper we want to share with you our experiences with modern communication technology in order to find useful tools for facilitating the co-operative work and the contacts of all the participants. The international IPD teams interacted on the Internet using various capabilities, including, but not limited to, ICQ, web phones, file servers, chat rooms, Email, etc. Video conferencing using ISDN-Vspan technology is used at critical points throughout the semester for doing plenary reviews on major milestones. Program evaluation focused on the technologies used, the processes involved, the nature of human interactions and the effectiveness of communications while the teams were engaged in product development.

2. START OF PROJECT

At the start of the project, each team was asked to design and create a team logo. This standard team building technique requires the members to specify the objectives of the logo, quantify the evaluation parameters, create multiple ideas and select one (or more, or in combinations) to end up with a single team logo. For the Lehigh-Fontys-Magdeburg **LFM** International team graphics and industrial design assistance was provided by Lehigh's Adjunct Professor of Industrial Design, Drew Snyder. In addition to **LFM** International (3 locations), there was team Lehigh-Fontys **LEFON** (2 locations) and the single site Lehigh team **GFJ Integrated**.

Next, the teams were introduced to a potential new product by the sponsor, Philips: an automatic lamp switch for TV back light in the living room. The first stumbling block of the project was the project definition itself. The teams were unclear on exactly what the sponsor wanted; it took e.g. FOUR weeks before the Lehigh teams could imagine how a 'cosy' European living room looks like! Only after sending some pictures of typical European living rooms it became clear. But also towards the design there were problems: was it a light control or a light and a control that was asked for?

While the teams were using the same text for the project start, the same textbook [5] and following the same schedule, the overall planning of the project was not clear. For example, the due dates and milestones were very unclear. There was nothing such as a total project planning. So it happened that teams didn't know the other teams had holidays or exams and therefore couldn't contribute. As a result team members failed to plan and work effectively.

They did not have the same stake in the outcomes since some were taking the course for credits and some were not. Interdependency of the team members was not well understood. Members were not able to recognise the importance of providing the information requested in a timely manner.

In addition, team tasks were ambiguously assigned from the inception of the project. The German and American engineers of the LFM team had the same mechanical engineering background to therefore worked at the same task individually until the middle of the project. At that point the design became two parts – namely, control box and lamp part – so that these teams were able to attack the different engineering problems in a parallel manner.

3. COMMUNICATION

The second problem arose when the teams in different places had to communicate about the project. Since the teams were separated by 6 time zones, communication was mostly limited to employ the Internet such as: Email, Videoconference and the **B**asic **S**upport for **C**o-operative **W**ork (BSCW) server. Real time communication (telephone and videoconferencing, chatting was only possible in a small time frame between 15.00 – 17.00 hr CET.

3.1 Email

Email is an excellent method to communicate with people in different locations because one can instantly send a message whenever problems or questions arise. However, during the project, many confusing or irrelevant messages flooded the server. The response time was also low: it took in average 2 days to get an answer on an Email sent! The sheer number of messages from multiple team members with no particular priority resulted in members not knowing if answers were needed or when. The volume of messages made some of the members numb to answer any messages. It was determined that Email messaging required a protocol: was the message received, was it read and understood, are the action items to be acted on, if so when, if not what alternative actions are to be taken. Because of the time zone differences, protocol included when messages should be sent and when a response would be expected.

3.2 Video Conference

With the video conference software package, iVisit, team members communicated directly over the Internet. This communication method is very similar to face-to-face meetings however, some technical obstacles were encountered due to the web cams and audio transmission. Very often the quality of the sound was not good enough to understand each other at the different sites. At the beginning of each conference, it usually took a long time to adjust each system. The causes of most problems were unknown, and the teams were forced to use the chat window, which allows users to write anything such as questions, answers and concerns on a computer screen. For smooth conversation, a better conferencing software tool should have been selected – that is critical for better understanding of each other, particularly during critical team-decision making sessions. Unlike face-to-face meeting, participants were not able to have an eye contact or read body language so that the conversation was rather awkward. Additionally, when one participant was speaking, others were not sure to whom the participant was talking so that it was difficult to know whose turn it was to speak or when was the right time to speak.

3.3 Computer Supported Co-operative Work

We have seen that Email was useful not only for the exchange of plain text messages but also to carry attachments with various types of information. However, it turned out that the number of such objects grew extensively during a project and that the attachments grew bigger in size also. The team members were often not in the possession of the needed objects or in possession of an old version. We apparently needed something which is commonly referred to as "Computer Supported Co-operative Work" or CSCW. Fontys University in Eindhoven was enthusiastic about the toolkit "Basic Support for Co-operative Work" or BSCW [5,6]. This server is used as a common, shared workspace for the groups.

The teams communicated the following of their experiences:

- Only the **LFM** team (the team with the three countries involved) used the facility extensively. The teams also indicated that the BSCW is a essential tool to make IPD-CE development possible.
- The most important use was the communication of text documents; secondly the sharing of programs and thirdly the planning of meetings.
- Positive experiences were: ease of use and convenience of central information storage being available anywhere.
- Negative experiences were: varying response time, additional effort to invest in working with this tool and since the server is open to every participant, folders and files are likely to be disorganised.

As the project went by, the number of files and folders became larger and larger such that it was sometimes very confusing to find the exact location of the needed file amongst a lot of others. Therefore, there should have been a responsible person who could always keep his or her eyes on the server's organisation.

4. QUESTIONNAIRES - THE TEAM'S OWN VIEW

Can Information and Communication Tools (ICT) be used effectively by an international product development team? In order to get some 'objective' answers we measured the 'view' of the student teams on several important aspects in the context of multi-side working (see Table 1).

Table 1 Measured issues of multi-side working.

1	*Customer focus*	This is about getting clear on needs, expectations, and priorities of those who receive the work. Pulling together an understanding of client needs and priorities is not easy for virtual teams and demands the utmost clarity of communication.
2	*Direction*	Direction defines the unique contribution of the team, from its broadest purposes to its specific actions and activities. When a team is dispersed getting this alignment and commitment is a real challenge.
3	*Understanding*	Understanding deals with learning from team members. It is more difficult to really understand fellow team members when using ICT tools. Understanding is also about being fully aware of the constraints and difficulties that involve multi-site working.
4	*Accountability*	This is the process of mutually agreeing on what results the team is expected to achieve with specific plans and activities, and a sense for how the team will be responsible to all involved organisations and to one another.
5	*Communication services*	The communication services cover the addressed exchange of information (in all kinds of formats). The word *interaction* (the ability to quickly react on one another) characterises communication services.
6	*Co-ordination services*	Co-ordination services focus on the way a (product creation) project has been organised, i.e. its working methods and procedures. *Sharing* (easy access to the right information) is the key concept here.
7	*Usage and Way of Working (WoW)*	It is not enough to have communication and co-ordination services in place! This key area measures how well the ICT solution is really used by the team. Does everyone know how to use it properly? Is sufficient support available? Etc.
8	*Attitude*	This is about the team's awareness, perception, eagerness, knowledge, etc. of today's ICT possibilities for dispersed (international) product development teams.

All these issues were covered by two questionnaires which we sent around (via Email) in the course of the project to all members of both teams. The first questionnaire concerned the soft issues (1 - 4) and was labelled the "Team Fitness Meter"; the second questionnaire "ICT Index" dealt with the more tangible issues of collaborative working (5 - 8).[1]

Per issue approximately 10 questions were posed. All questions were posed in the form of propositions. All students participating in the collaborative projects were requested to indicate to what degree they agree with each proposition.
The results of the questionnaire are summarised in Table 2. The overall response was 81% (9 out of 11 project members) for the **LFM** team and 71% (7 out of 9 project members) for the **LEFON** team.

Table 2 Results of the questionnaires

	LFM USA	**LFM** Netherlands	**LFM** Germany	**LEFON** USA	**LEFON** Netherlands
Customer focus	5.6	5.2	5.4	5.3	7.2
Direction	6.8	5.6	6.1	5.5	7.2
Understanding	6.3	5.2	6.5	5.6	7.0
Accountability	5.8	4.8	5.5	4.9	6.3
Communication services	5.4	5.0	5.4	5.4	6.1
Co-ordination services	4.4	5.5	5.8	6.0	6.2
Usage and WoW	5.0	4.5	5.4	5.5	6.4
Attitude	7.7	5.6	6.7	6.2	7.7

The ratings (x) in Table 2 must be interpreted as follows:
- $x < 4.5$: Infancy level, i.e. insufficient means for collaborative engineering.
- $4.5 < x < 6.5$: Beginner level, just started to build-up know-how.
- $x > 6.5$: Expert level, i.e. well-prepared for doing multi-side projects.

At first sight the ratings in Table 2 do not show much variation. Both teams are on beginners level. The average score of the **LFM** team ($\bar{x} = 5.6$) is somewhat lower than that of the **LEFON** team ($\bar{x} = 6.1$). Perhaps this simply reflects the fact that working together with 3 parties is more difficult than working together with 2 parties. Further the Dutch part of the **LEFON** team is well above average; with ($\bar{x} = 6.8$) this group even claims to be on expert level.

However, a closer look at the statistics reveals some interesting findings. First of all the above-average rating for the issue "Attitude" ($\bar{x} = 6.8$) clearly suggests that all students were highly motivated to experiment in international collaborative-engineering type of projects. Moreover, it can also be seen that the issues dealing with "Accountability" ($\bar{x} = 5.5$) and especially the "Way of Working" ($\bar{x} = 5.4$) are the most weak ones.

A detailed look at the questions concerning the "Accountability" issue shows the following main problems:
- The role of each team member is not clear, that is, there is unnecessary duplication of efforts or there are things falling through the cracks.
- Not all team members strive to live up to previously defined operating agreements.
- There is no mechanism of how to deal with failure to live up to these operating agreements.

[1] The Team Fitness Meter was a modified version of the one used by Henry & Hartzler [8] ; the ICT Index was developed during this project.

The low score of the "Way of Working" issue is caused by questions dealing with security and authentication. May be this is in the industry of more importance than in an educational environment. Also, difficulties in the usage of shared workspaces are mentioned, in particular configuration management (who's working on which version) and authorisation roles (viewing rights versus edit rights) are not well implemented. All these findings comply with remarks and statements that are made in previous sections of this paper

5. WORKSHOP AS AN ALTERNATIVE TOOL

For both teams the project leaders were Dutch. So, for the Dutch team members we experimented with a workshop titled "Multi-side Project Management". This workshop was held halfway the development project. In the workshop an atmosphere was created that enabled a critical reflection on "how things went till now". It appeared that such a workshop helped in team building and solving potential conflicts. The result of this workshop was that a number of action points and operating agreements (e.g. concerning the planning) were not only formulated but also prioritised. Of course, it is a pity that the German and American students were not involved.

Because of the potential benefits (it really accelerated and focused the development effort) it is recommended to organise in future projects such a workshop in an international setting, e.g. by using high-quality video conferencing. The real challenge is to reach the high level of interactivity and the 'openness' that is requested for this with the aid of electronic means.

6. EXPERIENCES

Each team had varying degrees of success meeting the project objectives. All members gained valuable experience in the process of global collaboration. The team members were frustrated by the limitations of the Internet to really communicate as opposed to just exchanging information. Trust and team cohesion remained low throughout the project.

All three teams carried out a market research. The German team also focussed on a perception research on what type of light has to be chosen and where to put the lamp nearby the TV set to have the best possible comfort and quality in watching TV. All three teams were able to complete some aspect of a prototype as shown below. It is not surprising that the single site **GFJ Integration** team completed the project on time and on budget. This team was able to take advantage of the other teams' marketing research and managed to meet two to three times each week and to be very productive. The international teams were lucky to have only one productive team meeting each week.

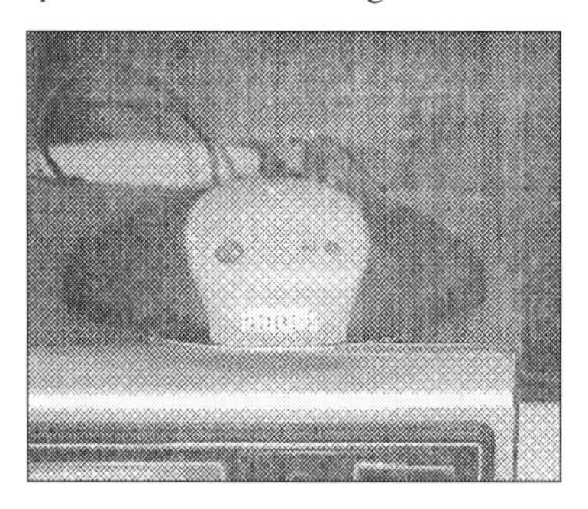

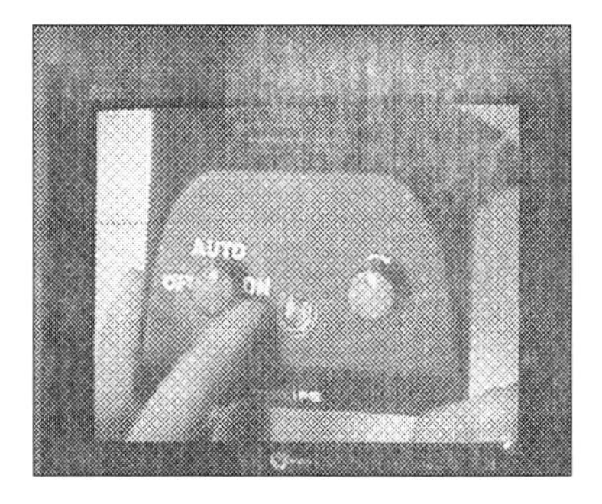

Fig. 1 Two prototypes of a switching light control box

The **GFJ Integration** team's prototype (Fig.1 , left) integrates well into Philips' product aesthetics while maintaining full design functionality. Team **LFM** International produced the compact design seen on the right. Team **LEFON** created a prototype controller package and the German team also created a prototype for the lamp and the light.

The Dutch students also developed a handbook as a guide for IPD-CE groups in future. This handbook will help groups with the use of the BSCW software, the internet videoconference software, MS project planning software and has a checklist for starting up new IPD-CE projects [9].

7. CONCLUSIONS

We experimented with modern communication technology in order to find useful tools for facilitating the co-operative work and the contacts of all the participants. Program evaluation focused on the technologies used, the processes involved, the nature of human interactions and the effectiveness of communications as well as the product are produced.

It turned out that group cohesion stayed low (students did not meet in real life), and that the Internet is not mature enough yet for desktop video conferencing. Chatting and Email were in this project by far the most important communication media. We also found out that the use of a Computer Support for Co-operative Work (CSCW) server is a possibility for information interchange in addition to Email attachments. The server can also be used as an electronic project archive. In future we expect the CSCW-server will be an essential tool for project support and traceability. Improvement must come from a better project organisation and a better team understanding of the deliverables and milestones earlier in the project phase. The use of MS project planning software is recommended. An improvement of video conferencing via the Internet could enhance the group performance enormously. Otherwise the (expensive) V-span ISDN video conference is necessary at the start up, the milestones, the workshop and the final presentation of the project.

Collaborative engineering is relatively a new approach to develop products. So it is not surprising that there are still many problems, which are new and unfamiliar to designers and engineers as seen in our project. As a result, it is strongly felt that a project must be well structured before its start. This implies that it is very crucial to first establish organisational matters and communication means before a project starts. Otherwise, the work carried out by each geographically dispersed team would be isolated and wouldn't properly interface with the parts developed at the other locations. In this way it is impossible to achieve the fastest product development cycle. In the worst-case scenario, the project might become "world-wide over-the-wall" engineering rather than collaborative, concurrent engineering. These problems should not be ignored and reflected on the next project.

Despite the problems encountered, the authors believe that the experience for the team members was worth the effort and therefore another project will be attempted in the fall 2000, with other students, but with the same universities and coaches.

REFERENCES

(1) Turino, Jon, *Managing concurrent engineering*, Van Nostrand Reinhold, New York, 1992.

(2) Peter A.M. van Kollenburg, George Punt *"Education in Concurrent Engineering is a Must"*, Proceedings of the 2nd International Symposium on Tools and Methods for Concurrent Engineering (TMCE98), Manchester, April 1998.

(3) Lehigh University Annual Industry Advisory Board meeting December 1999
(4) Ochs, J.B. proposal to National Aeronautics and Space Administration (NASA) via the Community of Agile Partners in Education (CAPE), April, 1999
(5) "Product Design and Development," Karl Ulrich and Steven Eppinger, Irwin McGraw-Hill, ISBN 0-07-116993-8
(6) Peter van Kollenburg, Herbert Veenstra, Dick v. Schenk Brill, Harald Ihle, Krijn Kater, *Integrated product development and experiences of communication in education,* Proceedings of the 3rd Int. Symposium TMCE2000, Delft, The Netherlands, April 2000.
(7) BSCW – GMD Research Center in Darmstadt (http://bscw.gmd.de/)
(8) Meg Hartzler, Jane E. Henry (1994). A How-To Manual for Building a Winning Work Team. ASQ Quality Press, Milwaukee, Wisconsin.
(9) S. Bonifacio, C. Treffers, R. Joosten, A. Rippe, P. Copier, P.I Sturme, R. Domburg, R. v/d Borne (1999), International Projects, manual for a successful project, Fontys University of Professional Education, The Netherlands

CSCW in industry and education – cross sectoral lessons

A I THOMSON, S P MACGREGOR, and **W J ION**
DMEM, University of Strathclyde, UK

ABSTRACT

This paper describes the findings of three recently completed research programs conducted at the University of Strathclyde. Each of the projects involves the implementation and usage of Computer Supported Co-operative Working (CSCW) technology within the design process. Two of the projects described are education based whilst the other was conducted in collaboration with industry based on live projects. The paper presents the findings of each of the projects and then explores opportunities for cross-sectoral lessons.

1. INTRODUCTION AND BACKGROUND

In recent years, extensive research and experimentation has been conducted in the use of Computer Supported Collaborative Working (CSCW) tools within both the industrial and educational arena [1-8, 9-14]. Opportunities exist for lessons to be learned through the identification of similarities and differences in the findings of such research. Allowing, recognition of generic findings and possible transferable lessons. This paper presents the findings of three recent research projects carried out at the University of Strathclyde. Two of the projects, ICON [9, 12] & ICON2 [10, 11, 12] investigated the usage of CSCW tools by disparate engineering design students. Whilst, a Design Council funded research project investigated the introduction and usage of shared workspace technology within the design process of three companies and their supply chain [16,17]. This paper presents an overview of the approach and findings of the ICON and the Design Council funded "Integration of Design Specialists Through Shared Workspaces" projects. This is followed by a comparative reflection of the results allowing identification of generic findings together with opportunities for cross-sectoral lessons.

2. ICON

2.1 The ICON concept

ICON (Institutional Collaboration Over Networks) began in June 1997 with a week-long collaborative design project involving four pairs of students. The students came from the Product Design Engineering course at Strathclyde University and the same course run jointly between Glasgow School of Art and Glasgow University. The main objective the ICON project being to ascertain the practicality of facilitating Internet based remote collaborative design projects for students. The students tackled different design briefs and were asked to provide a solution and present their findings at the end of the week. They were restricted to the use of network technologies such as audio/video conferencing and chat tools and were prohibited to communicate by any other means. Although difficulties were encountered through the week, the project was considered a success in that it proved virtual collaborative design projects were feasible. The difficulties encountered were also examined and resultant changes made to the system for the second project, ICON2, which ran for eight days in September 1998. This iterative process, illustrated in figure 1, resulted in modifications being made to the system in terms of the methodological approach and project organisation, in the pursuit of ensuring future success. By identifying the barriers to effective communication and work, benefits of adopting CSCW in a design context can be maximised and remedial strategies and guidelines can be developed for implementation by the design community.

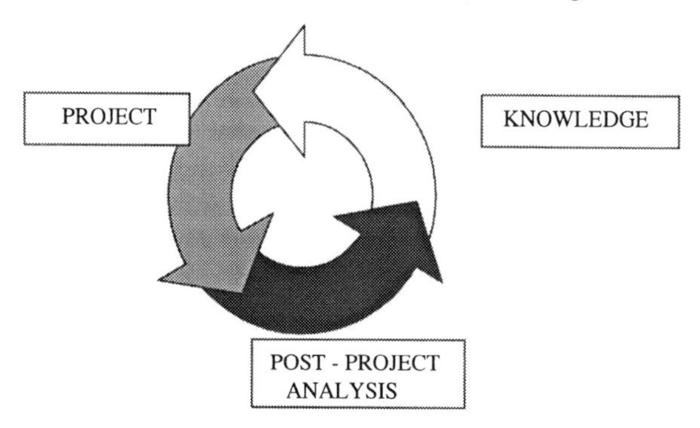

Fig. 1 Iterative development process

The underlying philosophy to the ICON projects involved improving the accessibility CSCW tools in order that policy makers in academia could implement similar projects easily, efficiently and with as little as possible start up. This included using as much freeware as possible and avoidance of ISDN.

2.2 Methodology

In the six weeks prior to the first ICON project three briefing sessions were arranged to introduce students to the project methodology, the technology and participating staff and students. Evaluation of the project took place through the implementation of pre/post project questionnaires and interviews, project diaries and video recordings of the final presentations. The design interface for the project came in the form of a project website incorporating design briefs, project schedules and technical support. The briefs, which were disclosed on the morning of the first project day varied for each team and were allocated randomly. True to the ICON philosophy participating students were restricted to using the tools and technologies made available to them. The computers used for ICON were cross platform PC/Mac, due to the existing facilities at each institution. CuSeeMe provided a single video link between the institutions with audio conferencing being provided for all teams. Microsoft office, BSCW (Basic Support for Co-operative Work), Netscape Navigator, Paintshop Pro, Adobe Photoshop, AutoCAD, and Peoplesize were provided to aid the completion of the briefs.

2.3 Findings

In general, the first ICON project was a success. Although technical difficulties were encountered throughout the week the students enjoyed the experience and reported a significant improvement in their computer competence. The students were content with the project, mainly due to low levels of expectation. Despite the positivity participants did not envisage such methods of working as constituting a replacement for conventional practices. Perceptions and levels of expectation were found to be central to the success of implementing such projects and this aspect will be discussed later in the paper. Participating students made good use of the tools, BSCW being the most successful. The use of audio was found to be a critical factor in effective communication, concurring with other research [13]. The following guidelines were distilled from the first ICON project. These were used as the basis for designing the system for ICON2 project:

• Agree on the technologies in advance and ensure that they can be used at each site;
• Wherever possible, standardise the hardware and software. Many difficulties were experienced in this project because communications technologies had to be cross platform i.e. PC & Mac;
• Allow enough time to set up and test the technologies. Attention to details such as student shared workspace registration and configuring email addresses on machines to be used for the project may also be necessary;
• Provide a variety of technologies so that if one communication channel fails another may be adopted;
• Allow students time to familiarise themselves with the technology before the project commences;
• Ensure staff with the necessary expertise are available to assist students having technical problems;
• Be realistic about what can be achieved. Eight students, four from each institution, was considered a manageable size for this project.

3. ICON2

3.1 ICON2

Similar to ICON, ICON2 involved the partnership of four pairs of students from each academic institution. Three teams were restricted to the use of network technologies to complete the brief whilst, a control team was also nominated who could meet as they wished and were allowed to use any means of communication with the notable exception of audio and video conferencing. The project lasted for eight days and was split into two main phases. Phase 1 comprised a sacrificial project that allowed the students to get used to the technology at the same time as conducting research for phase 2. The required deliverables from this part of the project comprised a Product Design Specification and a Theme Board. Phase 2 started with the participating students being presented with a design brief. Each team was to develop a product to meet the brief and to present their chosen concept using a CAD produced product layout drawing together with a rendered presentation graphic. The Clyde Virtual Design Studio (CVDS) [14], which was developed in the wake of the first ICON project, was used as the basis for tackling the brief. The CVDS integrated the set of tools/facilities that were required for the successful completion of the project. The CVDS consists of four main components:

Data Management - *storage space for project work:* In the form of BSCW which was downloaded free from the Internet and configured for use with the CVDS.
Communications Suite - *local audio, video, chat and whiteboard facilities:* Microsoft Net Meeting, Netscape Communicator and Ewgie chat were provided.

Local Applications - *relevant software programs:* AutoCAD R13, 3D Studio MAX, Microsoft Office, Paintshop Pro, Adobe Photoshop and PeopleSize were provided.
Reference Area - *General and project specific links and information.*

As with the first ICON project an effort was made to record as much information as possible. For ICON2 the following methods were employed:

- Daily project videos;
- Pre/post project questionnaires;
- Support staff logs;
- On-line diaries;
- One to one post project interviews;
- Presentation video.

3.2 Findings

The main findings on ICON2 can be classified in three areas as follows:

System and Tool Usage

There is much debate with regards to the effectiveness of synchronous V's asynchronous working practices [2, 6 & 9]. The overall split for all the teams in ICON2 was 68% asynchronous V's 32% synchronous. This level of synchronous work is relatively and may be attributed to the fact that the project was of short duration and that the students didn't know one another personally. Therefore, a high level of synchronous work was required in order to ensure the design was moving in the right direction, especially with tight deadlines looming. Other factors may include the ability of the tools to facilitate synchronous communication and the perseverance of the students, an attitude that was, perhaps, expected of them. Each of the three networked teams displayed very different patterns in communicating. Fig. 2 shows how team 3 collaborated over the course of the project.

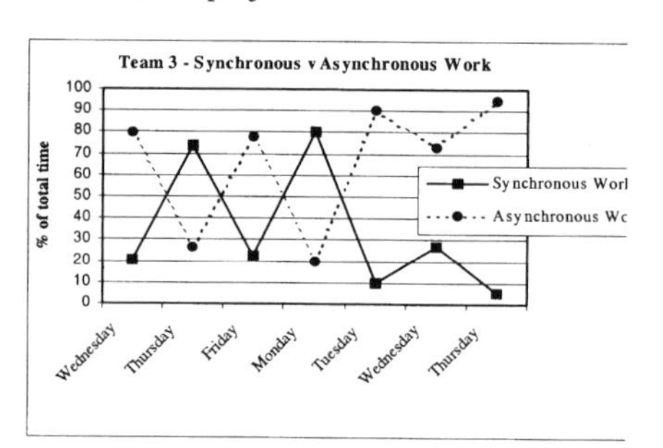

Fig. 2 Team 3 Daily Synchronous/Asynchronous split

Work Modes and Habits

From examination of the personal log and on-line diaries it became apparent that the participants generally, displayed the same work mode at each of the various stages. By examining this it may be possible to maximise benefits and ease teething trouble in future projects. To this end, figure 3 is proposed. This figure is basically a summary of the work practices and attitudes of the ICON2 participants. The students picked up the technology relatively quickly and used only what they found useful after a while. Other results included a decrease in the need for technical support although most students agreed that constant support would always be welcomed. One aspect of the model that could be improved is the participants' tendency to become less tolerant as the project goes on. As one evaluator mentioned, " . . . *the more it goes well, the more annoying it is when the odd thing goes wrong.*" These frustrations can be attributed, partly, to the tight time-scale of the project.

Additional training and experience of using such a system should result in minimal frustrations and less detraction from the design process. We may not be able to change the fact that participants become less tolerant as the project, but if we can manage the project effectively this may not matter as much. Additionally, other measures such as on-line tutorials may substitute the need for software support.

Stage	Induction	Familiarisation	Expert
Attitude	Keen	Content	Frustrated/ Stressed
Tools Usage	Using a wide variety	Narrowing usage	Whatever appropriate to deliverables
Attitude to tools	*"open to anything"*	*"open to what works"*	*"nothing works"*
Level of back-up	Tools instruction needed	Software support only	Moderate software support

Fig. 3 Work-modes and Attitudes During Icon2

Benefits
The main benefits of the ICON2 project can be summarised as follows:

• A steep learning curve for students resulted in significant improvements in their computer skill and confidence levels;
• Greater experience and knowledge for staff for eventual input into a new system design;
• Widening of scope of students knowledge and experience through working with students from another institution;
• Strengthening of inter-institutional links;
• Student perception of certain elements of the design process being improved through practical project;

Barriers
The barriers encountered in ICON2 can be categorised as:

• Technology
• Educational Issues
• Social/Cultural
• Project Management
• Psychological/physiological

For a detailed discussion regarding the barriers, refer to [15].

3.3 Conclusions/Recommendations

Analysis of findings allows certain guidelines to be produced for future virtual collaborative design projects.

- A project manager or champion should be appointed who should be aware of the issues and possible outputs of the project;
- Any additional input to projects should be cleared by the project manager and its effects closely considered;
- Future collaborative design projects should closely resemble conventional semester long modules so that the technology is the only differing factor in the project;
- Students should be trained adequately in all software that is to be used in project. At least an introduction to the packages should be completed;
- Participating students should be trained in adopting the correct behaviour for such projects;
- Periodic contact between colleague(s) should be encouraged to ensure working towards a common goal;
- Depending on the specifics of the teaching methodology consideration should be made to letting design partners' meet physically before and/or after the project. Working on the start of the project in person and presenting findings afterwards are options;
- Encourage high level synchronous collaboration after long periods of non-contact. In a conventional project this will usually mean the start and end of sessions and the start of important phases;
- Materials should be developed which can support the detail design phase of the design project including online specifications and design methods;
- Design briefs should be similar but not identical to encourage the correct levels of inter-institutional and collocated collaboration;
- On site support at participating institutions is advised for the start of any project until students become familiar with technology;
- Software support should be considered for stages of the project when students are required to produce an output. Online support is an option;
- Adequate space should be provided for students to carry out other activities at their workstation, such as sketching and writing;
- An awareness of stages of behaviour in project participants should be evident in support staff;
- An awareness of possible effects of different personalities and how to respond is also advised;
- Students should take appropriate exercise to combat any physiological problems associated with long periods in front of a computer;

Another important need for such projects is that of *setting up* future projects, ensuring that any preparation is completed as economically and effectively as possible. The following stages have been identified:

IDENTIFICATION - *methodology, resources*
INSTALLATION - *technology*
PREPARATION - *briefs, students*
RUNNING & MONITORING - *support*
EVALUATION - *input to next iteration*

4. INTEGRATION OF DESIGN SPECIALISTS THROUGH SHARED WORKSPACES

4.1 Introduction and Background

This recently completed research, funded by the UK Design Council, investigated the introduction and usage of shared workspace technology within the design process of a number of small and medium sized enterprises (SME's).

Within the context of this paper the term shared workspace, has been used to describe a computer based collaborative working system typically consisting of the following functionality:

- video and data conferencing ;
- real time application sharing ;
- shared whiteboards;
- file transfer.

Development and use of such technology to date has been dominated by large multinational companies. Ford, for example, have used the latest collaborative technologies to allow their seven design centres, each of which specialises in different aspects of design, to communicate effectively across great distances and different time zones. Benefits demonstrated by such projects include:

- improvements in the flow of work allows companies to move and react faster;
- product development lead time and costs are reduced while maintaining or improving quality;
- time to market is reduced;
- relationships and efficiency of communications throughout the supply chain.

It is clear that current desktop data and video conferencing technology offers the possibility for companies to collaborate effectively and at relatively low cost over networks using personal computers. Furthermore, the benefits that collaborative working technologies can bring to the new product development process are apparent. However, their widespread use has been restricted by a number of organisational and technological issues. The main aim of this research is to address the key issues relating to the implementation and adoption of these technologies within companies and their supply chain.

4.2 Project Approach

The research approach adopted is best described as a series of industrial case studies involving a number of companies from a range of industries specifically Product Design, Construction and Electronics Manufacture.The general methodology adopted within each of the companies was to run successive case studies each building upon and testing the findings of the previous. Therefore, each case study followed a different methodology focusing on slightly different aspects. The first company case study commenced in May 1997 with a series of trials being carried out in Hulley & Kirkwood a mechanical and electrical building services consultancy. This was followed by a product design company, Devpro starting in December 1997. Finally, the Keltek electronics case study began in June 1998.

A variety of data collection methods each aimed at capturing specific types of information were devised and employed during the case studies specifically:

- Initial Structured Interview
- Post Demonstration Questionnaire
- Diary of Observations:
- On-Line Logging System
- Weekly Questionnaires:
- Final Interview.

4.3 Findings

The results of this research can be classified within the following areas:

- Barriers to the introduction and usage of the shared workspace;
- Typical system usage;
- Benefits obtained through system usage;
- Perceptions of the company throughout the introduction and usage of the shared workspace;

Barriers to the Introduction and Use of Shared Workspaces

Barriers identified during the introduction and use of the shared workspace can be classified under the following main areas:

- Management;
- Psychological / Perceptions;
- Technology Related;
- Training.

A full description of each of the barriers identified is provided in [16]

Benefits of Shared Workspace Usage

The main benefits achieved within the case study companies can be summarised as:

- Reduction in the time taken to carry out a variety of design activities due to less re-work, ambiguity, file transferring and paper chasing;
- Improved design quality;
- The companies design costs are reduced therefore, reducing the cost of the services to their clients giving them a market advantage over their competitors;
- Due to improved communication clients invest less time on design, reducing their costs further;
- Companies adopting the technology feel they are getting a better response from remote parties using the shared workspace than they would get adopting a conventional communication tools.

Typical System Usage

In general, the shared workspace technology was not used to replace travel. Despite the fact that prior to the introduction of the technology companies felt one of the greatest benefits they would achieve would be a reduction in frequency of travel and savings relating to this. In reality, there were certain activities that each of the companies felt could not be carried out remotely therefore, the reduction in travel and related costs was found to be negligible. Typically, the system was used to enhance design activities that were previously carried out using conventional asynchronous communication media such as the application of e-mail, telephone, fax etc. to carry out the following activities:

- introducing and discussing design changes;
- clarifying design details;
- presenting designs to clients for approval;
- discussing project progress.

The shared workspace was used differently in each of the participating companies. Hulley and Kirkwood initially found it difficult to co-ordinate times for synchronous use of the system. After a few months of barely using the system they devised a method of employing the system in an asynchronous manner, which was found to be advantageous. In contrast, Keltek adopted the shared workspace in a synchronous manner from commencement of usage. The main difference between the companies being that the key system users in Keltek were predominantly located at their desk whilst, the key system users adopting the system asynchronously spent a considerable percentage of their time out of the office.

For synchronous use application sharing was found to be the most useful tool, being employed almost 100% of the time. On the other hand, Hulley and Kirkwood who adopted an asynchronous mode of working found the whiteboard to be an extremely useful tool employing it more than 90% of the time. Usage of the shared workspace went through 'cycles' i.e. short periods when it was used synchronously almost on a daily basis followed directly by longer periods when it may not be used for several months whilst designers work alone of with co-located team members.

Company Perceptions

Company perceptions changed dramatically throughout the project. Initially, prior to the introduction of the technology all of the companies were enthusiastic at the prospect of using the shared workspace. Initial impressions were that the shared workspace would have a positive impact on communication between remote design team members. Preconceived benefits include:

- Reduced travel, savings in flight tickets and a reduction in the time spent travelling;
- Improved communication within the distributed design team;
- Closer working relationships within the company and their supply chain;
- Better quality products;
- Reduction in the time taken to execute interactive processes, fewer redesigns.

Once the system was introduced in each of the companies initial enthusiasm was thwarted by a number of barriers, primarily relating to:

- The location of the shared workspace;
- Confidence in using the system;
- Technical issues;
- Fire fighting.

Although over 90% of prospective system users had initially, expressed an enthusiastic interest in using the shared workspace when the system was installed less than five percent of them retained this enthusiasm. In instances where individuals overcame initial barriers and used the system to their advantage it was found that initial enthusiasm returned. The companies who adopted the system within their design process have either purchased addition systems or have made plans to do so in the near future.

Guidelines

The main output of this research was the development of guidelines for the effective introduction and usage of the shared workspaces within the design process. These guidelines follow five basic stages:

1. Recognition of need;
2. Preparation;
3. Plan training
4. Introduction
5. Adoption

5. COMPARITIVE FINDINGS AND CONCLUSIONS

5.1 Common areas

Through comparing the findings for the educational and industrial projects described in this paper it is clear there are areas of similarity between the project findings. These are discussed in the following paragraphs.

***Barriers*:**

The majority of barriers identified in both projects can clearly be classified under common headings, specifically:

Technology;
Social/ cultural;
Management;
Psychological/physiological;
Training.

Usage:

• In both the educational and industrial based projects it was found that the technology was employed in different ways by each "team";

• Majority of communication still asynchronous despite user preconceptions that synchronous communication would increase dramatically. This finding is common to other research projects [5].

• Industrial case studies show that users go through short phases of intensive synchronous usage followed directly by longer periods of asynchronous communication. A similar pattern is apparent in the ICON2 project illustrated in figure 2;

• Both projects show that CSCW tools cannot replace face to face meetings, in their present form;

Technology:

• A lack of sufficient training/preparation can lead to a severe lack of confidence in the system and ultimately, non-use;

• Good quality audio was found to be absolutely critical in both educational and industrial usage with video taking the role of a secondary communication link. This result is common to other research findings [13];

• Both projects show that a lack of hardware / software standardisation can render the system unusable;

Management:

• As CSCW involves the collaboration of people and institutions conflicting objectives can sometimes mean that ventures fail to become reality. Both the educational and industrial projects showed that effective management is crucial to compromising on different agendas such as policy, timetabling and resources;

• A project champion is essential in both industrial and academic projects to ensure successful management of the system.

5.2 Transference of Lessons Between Sectors

The majority of findings from both projects are common. However, both projects provide scope for lessons to be transferred from industry to education and vice versa.

Industry to Education:

• More barriers were identified in the industrial based projects, due mainly to the almost artificial "sheltered" nature of the ICON projects. These additional barriers have been developed in to guidelines which could prove useful in future educational projects;

• Industrial case studies can be used as stand alone teaching material within educational environments in order to provide students with a realistic overview of "the real world";

Education to Industry:

• Both ICON projects were well managed and prepared for well in advance of commencing. As a result, system usage was smoother than within the companies. Industrial case studies show that companies are keen to commence usage in order to achieve the perceived benefits and tend to gloss over the preparation and management stage, often to the detriment of successful technology implementation;

• In ICON2 a variety of technologies were provided to ensure a back-up was available in case of failure of other communication media. In addition, the continuous availability of technical support eased potential problems. This approach would prove beneficial in industry where key system users become very frustrated when communication technology fails, often becoming annoyed to the point where they may not use the system again ;

• The ICON projects showed that students adopted high levels of synchronous work to cope with short project deadlines. Results from industry showed key system users tended to "back off" employing technology when tight deadlines were looming often resorting to conventional asynchronous modes which are more time consuming.

• Throughout the ICON projects, particular attention was paid to the learning process that the students went through in the course of the project, both in terms of the new technology and the core material. As CSCW technologies are new to most people in education and industry, the latter can learn from the former in methods that maximise quick and efficient uptake of the new systems - all engage in the learning process.

5.3 Conclusions

It is evident from the research findings presented in this paper that lessons can be transferred between education and industry sectors. Furthermore, common guidelines can be developed in the form of procedural stages to facilitate the successful implementation and usage of CSCW technology within both arenas.

REFERENCES

(1) C.E.Siemieniuch & M. Sinclair, "Real-time collaboration in design engineering: an expensive fantasy or affordable reality ?, Behaviour & Information Technology", Vol. 18, No. 5, 361-371, 1999.

(2) S. Nidamarthi, R.H. Allen, S.P. Regalla & R.D.Sriram, Observations from multidisciplinary, internet-based collaboration on a practical design project", International Conference On Engineering Design ICED 99 Munich, P709-714, August 24-26,1999

(3) Sandkuhl, K & F, Fuchs-Kittowski, "Telecooperation in decentralized organisations: conclusions based on empirical research, Behaviour & Information Technology", Vol. 18, No. 5, 339- 347, 1999.

(4) R, Fructer, 1996. "Interdisciplinary Communication Medium in Support of Syncronous and Asyncronous Collaborative Design". *Proceedings of the First International Conference of Information Technology in Civil and Structural Engineering Design (ITCSED).* University of Strathclyde, Glasgow. 14-16th August.

(5) A. May, C. Carter, S.M.Joyner, W. McAllister, A. Meftah, P. Perrot, P. Pascarella, H. Chodura, M. Doublier, P. Carpenter, P. Caruso, C.Doran, V D'Andreas, P. Foster, J. Pennington, V. Sleeman & R. Savage, "Team based european automotive manufacturing: final results of demonstrator evaluation" (Rover/ Team/ WP5/DRR007) Loughborough, Leics LE11 1RG, UK: HUSAT Research Institute, Loughborough University, pp1-75

(6) A.Milne & L.Leifer, "The Ecology of Innovation in Engineering Design", International Conference On Engineering Design ICED 99 Munich, Vol 2, P935-940 August 24-26,1999

(7) Petrie, C. "Madefast" Home Page. http://www.madefast.org/

(8) "Suppliers and Manufacturers in Automotive Collaboration (SMAC)" Home Page, http:// greenfinch. analysis. co. uk /race /p17/present/smac/default.htm

(9) M. Sclater, N. Sclater, L. Campbell, "ICON: evaluating collaborative technologies", Active Learning, vol 7, CTI, Dec 1997.

(10) S. P. MacGregor, W. J. Ion, "Introducing and developing virtual design environments - an effective platform for collaborative design projects in academia", Proceedings of EDE '99, University of Strathclyde, September 1999, pp243-247.

(11) S. P. MacGregor, "Investigation and development of guidelines for the implementation of collaborative virtual design projects in academia", MEng dissertation, DMEM, University of Strathclyde, 1999.

(12) L. Campbell, I. Ali-MacLachlan, A. I. Thomson, W. J. Ion, A. S. MacDonald, "Institutional collaboration over networks - ICON, A comparison of two collaborative design projects", Proceedings of CADE '99, University of Teeside, April 1999.

(13) J.S.Kirschman & J.S. Greenstein, "The use of computer supported cooperative work applications in student engineering design teams: matching tools to tasks", International Conference On Engineering Design ICED 99 Munich, P1299-1302, August 24-26,1999

(14) W. J. Ion, A. I. Thomson, D. J. Mailer, "Development and evaluation of a virtual design studio", Proceedings of EDE '99, University of Strathclyde, September 1999, pp163-172.

(15) N. Sclater, H. Grierson, W. J. Ion, S. P. MacGregor, "Online collaborative design projects: overcoming barriers to communication", to appear in special issue on virtual universities in the International Journal of Engineering Education, Summer 2000.

(16) A.I.Thomson & W.J.Ion, "Integration of Design Specialists Through Shared Workspaces" Design Council Final Report,

(17) A.I.Thomson & W.J.Ion, "Introducing and Using Shared Workspaces in the Design Process: From Perceptions to Effective Widespread Usage, International Conference On Engineering Design ICED 99 Munich, P1829-1832, August 24-26,1999

Internet learning

M HOWARTH
Faculty of Technology, Bolton Institute, UK

ABSTRACT

Bolton Institute is a University Sector provider situated in the North West of England. The Institute has over 3000 full-time students studying a wide variety of courses. The Faculty of Technology is based on the main campus in the centre of Bolton and is a leader in Remote Access Learning. Courses from Foundation to Postgraduate can be studied in a range of disciplines such as Product Design, Automobile & Mechanical Engineering, Electronics, Leisure Computing and Textiles. The Faculty is home to a centre of excellence in Information Technology and was recently awarded over £1 million to develop courses to be taught using the World Wide Web. Many of the courses are accredited by professional institutions. This presentation will discuss the use of the Internet for the delivery of graduate and postgraduate courses.

1. BACKGROUND

Educators have seen a real growth in the use of information sources and learning materials using the computer and associated software. Many higher education institutions are positively embracing the technology for learning, available on the Internet. Indeed some would argue that it is the correct and only way forward for the next millennium.

Remote Access Learning, previously known as Distance Learning has become the buzz in many circles, and Financial Managers see the use of such methodology as a means to drive for further efficiencies within the HE system.

The paper will seek to argue the advantages of such provision, by reviewing a pioneering project taking place at Bolton Institute. It will also look at the future and some of the possible advantages of embarking fully on this methodology of learning.

2. THE BEGINNING

The rapid advance in computing technology, and its availability, allows educators to use the medium to help advance and professionalise their subject knowledge and presentation.

A beginning saw the use of simple programmes in laboratory experiments, followed by bought-out software and increasing use of CD-ROM. The development of the World Wide Web has opened up a vast amount of information and new methodologies, which can help the learning process and also develop the necessary computing skills required in the technology age.

Distance learning has been around for many years. Open University degrees, universities and other HE providers have developed their own distance learning courses since the 1960's, as well as private organisations who specialise in training as opposed to education.

3. THE AMI PROJECT

The Faculty of Technology, along with the University of Northumbria, has pioneered one of the UK's first courses at postgraduate level to be delivered purely over the Internet. The MSc Advanced Microelectronics for Industrialists (AMI) fully embraces the new technologies available. Key features:-

- New concept in delivering a postgraduate course.
- Teaching materials delivered across the Internet.
- Direct communication with specialist tutors via e-mail.
- Access to industry standard design tools.
- Modules approved by both Academics and Industrialists.
- State-of-the-art computer hardware provided to students.

The UK government has set out principles for the learning age which seek to ensure the following:-

- Investment in learning to benefit everyone.
- Lifting barriers to learning.
- Putting people first.
- Sharing responsibility with employers, employees and the community.
- Achieving world-class standards and value for money.

Quite clearly, the use of Internet learning will help towards the learning age. One of the major benefits to be gained from Internet-based learning, is that it can be provided at a time and place to suit the individual or organisation without disruption tot he day-to-day operation of the business. The flexibility of being able to go on-line at all times has to be a major step forward in learning provision.

Many of the participants in the AMI programme are seeking to update skills and to comply with the wishes of professional bodies for Continued Professional Development (CPD).

The AMI programme has received funding of £0.5 million to develop the learning materials. The money has been generally allocated to relieve academic staff and employ Technical Authors.

4. STUDENT'S VIEW

The programme allows the student to:-

- Access all study material on the Internet.
- Study at home or place of work.
- Access CAD tools. (First of its kind).
- Be continually assessed by submission of assignment work.
- Learn at their own rate.
- Acquire a postgraduate MSc qualification.

5. COURSE STRUCTURE AND OPERATION

The course is structured to be delivered over a period of three years, with years 1 & 2 coursework and year 3 a major project.

Each student is supplied with:-

- High specification PC.
- Subscription to an Internet Service Provider.
- ISDN communication network.
- Software Windows NT, Microsoft Office 97, X Windows emulation, Intel Pro-Share video conferencing.

6. INFRASTRUCTURE

fig 1LINK http://www.bolton.ac.uk/staff/mh7/servers.gif

Each of the taught modules comprises of:-

- Formal study – chapters downloaded from Internet books – journals – interactive problem solving.
- Directed Learning – study of material from other websites and library resources.
- Self-Assessment Questions – specimen answers available on web site.
- Assignments.
- Projects.

Support is available to all students via:-

- E-mail.
- Video conferencing.
- Telephone.
- Worked solution to self-assessment questions.
- One-to-one contact, with pre-arranged meetings with tutors.
- Contact with other students via e-mail.

7. COURSE FEES

- £840 per 10 credit module, plus a registration fee of £60.
- Faculty registration of £940 as a contribution to the PC.
- Total charge for full course £9520.
- Bursaries, grants may be available.

8. NEW INITIATIVES IN DESIGN

The Faculty operates a highly successful course in Computer Aided Product Development. This MSc is presently fully funded by the European Union. The course will be developed during 1999-2000, to be offered on the same basis as the AMI programme. Some of the learning materials are presently available on the web site:-

http://www.bolton.ac.uk/staff/mh7

The course operates full-time over one year, and offers specialisations in:-

- Computer Aided Modelling.
- Product Development Management.
- Creative Studies.
- Major Project (student choice).

Optional modules are available in:-

- Manufacturing.
- Materials.
- CAD.
- Finite Element Analysis.

Market research has shown that Industrial and Product Designers need to have the necessary computing skills, together with the more traditional methodologies such as Sketching, Rendering and Model Making. The course has seen students graduate into a number of design related positions such as Jaguar, Ford and Consultancy.

The initiative, under the guidance of the author, will seek to enrol students from both the UK and overseas. Difficulties may be encountered in ISDN provision and Service Providers, but the Faculty remains confident of success.

9. CAD DEVELOPMENT WORK

Major development work is taking place to allow remote access to CAD modelling software. There is a strong need for education and training in the are of 3D modelling and Finite Element Analysis. The Faculty will soon be in a position to benchmark access to in-house software which is licensed from SDRC. Employees and potential students will be able to access the Faculty server and run the software in real time from home and at work. The learning materials will consist of guidance notes and specific examples together with dedicated assignments to be completed by the student or trainee. Examples of the types of visuals can be seen in Diagrams 2-9 below.

Diagram 2 Building a simple model

Diagram 3 Setting the view

Diagram 4 The finished model

Diagram 5 Generating a plastic bottle

Diagram 6 Setting the lofted shape

Diagram 7 Lofting the bottle

Diagram 8 Final rendered view of the bottle

Diagram 9 Setting a machining cycle

Diagram 10 A Detail Drawing

Diagram 11 An exploded view

Diagram 12 A prosthetic Hand

Diagrams are links to a web site and the author will demo on the day of presentation.

Further work on animation will take place to run through exercises and use of Real Player or similar tools.

10. CAUTIONARY NOTE

The author feels, along with other academics, that Remote Access Learning should not be seen as a replacement for the traditional attendance at a university, where the interaction of people is so important. The development of young people into adulthood remains a key, to attendance at university, where the important communication skills and recognition of individual character from an identifying stamp on a person's life. The vast majority of students who attend university remain committed to the ideal, and the use of new technologies should be seen as a welcome addition to the completion of a graduate and postgraduate education.

NOTE: This is an internet based presentation and reliant on the protocols being set up to show a live demo of remotely accessing Bolton Institute CAD software. The paper is presented as background to the live event. If the protocols are not ready for the day then the author will refer delegates to appropriate web based information.

Delegates should visit web site for a preview:
www.bolton.ac.uk/staff/mh7/internetbrighton.html

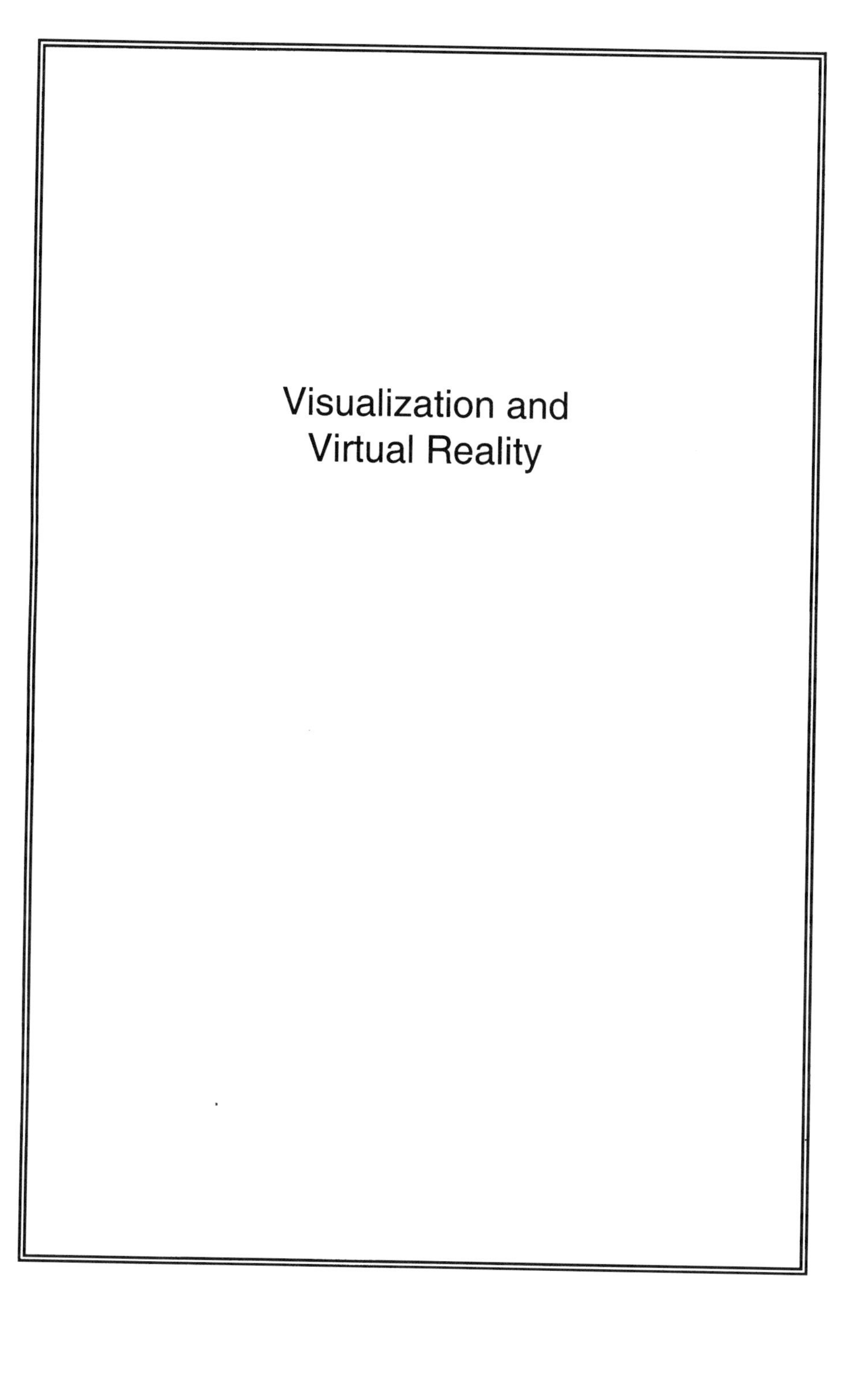

Visualization and Virtual Reality

The Virtual Design Educator: realistic or optimistic? An approach to teaching design virtually

E JAY, M WHITE, and **P F LISTER**
Centre for VLSI and Computer Graphics, University of Sussex, UK

SYNOPSIS

In the age of global communication, gathering information is becoming easier and easier by the day. In a teaching context the problem does not lie in getting information but in how to teach with it. Multimedia tools provide interactivity and often speed up the learning process. Newer tools enable the user to be totally immersed in a three-dimensional world. These *virtual worlds* were first introduced to the design community as rapid prototyping tools or Virtual Prototyping tools. In this paper we propose a view of how virtual worlds could be used as pedagogical tools. Our approach comes at a very high level, which encompass the whole design process, and will include off-the-shelf technologies.

1. OVERVIEW OF STATE OF THE ART DESIGN TECHNOLOGIES AND METHODOLOGIES

This paper is designed to show the potential of Virtual Reality technology in the context of design education and industrial design. Our aim is to show that a Virtual Environment (VE) can be used to teach design either generically or within a specific domain e.g. engineering. We adopt a top-down (1) approach whereby we define a generic design model and show that most existing methodologies can be derived from this model.

This section is an overview of design methodologies and technologies currently used in industry and by the teaching community. Note that design will be addressed in its most generic sense with some occasional examples.

1.1 Design methodologies

Design methodologies try to define and relate the different activities and tasks, designers may use to define a design process. These activities are often formalised and express aims and objectives at a specific stage of the design. These methodologies are usually designed with a particular emphasis on one part of the design. For example, creative methods put the emphasis on the development of creative thinking and will therefore tend to extend all phases involving a creative process. Total Quality Design focuses on customer need, and introduces design interaction within the team early in the process. In other words, the success of a design methodology often depends on several factors such as team cohesion, interaction.

" Design is an activity that shapes its objects- create their forms- in accordance with the goals or purpose of those objects"

This definition (2) identifies key aspects of a design process. Design is an activity that needs:

- Identification of requirements
- Definition of goals, purposes and properties
- Specification of a work environment
- Definition of an entity, or entities

These issues are usually well embodied in every design methodology. They contribute to the success of any design, and constitute key parts of any teaching method.

1.1.1 Common attributes to design methodologies

Industrial design often follows an in-house methodology that is a variant of a generic methodology and addresses the particular needs of a market at a particular time. Here, we identify the factors that will influence a design, regardless of the methodology used. This will help us define the key areas that need to be focussed on for teaching in our virtual approach.

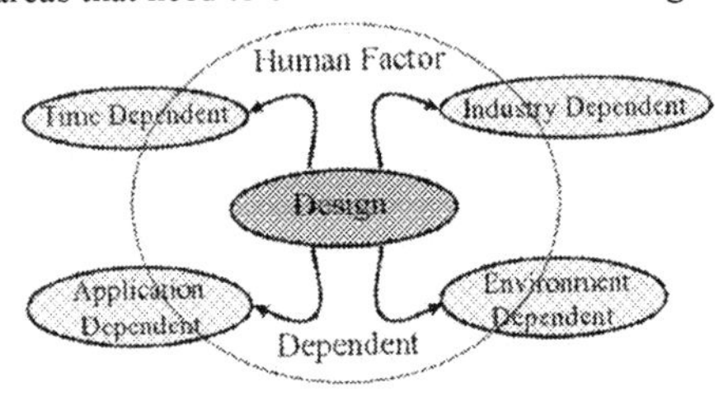

Fig. 1 Generic Design Dependencies

Futuristic and old-fashioned design may fail because of inadequacies in their functionality or complexity. A design will depend on several key factors (see Fig. 1). A design is:

- **Time dependent**: it has to fulfil specific needs at a particular moment in time with the available technology, and within cost.
- **Industry dependent**: it is very much different for each industry and has to fulfil the needs of a specific market.
- **Environment dependent**: in an educational context this attribute will relate to the teaching method used e.g. studio-based teaching (3), and also to the resources used. In an industrial context "environment" can relate to the place of work, but also to a strictly environmental constraint imposed on the design, e.g. the construction and mechanical industries. Moreover environment also refers to the milieu of the team.
- **Application dependent:** the application is what the design is "designed" for.
- **Human Factor dependent:** human factors influence everything we do and will therefore influence all of the previous dependency attributes.

1.1.2 A generic design model

The generic design model we will use throughout this paper is shown in Fig. 2.

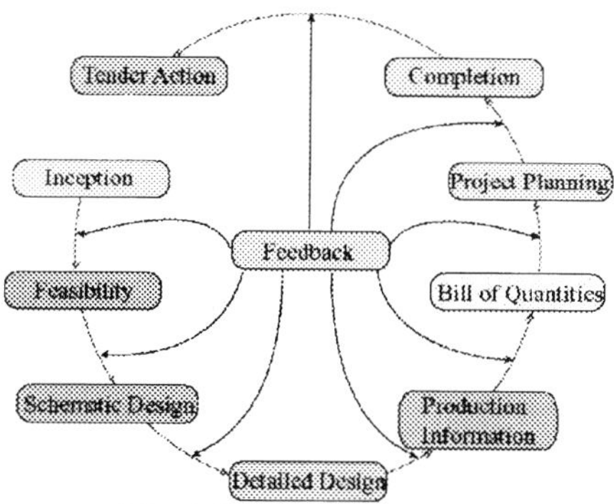

Fig. 2 Generic design flow

A great emphasis is put on continuous feedback at all stages. This model is flexible enough to encompass most existing design methodologies. It provides the student with a global view of design by treating each module as a defined black box. It is moreover, essential in our approach, because every design procedure will be treated as a variant of this model. The nine phases of this design flow are:

1. *Inception*: definition of client (company, individual or market) requirements and needs.
2. *Feasibility*: exchange of ideas between the design team members on the possible solution to the problem posed. This could consist of brainstorming.
3. *Schematic Design*: definition of rough sketches as a first solution to the problem.
4. *Detailed Design*: feedback enables designers to draw/prototype an advanced solution to the problem.
5. *Production information*: definition of drawings/prototype aimed at the production team, and based on their procedures.
6. *Bill of Quantities*: defining the nature of materials to be used.
7. *Project Planning*: define the order of the construction/building phases, specify deadlines, using charts (Gantt, Pert).
8. *Completion*: writing of the final report containing all outcomes of the previous phases.
9. *Tender Action*: interaction and contracting of external entities.

A key part of any design process is to define objectives, aims and purposes. This definition process requires a high level of interaction between team members at an early stage. Members, who will almost certainly constitute a cross-disciplinary team, need to share thoughts, expertise and build on a set of requirements based on market needs. Constant feedback from clients, potential users and third parties will help during this process.

1.2 Advanced design technologies

Design technologies are often used during the detailed design stage, which can consist of CAD (Computer Aided Design) or other computer based technique. Three-dimensional CAD tools e.g. AutoCad or Design Workshop, are mainly used for their prototyping function:

- *CAD*: has been used for many years by designers and consists of a software based 3D graphics engine that enable designers to build and prototype objects.
- *Augmented Reality*: is an approach that is half way between reality and virtual reality. It consists of superposing computer-generated images onto real objects by using special glasses. This technology is being used by Boeing on their construction site (7). Designers

could benefit from this technology in the sense that they could "virtually" add value to an object and test its aesthetic properties in real time.

Fig. 3 Virtual Reality at Boeing (Source: The Boeing Company)

- *Hollographic Design*: can enable the designers to physically see a life size model of their new design. New advances in holographic design have produced impressive results (5).
- *Virtual Testing*: is used to test concept and product virtually, prior to manufacturing (10).
- *Rapid Prototyping*: is an advanced technology recently introduced to designers. It consists of a software-based engine that enables designers to quickly model ideas and concept prior to manufacturing.
- *Cybernetics*: technology could one day be used by designers to aid the rapid prototyping phase, or indeed when coupled with artificial intelligence could aid the creation process.
- *Very high-level languages*: are mainly used for hardware design purposes, and enables designers to define hardware components with a very high level of abstraction.
- *Internet*: is a wonderful tool for learning purposes, and is therefore significant in an educational context.

1.3 Teaching design

Design is one of the most difficult subjects to teach and is often taught as part of another course e.g. engineering. Teachers structure their lectures around particular applications. Design education is now focussed around a more technology-centred approach as opposed to a human-centred one. The danger posed by technology-centred approaches has been outlined in (6). This approach may result in too much attention being focused on how to solve a problem rather than on how to define it. The effect of technology on design education is determined by the way it is introduced in design courses. A quick, unsubtle, introduction of technology into an already overcrowded curriculum may well slow the teaching process, and design education. Technology needs to be an integral part of the course for its use to be perceived in a positive way. Virtual technologies may be the solution to such issues. Teachers and students will be part of the same virtual experience from the beginning of the design learning process. Inside this environment they will interact, communicate and design. An approach to using VE in the classroom is already being tested (8). This approach called MUD (Multi-User Domain) has shown great potential but it has proven difficult to keep order within the environment. A mix between human-centred and technology centred approaches implies that design educators should teach with an industrial perspective including, as appropriate, industrial practitioners. Teachers should then introduce interactive schemes enabling the design team to interact with other teams e.g. the construction team.

2. THE VIRTUAL DESIGN EDUCATOR (VDE)

Section 1 identified a generic design methodology, as well as novel technologies that may help designers during some phases of the design. One key generic dependency defined in section 1.1.1, is the environment related effect on the design. This attribute now corresponds to a virtual environment. Virtual Environments (VE) are three-dimensional (3D) computer-generated scenes in which individuals exist as computer-generated objects known as avatars, and communicate in a virtual reality context through haptic interfaces (n-dimensional human-computer interface). These are therefore totally immersive or can be made semi-immersive for cost reasons. The potentials of VEs are enormous, especially in an educational context. This section explains the whole concept of the VDE, and shows how it can be used in a design education context as well as in an industrial design context. The VDE is *not* an Internet virtual classroom or a virtual video conferencing tool. It encompasses many existing technologies. Note that at present, the VDE is still a concept idea and has not yet been implemented practically.

2.1 Definition of the VDE

The VDE is a design integration platform, which brings all the participating design entities to a common level. The non-exhaustive set of requirements for such environment is as follows.

2.1.1 Educational requirements

In VDE terms we have a *design manager* and a *design team* made of designers and most probably other professionals. The educational requirements are as follows:

- Simplicity of use
- Use of pre-loaded design examples e.g. industrial designs
- Design by immediate practice
- Development of team and leadership skills
- Full interaction between the design team
- Possibility of interaction with the design e.g. changing the design map (see Fig. 4)
- Interaction between the design manager and the design team (students)

2.1.2 Technological requirements

A virtual tool aimed at education and design should be simple to use. The technological requirements are:

- Advanced n-DOF (Degree Of Freedom) haptic interface enabling total control of phases 3 and 4 (see page 3)
- Complex underlying procedures are transparent to the user and can be seen as black boxes
- Hierarchical relationships between avatars
- Full interaction between all members of the design team
- No geographical constraints
- Should use off the shelf components to reduce costs
- Modularity and realism

Design can be a very abstract idea, which evolves, and results in a very practical solution for a particular problem. In an educational context this makes design one of the most difficult subjects to teach.

2.2 Design of the VDE

The VDE focuses on all the phases of the design and not only on scheme and detailed design. It is not only a Virtual Prototyping tool.

2.2.1 VDE Design map

The design configuration presented in Fig. 2 is mapped into the VDE design plan manager to give a design plan (see Fig. 4). This allows the design manager to define important characteristics of each phases of the design flow. These are represented as cylinders. This schematic enables visualisation of essential properties of each design task in 3D, and can be customised. This concept can easily be extended to the whole management of the project, and can support multiple design representation. Indeed a design plan can have several degrees of abstraction, and can be used to represent progress, resource allocation, and could be used to dynamically re-allocate resources. It is also an important concept to facilitate design re-use.

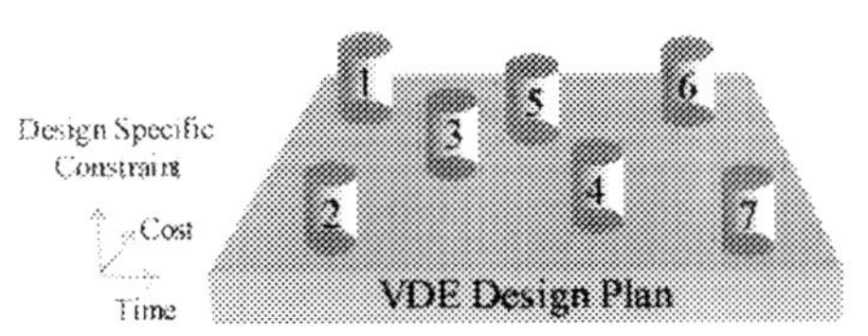

Fig. 4 VDE Design Plan

For educational purposes the design map is an interesting concept in that it enables students to see the impact of a particular phase on the whole design. The modularity of this map will enable them to modify the structure of the design interactively and visualise its consequences. In an industrial design context it will facilitate definition of milestones.

2.2.2 The VDE system and technology

The head of the design team, the design manager (see Fig. 5), is the major person that takes decision and analyses feedback. His or her role is to take into account each entity of the design team, who will communicate their information via the Internet or an intranet and through a VDE sub-terminal. The design manager then takes decisions and may or may not decide to update the design that sits on the VDE main server. Only she has constant read/write access to the VDE main server. Each VDE sub-terminal has constant read-only access to the VDE server. There are no geographical constraints imposed on this system and VDE sub-terminals can interact with each other. As phases 1 and 2 of the design procedure consist of brainstorming and verbal interaction, they will be implemented in the VDE as voice e-mail or video conferencing. As the users will be immersed in the VDE, this involves the view-dependent mapping of video streams onto virtual walls. Phases 3 and 4 will be implemented in such a way that they provide the user with maximum flexibility to translate thoughts and concepts into objects. A volume modeller allows the user to interactively model volume elements and 3D deformable surfaces using a haptic interface e.g. a glove. A reshaping tool will enable constant refinement of the design and enable the definition and implementation of properties. The team responsible for the aesthetics will generate their first version, present it to the design manager who in turn will make it available the rest of the team. The advantage of this procedure is that it enables each team to communicate their work as they do it, and can therefore speedup design. Testing the design in context can be done using *application scenarios*. These are virtual sub-environments within the VDE that are used to test, simulate and model the behaviour of a specific design in a particular pre-defined context or scenario. DODE or Domain Oriented Design Environments (9,11) is a model that allows users to define, refine and reformulate a problem by constant interaction between domain specific experts and software expert. It is very appropriate for learning in a design education context. This enables a design environment that is tied to the needs of designers for a specific problem. "Seeding, evolutionary growth and reseeding" allows constant interaction and incremental evolution of the environment during its design. This makes such environment totally flexible toward tasks that have not been thought of by the creators of the design environment. The Toolbelt (12) model is based on a DODE and is built from off-the-shelf components. It raises the need for design models supporting multiple design representations and their interdependencies. Our approach is very much a virtual extension of the DODE, which gives sufficient flexibility for automatic update of multiple design representation. Indeed, our design plan enables multiple design methodologies that can be interdependent and parallel. The generic design methodology chosen enables this representation as each phase can be seen as a domain. The sub terminals are domain specific entity, which works toward the development of the design and the environment.

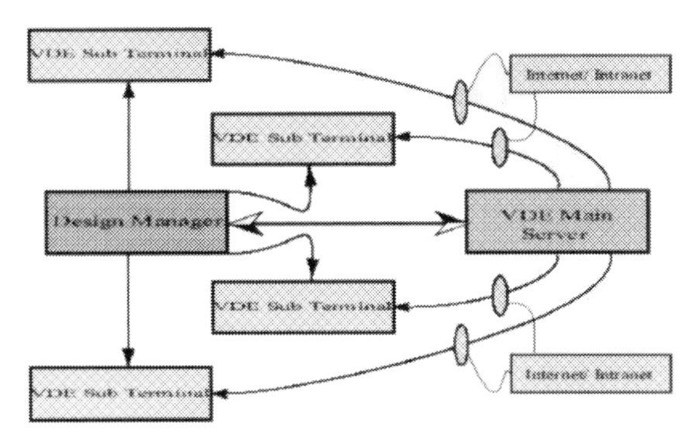

Fig. 5 The VDE Interaction System

2.3 Pro and cons of the VDE

Few schools and universities will have the financial support to accommodate a fully immersive virtual environment, typically consisting of a CAVE (CAVE Automatic Virtual Environment). A CAVE implementation of the VDE is therefore much more feasible in an industrial environment. When used in a teaching environment the VDE may require a phase of training and especially for student use of haptic devices. These tools can be difficult to use and much research is currently in progress to make them more human-like.

The VDE does not replace in any way the teacher and forces the members of the team to communicate from the beginning of the design. Interaction is paramount in this environment. It enables multiple structural approaches to design that can be modified. This is particularly interesting in a design education context to show how designs are influenced by the methodology used. Moreover the VDE can be built from existing technology despite a full implementation requiring the use of advanced tools e.g. a volume modeller. Finally, the VDE offers industrial designers the possibility of doing "distance designing" while still having the feeling that everything is done in-house.

3. REAL LIFE EXAMPLE APPLICATIONS OF THE VDE

A typical design study laboratory using the VDE will consist of students, a lecturer and a network of computer (or a CAVE) and some haptic devices. Students will register in the environment as design team members and create a profile defining their role inside the team. They will communicate with each other inside the environment through avatars. Brainstorming sessions to define the problem may take place inside the environment or outside. Depending on their role in the design each sub-team will use different tools of the VDE. In the specific context of electronic design, it is important to have the opportunity to test several designs. For example, a motherboard requires the integration of many experts to design each component, and also to integrate them on the board. Each team of experts will provide individual contribution on their own specific components. In such a complex application a change in one component may induce a series of dramatic changes for the other teams. Our approach can cope with such changes in that in the first phase of the design the design manager could define several design plans into which several design attempts can evolve in parallel.

4. CONCLUSION: REALISTIC OR OPTIMISTIC?

The concept of the VDE is realisable using mostly technology currently available. Some research still needs to be done in the area of the volume modeller in order to be able to model surfaces in real-time. It enables total control of the design and the realisation of application scenes and design maps. It does not replace the teacher but rather puts her in a different position where she can still provide the students with all the information they need, and give them virtual assignments. The VDE is therefore realistic in term of education but slightly optimistic in terms of the technology to provide real-time design.

REFERENCES

(1) Jacques Giard, "Industrial Design Education: incompatibility with education in art and architecture", IDSA Design Education Conference Proceedings, 1999, IDSA Design Education Conference Proceedings, 1999.

(2) Akiyama K., "Function Analysis: Systematic Improvement of Quality and Performance", Productivity Press Inc., Cambridge MA, 1991.

(3) Jim Budd, Surya Vanka, Andy Runton, ""idesign" Collaboration in the Digital Design Studio "The Global Design Community at Your Fingertips"", IDSA Design Education Conference Proceedings, 1999.

(4) "Research Opportunities in Engineering Design", NSF Strategic Planning Workshop Final Report, April 1996, http://enws121.eas.asu.edu/events/NSF/report.html

(5) Zebra Imaging web site: http://www.zebraimaging.com/

(6) William Bullock, Lorraine Justice, "A Future Role for Research Universities", IDSA Design Education Conference Proceedings, 1999, IDSA Design Education Conference Proceedings, 1999, http://www.idsa.org/whatsnew/99ed_proceed/paper006.htm

(7) Boeing Company web site: http://www.boeing.com/assocproducts/art/tech_focus.html

(8) Tari Lin Fanderclai, "MUDs in Education: New Environments, New Pedagogies", Computer-Mediated Communication Magazine, Vol. 2, no. 1, p. 8, January 1, 1995.

(9) Gerhard Fisher, "Domain-Oriented Design Environments: Supporting Individual and Society Creativity", http://www.cs.colorado.edu/~gerhard/papers/cmcd-99.pdf

(10) Sankar Jayaram, Yong Wang, Uma Jayaram, Kevin Lyons, Peter Hart, "A Virtual Assembly Design Environment", IEEE Computer Graphics & Applications, Vol. 19, No. 6, November/December 1999.

(11) Gerhard Fisher, Ray McCall, Jonathan Ostwald, Brent Reeves and Frank Shipman, "Seeding, Evolutionary Growth and Reseeding: Supporting the Incremental Development of Design Environment", Human Factors in Computing Systems, CHI'94 Conference Proceedings 1994, pp 292-298.

(12) Tamara Sumner, "The High-Tech Toolbelt: A Study of Designers in the Workplace", CHI'95 Conference Proceedings 1995.

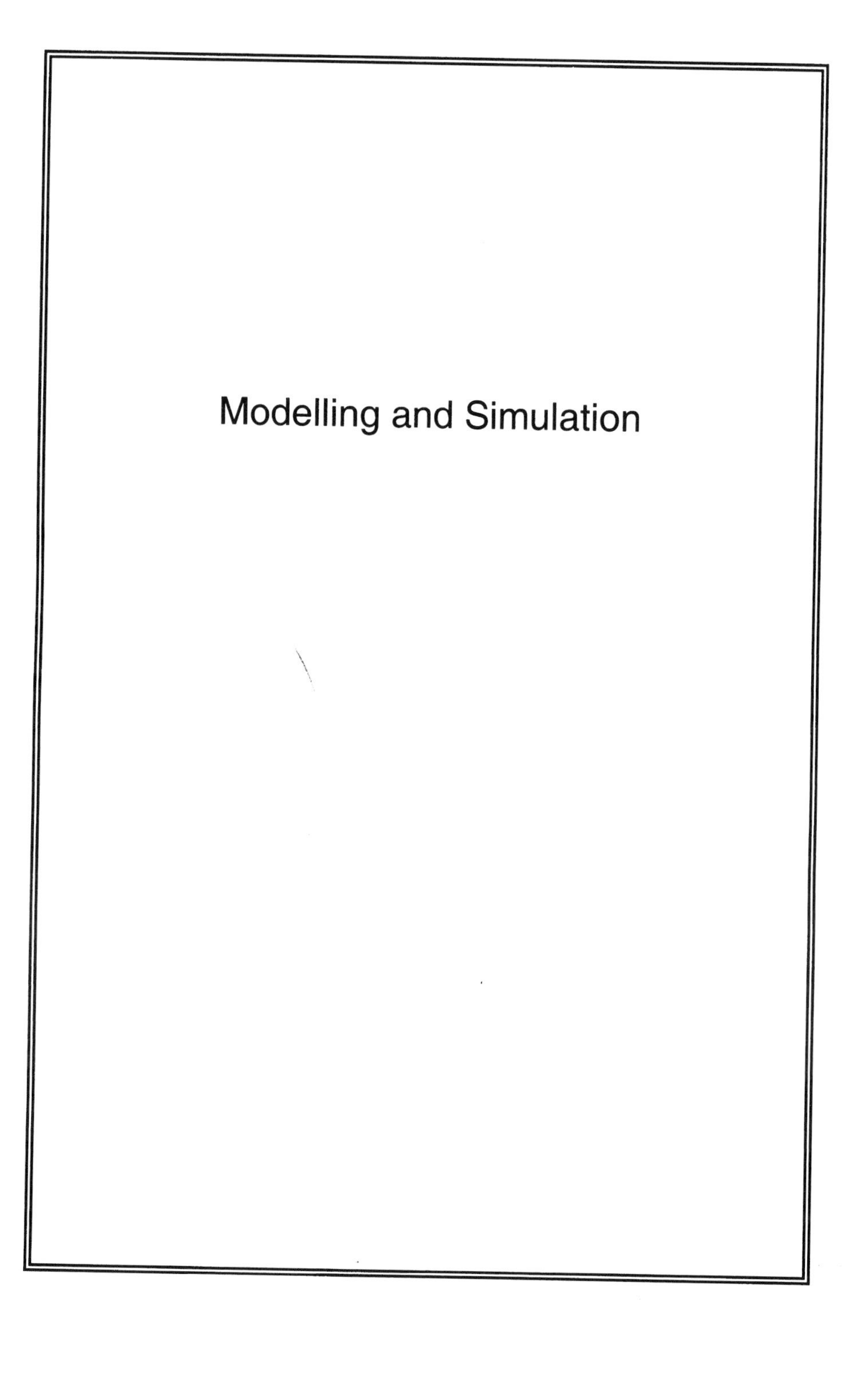

Modelling and Simulation

Modelling and design intent

W F GAUGHRAN
Department of Manufacturing and Operations Engineering, University of Limerick, Ireland

ABSTRACT

Computer modelling software endeavours to generate technically accurate and sometimes photo-realistic models of design intent. Significant time investment is devoted to becoming proficient in the use of one or more CAD software packages, in the design, production and modification of products from concept to manufacture. Creatively using CAD software, depends greatly on the users cognitive ability to visualise the design intent and to interact with the developing model of the product. Research shows that the cognitive modelling capacity of the engineer has a significant impact on the process, and allows the designer to build 'cognitive rapid-prototypes'. This ability can be developed by appropriate intervention strategies. At the University of Limerick, interactive tutorial software in conjunction with other strategies have been developed to improve the users spatial modelling capacity. Significant improvements are evidenced, in the users spatial ability and visualisation skills - resulting in greater creativity and more productive use of CAD modelling systems as a design tool.

Key Words: *Design Intent, Visualisation, Cognitive Modelling, Creativity*

1. DESIGN VISUALISATION

One simple dictionary definition of design is: 'to fashion after a plan' or 'an outline sketch of something to be executed or constructed (Wester) Another is: 'a preliminary plan or sketch for the making or production of....' (Oxford). While the foregoing all have to do with designing they omit a fundamental element of design; that is to create something that has not existed before. The 'definition' covers a wide range of activities from writing, to bridge design, to sculpting, as well as applying to many other creative people in society.

It has been said that *'design is the essence of engineering'* (Deiter, 1991). Design is such a diverse activity that it is difficult to assign to it a concise definition and generally it is generically described. However, the following definition is quite satisfactory in relation to design in engineering: *'design establishes and defines solutions to and pertinent structures for problems not solved before, or new solutions to problems which have previously been solved in a different way'*. (Blumrich, 1970). This definition gives a very broad brief to the engineer.

There are numerous models (e.g. Fig. 1) of design processes but whichever approach is taken the *visualisation skills* of the individuals involved will figure highly in bringing a 'product' from concept to successful manufacture.

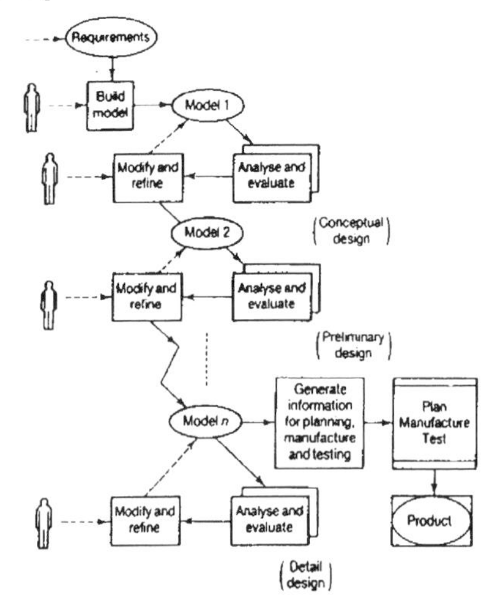

As can be seen in Ohsuga's model the act of modelling itself plays a vital part in the design of any product. These models can in turn have a variety of forms, such as mathematical models, two or three-dimensional CAD models, parametric models, scaled or full-size tactile models and so on.

The 'externalisation' of the model requires that refinement and modification be an ongoing part of the process. As an example Opel/Vauxhall Motors in developing the 'Astra' made no fewer than twenty eight one-third scale models from countless original sketches and computer models. The model is used to define as well as refine the designer's intent.

Fig. 1 Ohsuga's design process model

Would better visualisation skills, the ability to build more accurate cognitive models, have compressed the process and saved a great deal of money and very valuable time? There is no doubt as to the value of a concrete model. It clarifies the design intent and refines the original ideas. It interacts not just with the visual senses but also with the tactile. Figure 2 shows an example from Opel/Vauxhall Motors.

Fig. 2 Scale Models refine the Cognitive Model and visa-versa

2. BUILDING COGNITIVE MODELS

The cognitive model is at the core of creative activity. *The cognitive model, is the mind's image of an artifact, system or product.* It may be generically described as: *the ability to build and manipulate images and ideas* so as to define, refine and communicate design problems, ideas and solutions. It is important that each modelling stage should be as accurate and as close as possible to the end objective. The machine will only do what the computer directs. But what of the quality of the computer model – how close is it to the design intent and how clear is that intent in the mind of the design engineer? The CAD model will have taken considerable time and effort to generate and will require modification and development of the initial representation. Can this time be compressed and can the initial models be closer to the 'ideal' at an earlier stage?. Minimising human intervention in the production process is generally advantageous but the input of human intelligence and particularly human creativity is essential at the embryonic stage of development or designing. This is why the *cognitive model* must precede any other system or method of modelling. Contributing to the building of the cognitive model are abilities such as spatial abilities, graphicacy, graphic ideation and overall visualisation skills.

The cognitive model allows the designer to analyse and interrogate the initial concept as it finds expression through the freehand sketch or is cognitively manipulated in other ways.

3. DEFINING THE DESIGN PROBLEM

Following the cognitive model a simple or outline sketch may be used to externalise the initial idea. This sketch is in itself a 'model'. To express or 'model' a problem, even when it may be poorly defined, the act of drawing or outline sketching is usually employed. In *'ideas sketching'* the lack of any clear detail may imply a lack of understanding on the designers part or it may be deliberate so as not to clearly define or to focus too early on a single solution or perhaps on a definition. This allows the designer to explore the possibilities not only in relation to the problem solution but also of the problem itself. This exploratory strategy is not new – it has been employed for centuries. Many of DaVinci's design ideas are examples of this exploratory strategy (see Fig 3).

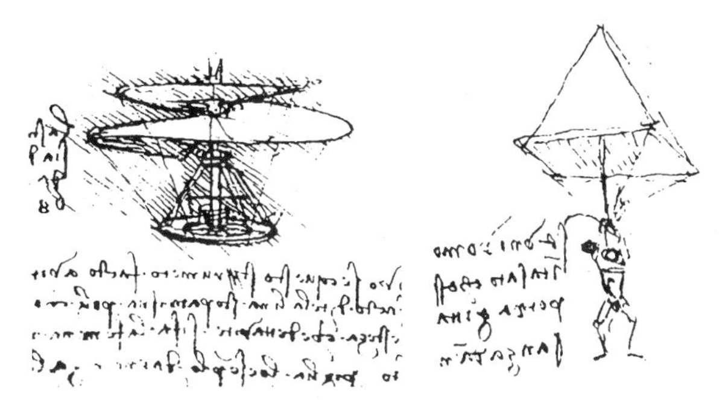

Fig. 3 DaVinci explored his ideas through graphic ideation

A characteristic of such outline drawings and sketches is that they will leave vague parts of the problem or solution which are not of immediate concern. They may also be vague because the designer is struggling to define or endeavouring to shape the overall problem or its solution. The sketch is therefore a vital element at the early stages of designing, it represents an external model of the cognitive image, it is the first visible model of either the problem or its solution. *The sketch is representation of the cognitive prototype.*

4. GRAPHICACY AND MODELLING

The ability to draw and to encode the meanings of drawings is termed 'graphicacy'. Graphicacy is defined as: *'the ability to encode and communicate graphical information'* (Gaughran 1996, 1997). As part of a design strategy there are a number of stages in modelling through drawings. The initial stage is the expression of the embryonic image of the idea through outline sketching. This activity is described as *'graphic ideation'*, (as illustrated in Figure 3 and the modern example in Figure 4), and Mc Kim 1980, suggests that this is: *'the activity of using drawing and sketching to assist the thinking process'*. This externalisation acts to stimulate 'cognitive ideation', which is the capacity of the mind to form, entertain and relate ideas.

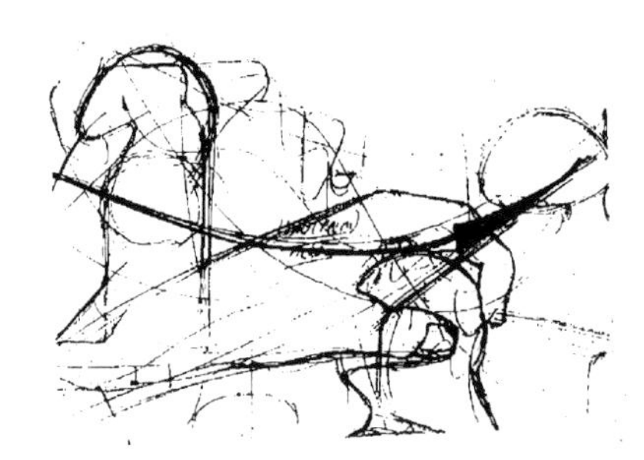

Fig. 4 Exploring ideas through 'graphic ideation'

The graphics in acting as a stimulus to building the cognitive model become more refined and detailed as the initial idea or problem is developed and focused. The graphic ideation stage should allow ideas to flow and images to be built and interrogated. This requires a further capability unique to human intellect namely *'spatial ability'*

5. COGNITIVE MODELLING AND SPATIAL ABILITIES

The graphical or cognitive model is concerned with representing entities in space, either two or three-dimensional. In order to visualise or represent spatial elements effectively it is necessary to understand spatial relationships and their functions. This cognitive activity has been identified as *'spatial ability'* and has a close association with the skills of graphicacy. It also is the underpinning and necessary prerequisite skill which will allow cognitive models to be built and developed. Spatial ability may be defined as: *the ability to visualise, manipulate and interrelate real or imaginary visual images.*

Drawing or graphics, whether freehand, formal, or computer generated, deal with the representation of spatial elements. This graphics activity helps to develop a 'graphic intelligence' which is closely related to spatial intelligence.

Gaughran, (1990, 1996, 1998) identified five hierarchical sub-factors of spatial ability viz: *image holding and comparing, planar rotation, orientation, kinetic imagery and dynamic imagery.* Each and sometimes all of these sub-factors will be brought into operation as required in the drawing or design problem to be solved or expressed. In designing intervention strategies to develop cognitive modelling abilities the Gaughranian sub-division of spatial ability into the five sub-factors above, may be further sub-divided according to factor-of-difficulty, e.g. basic, intermediate and advanced.

Individuals with a developed sense of spatial ability will perform at a significantly higher level in, engineering, design and in technology and science. (Lord, 1984; Smith, 1964; McGee, 1979; Gaughran, 1996). This is also supported by the performance of left-handers in the areas of visual/spatial memory and intelligence where they appear to have an advantage over their right-hand counterparts. Left-handedness appears to indicate right brain dominance.

6. TAKING COGNIZANCE OF LATERALISATION

The ability to visualise or spatialise systems or configurations is in nearly all cases a right-cerebral function. Both the left and right hemispheres of the brain contribute to a given task. The mathematician with visualisation ability will be better at math and the engineering designer with this ability will be a better designer. Research suggests that the cerebral dual-processing model is best suited to designing and that particular elements of a process will call for left-right shift in arriving at a design solution. Anita Cross suggests that there is a thinking peculiar to design which is clearly linked to non verbal codes which designers acquire and use.

The designing of tutorial media is complex in that it requires taking cognizance of the variety of intellectual levels and the type of cognitive processes involved. In 1990 the author designed and tested a series of computer based tutorials to determine whether this mode was effective. While the positive results confirm that there is cerebral specialisation in connection with visuo-spatial processing, the significant evidence of cerebral lateralisation is quite interesting.

A cohort of about 120 students were involved of which 12.5% were left-handed. The complete cohort undertook the same pre-test and were later divided for tutorial sessions. They were given exactly the same series of tests on the post-test and the improved score of the control half of the cohort was used as the 'familiarity-factor' and subtracted from the raw scores of the experimental group. This produced an average improvement in score of just under 7% for the experimental group and this, when subjected to statistical analyses proved to be significant.

Points of note:

- In the pre-test the sinistrals scored 12% better than the dextrals
- In the post-test the sinistral's score improvement was nearly double that of the dextrals.
- The top seven scores in the experimental cohort scored above 75% - three of these were sinistrals.

Recent studies at the University of Limerick (since 1992) on whether previous drawing experience affects the computer aided design scores of undergraduates are also interesting. There was a 17% difference in scores by those students with previous advanced drawing experience. The results at the top of the group were particularly interesting. Seven of the top eight performers had additional drawing experience. The eight member of this cohort had an 'A' in higher level mathematics and he was left-handed. Two of the top five were left-handed (sinistrals constituted 12% of the cohort of 75 students).

The foregoing is not intending to say that we need more left-handed engineers but it does illustrate that where handiness is considered it is an indicator of cerebral specialism and lateralisation. It is also worth noting that sinistrals are less specialised cerebrally. As the corpus callosum is the nerve mass which communicates between the cerebral hemispheres the interaction between left and right appears greater in sinistrals. Research has also shown that where lateral specialisation is concentrated on one side that hemispherical asymmetry occurs (the frequently used hemisphere will be larger).

Lateralisation involves the two hemispheres in different aspects of cognitive processing. In order to improve engineering performance, cognizance should be taken of the necessity to stimulate right-brain activities such as holistic reasoning and image building or *cognitive modelling* and the important human intelligence of *spatial ability*. While many contend that individuals either possess this ability or they don't (that it is innate), research shows that with *appropriate tutorial intervention*, this ability can be developed and enhanced.

7. INTERVENTION STRATEGIES

In order for tutorial intervention to be effective cognizance must be taken of individuals with quite limited spatial experience. Considering the speed of perception in relation to the angle of rotation, Shepard and Metzler (1971) found when using cubelar arrays in perspective that the average rate of rotation is approximately 60° per second. They (Shepard and Metzler) presented perspective view in three different ways: (a) pair that differs by an 80° rotation within the picture plane; (b) differs by rotation in depth and (c) cannot be brought into congruence by any rotation, they are enantiomorphic (mirror images). The mirror image causes difficulty for most learners due to cognitive confusion with left-right orientation. These examples illustrate how a progressive strategy may be employed to increase the factor-of-difficulty and how in three-dimensions the process of encoding, comparison, search, rotation and decision can be followed. The individual with a more advanced spatial ability will quickly reduce the set of alternatives in processing correct solutions.

The self-paced interactive computer tutorial, developed at the University of Limerick, caters for all Gaughran's spatial sub-factors and is linked to a computer based test bank. Before embarking on the computer based the test-bank, the animated self-paced tutorials are employed. In addition to the tutorial taking cognizance of all five spatial sub-factors it allows participation at different levels in each. Where the learner has difficulty in deciding on a solution to the modelling problems presented, then they can select the 'solution' button to witness an animated demonstration of the solution. The computer based tutorials are supplemented by other tutorials and tests in an earlier module which is part of the general preparation for design and CAD. The impact of the early module as well as the purpose-built visualisation package is quite encouraging. Time spent on the two-dimensional CAD and in parametric and solid modellers is used with greater efficiency and an improved learning curve is achieved. The activities involved in test-bank, in addition to measuring progress, appear to also contribute to the overall development strategy.

8. DISCUSSION

Usually 'visualisation' is associated with solid or parametric computer modelling and the sophistication of the software package in representing the product or component. The more accurate and 'real' the computer representation is, the better the visualisation is considered to be. This is especially true for individuals who are lacking in cognitive modelling abilities. However it is the cognitive model which allows the engineer to generate the design at conceptualisation of the idea. Cognitive modelling proficiency empowers the designer with the ability to modify and develop the design as it evolves and make intelligent evaluations at all stages in the development of any component or product.

A recent article in CADCAM magazine writing on the modelling package 'SolidWorks' says: '*Successful production modelling depends on being able to create models to the design intent*' (Colin Mathews, Mar. '98). The intent is difficult to realise if it is not clear in the mind of the designer/engineer at the outset. Very often it is in the formulation of the 'intent' that the skills of human visualisation are most called upon. It is with this in mind that working to improve the cognitive modelling skills of engineering designers is seen as very important. The model, whether it is the tactile three-dimensional model, the computer generated parametric model or the rapidly generated prototype, is preceded by the cognitive-model. All other models are externalizations of the cognitive model.

At the University of Limerick, the interactive self-teach modelling package should, together with the test-bank counterpart and early intervention strategies, improve performance in all Design/CAD activities. The design strategy of the intervention tutorial takes cognizance of 'lateralisation' of the desired cognitive skills. The improved visualisation skills are not confined to design and CAD activities. The ability to cognitively model (visualise) problems as well as solution affects most of the engineer's activities. These intervention strategies contribute to producing engineers who are more confident, competent and efficient in modeling the design intent before it is externalized for definition and development.

REFERENCES

(1) DEITER, G.E., 1991, Engineering Design, McGraw Hill, Boynes 1978.
(2) EDWARDS, B., 1979 and 1993, Drawing on the Right Side of the Brain, Collins and Sons, Glasgow.
(3) FRENCH, M.S., 1951, Conceptual Design for Engineers, Design Cornal London.
(4) GAUGHRAN, W.F., 1997, Intelligent Manufacturing and Lateralisation, FAIM Conference. Middlesbrough, England.

(5) GAUGHRAN, W.F., 1997, Engineering Design Strategies - Shifting the Focus, ICPR Conference. Osaka, Japan.
(6) GAUGHRAN. W.F. 1996 Design Intelligence in Engineering, Paper to the IMC-13 Conference, Limerick Sept. 1996
(7) GRAHAM, R,B., 1990, Physiological Psychology, Wadsworth
(8) LEVY, J., 1976. Cerebral Lateralization and Spatial Ability, Behaviour Genetics, T, No. 2, 171-188
(9) LEVY, J. 1974, Phychobiological Implications of Bilateral Asymmetry, in Hemisphere Function in the Human Brain, Wiley, New York.
(10) LOHMAN, P.F., 1979, Spatial Ability: A review and reanalysis of the Correlation Literature, Technical Report No. 8, Standord University.
(11) LORD, T.R., 1984 A Pleas for Right Brain Usage. Journal of Research in Science Teaching, 22, V, 395-405.
(12) MAMAW R.J. and PELLEGRINO J.W., 1984 Individual Differences in Complex Spatial Processing Journal of Ed. Psychology, V76, 5, 920-939
(13) MOSES, B., 1982 Visualisation - A Different approach to Problem Solving. School Science and Mathematics, 82-2, 141-147.
(14) ROSENBLATT, E. AND WINNER, e., 1988 The Art of Children's Drawings. Journal of Aesthetic Education, Vol.22, No. 1, Spring, 1988.
(15) SEDDON G.M., ENIAIYEJU P.A. and JOSOH J., 1984 The Visualization of Rotation in Diagrams of Three Dimensional Structure American Ed. Research Journal, Spring '89, 21, 1, 25-38
(16) SHEPARD R.N. AND METZLER J., 1971 Mental Rotation of Three-Dimensional Objects Science, Vol. 171, Feb. '71
(17) SMITH, I.M., 1964. Spatial Ability - Its Educational and Social Significance. N.F.E. R., Nelson, Berks.
(18) VERNON P.E., 1961 The Structure of Human Abilities Methuen, London.
(19) ZIMMERMAN W.S., 1954 Hypotheses Concerning the Nature of Spatial Factors Education and Physical Measurement Vol. 14, 306-400.

Time Compression

Teaching of concurrent engineering – a case study

J-F PETIOT and **F-O MARTIN**
Equipe CMAO et Productique, Ecole de Nantes, France

SYNOPSIS

The increasing competition in industrial world imposes a production increasingly faster, less expensive and of better quality. The traditional design methods cannot realise these requirements. So the concurrent engineering (CE) concept was proposed. This new methodology requires a tight collaboration of all the trades dealing with the product all along its lifecycle. It is then important to learn this new methodology to future engineers. In this article is described a design teaching session for engineers students. In this tutorial a group composed of four students has to design a product in collaboration. Finally, the efficiency of this teaching method is discussed.

1. INTRODUCTION TO CONCURRENT ENGINEERING

The definition of concurrent engineering given by the Institute for Defense Analysis (1) is as follow: "*Concurrent Engineering is a systematic approach to the integrated, concurrent design of products and their related processes, including manufacture and support. This approach is intended to cause the developer, from the outset, to consider all elements of the product lifecycle from concept through disposal, including quality control, cost, scheduling and user requirements*".

The practical application of this approach in industry requires the realisation of the following key points :

- Concurrent design of the product and its related processes (2) (3),
- Constitution of a multi-disciplinary project team, composed of experts for every stage of the product lifecycle (figure 1). A project manager synchronises the organisation (trade/function matrix (4)),
- Temporal overlapping of the tasks, in order to shorten the development time of the product,
- Use of data processing tools and simulation methods allowing to take into account downstream trades constraints :
    - Virtual prototyping
    - Fast prototyping

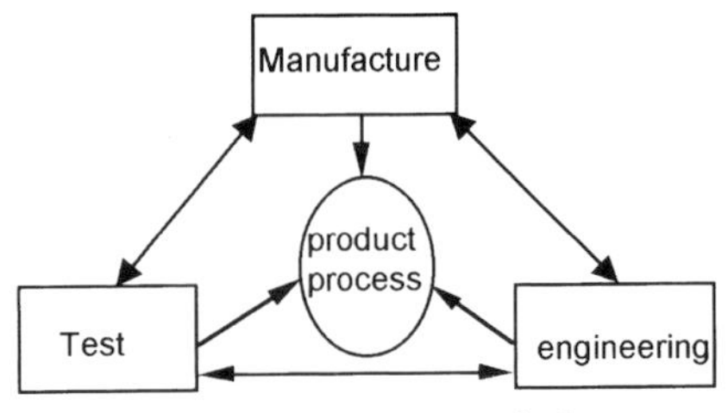

Fig. 1 The concurrent design

In addition to these changes related to the organisation, the implementation of CE requires abilities on a human level :

- Relationship skills and ability to work with team-mates(5) (6),
- Skills for mediation and negotiation (7).

We propose to focus our teaching session on these last two points. We think that engineer student courses had so far touched on the technical aspect only. Hence, it is essential to bring some knowledge about this side of the design activity.

The mechanical engineers that we train in our engineering school (Ecole Centrale de Nantes) will certainly have to deal with CE in their future jobs. So we believe that it is very important to teach them basic principles and human skills related to it. We want to particularly underline the following points :

- alternation between analysis and synthesis stages in design,
- research of criteria to evaluate the solution,
- existence of two types of information flows : technical information and organisational information,
- pre-eminent role of communication,
- role of experience and know-how capitalisation,
- needs to negotiate and to have a strategy,
- basic knowledge of compromise and robustness in design,
- need of assistance and simulation tools.

In the next paragraph, we present the experimental approach aiming at study the design process from an organisational point of view. There are several methods to teach design which are based on games (8). We present the game that we have chosen.

2. THE DELTA DESIGN GAME (9)

2.1 Description of the game

The principle of this game is quite simple. A team of four designers is set up, each one being expert in a particular field. Each team has one hour maximum to design a house (figure 2). But this house is not an ordinary one. The action takes place in an other world with only two dimensions. The inhabitants of this planet are not living on a sphere but on a plan. This means that the principles that rule this world are slightly different from the one that we know on Earth. Each expert of the team has thus new knowledge to learn before the design phase.

Fig. 2 Delta design game

The game is composed of a map and of about 40 deltas red and blue. The deltas (figure 3) are the materials that the players must use to build the house. These are equilateral triangles made of cardboard. The map is an A3 sheet of paper with a grid (figure 2). The hardware is quite simple. But the difficulty of the game stays in the relationship between all the players.

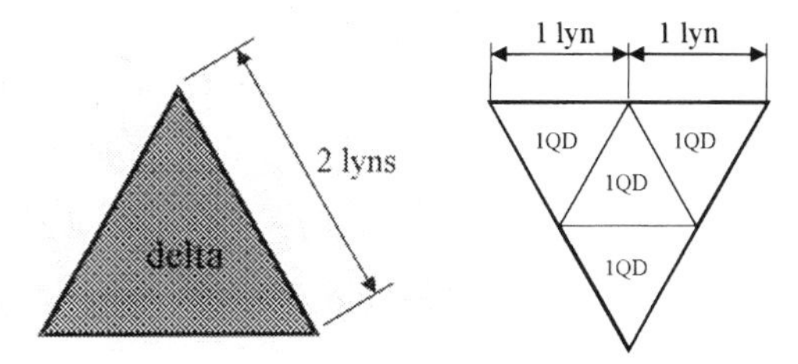

Fig. 3 Example of delta

All the subtlety of the game lie in the goals and constraints assigned to members of the team. The game has been designed in order to avoid obvious solutions. The players have to negotiate to impose their goals. The perfect design, i.e. every goals fulfilled and every constraints respected, can hardly be found in only one hour of work. But to reach an acceptable solution, the players have to think about a good way to work together.

Each team can choose its organisation freely. There are no other constraints aside the roles assigned to each player. The roles are as follows :

- The **project manager** main concern is cost and schedule, the interpretation and reconciliation of performance specifications, and negotiations with the contractor and client. He want to keep costs and time-to-build at a minimum, but not at the expense of quality. But he is not the leader of the team. He can't take decisions in the name of the team, unless all the players agree.
- The **structural engineer** main concern is to see that the design holds together as a physical structure under prescribed loading conditions. He must see to it that the two points at which the structure is tied to ground are appropriately chosen and that continuity of the structure is maintained.
- The **thermal engineer** want to insure that the design meets the "comfort-zone" conditions specified in terms of an average temperature. He must also ensure that the temperature of all individual deltas stays within certain bounds.

- The **architect** concern is with both the form of the design in and of itself and how it stands in its setting. He must see to it that the interior of the residence takes an appropriate form and that egress is convenient. He should also develop a design with character.

The design game can be see as the determination of **design parameters** satisfying **design criteria**, and of the **design strategy** aiming to reach the objective (figure 4).

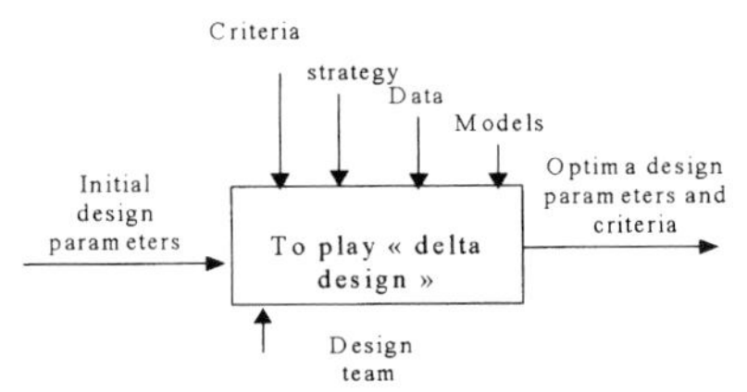

Fig. 4 SADT of the design activity

Glossary:

- **design parameters :** these are the parameters that have to be set to realise the design product (position of the bricks, colours...).
- **design criteria :** these are the evaluation criteria of a solution (like in functional analysis). Their quantitative value represents an idea of the qualitative value of the product.
- **design strategy :** this is a protocol used to organise the design process in the more efficient way according to the time and the problems met during the design.
- **Data :** these are the data of the problem
- **Models :** these are the models used by the experts to calculate the criteria related to their field of expertise. It has the same meaning as the scientific approach (it is an approximation of the reality that allow the user to forecast the reaction of a system).

2.2 Relation with concurrent engineering

The students confront a situation of innovative design in a CE context. All the characteristics of concurrent engineering are represented (except the design of the manufacturing process):

- There is a multi-field team of expert in different fields.
- There is a temporal overlapping of the task. This means that each expert is working at the same time on the same product (figure 5).

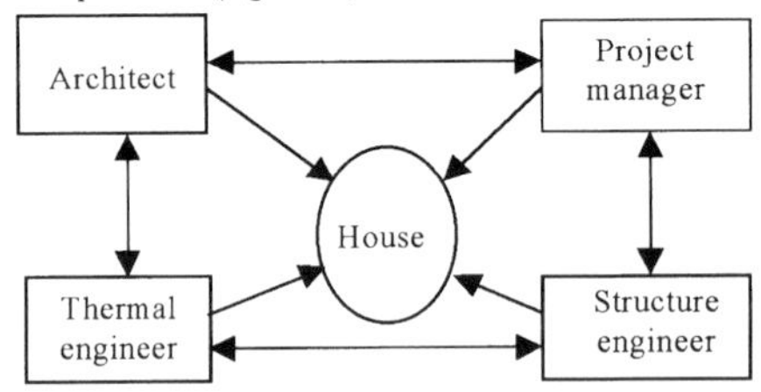

Fig. 5 Delta design game

However there is no leader in the team. This means that every decision has to be discussed and agreed by every player. This configuration emphasises the communication. Each one has to integrate gradually in his way of thinking the constraints from the other fields of expertise. The students learn that expertise in a technical field is important but this is just one part of the job. The designer must in addition have human skills linked to his work organisation.

The evaluation criteria for the solution are given in the rules. But they have no computational tools or simulation tools to help them in their tasks. This permits to make them think about the utility of these kind of tools in a more complex environment.

The context is clear enough to let students perceive relationship and communication side of the design. They can think about the design activity in a more general way than about technical details.

3. TEACHING OF CONCURRENT ENGINEERING BY EXPERIMENT

We describe in this paragraph the implementation of the means previously chosen, used to teach CE to our students. The tutorials are divided in two sessions of four hours each, devoted to the Delta Design game (10). The students are gathered in 6 groups of 4.

3.1 Description of the tutorial sessions

We decided to do two plays of the game, one by session. The two sessions are organised in a slightly different way. The schedule of the two sessions is as follow :

Table 1 Schedule of the tutorial sessions

SESSION 1	Training phase : 2h	Play #1 : 1h	Presentation of the results : 1h
SESSION 2	Analysis of the game : 1h30	Play #2 : 1h30	Presentation of the results : 1h

For the first session, the students only have the rules of the game. They have 2 hours to learn their trade. To improve the efficiency of the training, the students are gathered by trade for this phase. By this way, the experts in one field are the only one to know the kind of calculation they have to proceed and how act on the design parameters to fulfil their criteria. Next to this phase begins the play for one hour. Finally each design team present its work to the teacher.

The second session is different. Since the students are already trained in their expertise, they don't have to do training again. The first phase of this second session is devoted to an analysis of the game and a recall of the concurrent engineering basis. We give them a paper with the main points of this course. During this phase we underline the goals that we want to fulfil with this game, that is :

- the human factor is as important as the technical expertise,
- a high level of communication is required,
- the way to negotiate act upon the global result of the design,
- implementation of strategies are required to "act in complexity",
- to give an overview of every sides of the engineers trade,
- finally, to make them think about their future trade.

We give them also a glossary of the terms they will have to use to write a report. Thanks to this document, they have a common basis that allows them to communicate more easily. The glossary explains terms like Design Parameter, Criteria, Models...With this preliminary analysis and the experience they acquired on the previous session, we expect them to focus their attention on the way to work together more than on their own personal goals.

3.2 Results

We asked the students to make a report of the two sessions. From these reports we can notice several points.

For the first session, we gave to the students only the rules of the game. They were all alone to decide the way of working. They faced unexpected problems related to the communication. They could not meet their own requirements because of the constraints of their colleagues. The game that was apparently easy turned out to be tricky. During this first session they learned how manipulate the game to reach their goals. They also learned that the negotiation is an important part of the designer trade.

The structure of the design process concerning the alternation of analysis and synthesis stages was well understood. It is an intuitive notion that they already knew. But the game allows them to experience it. There are two trades (thermal and structural engineers) that need more time than the other to calculate their criteria. So the team had to stop the synthesis of the product to let them finish their calculation.

Concerning the research of the evaluation criteria of the solution, the game is not designed to teach this kind of thing. Indeed, the evaluation criteria of the solution are given in the rules of the game and never change during the play. However, the analysis of design process given in the beginning of the second session makes the students aware of the relations that exist between the criteria and the goals. They understood that it is important to have a goal to reach and that this goal is strongly related to the classification of the criteria. Most of the students tried to build a cheap house, but some of them explained that there are several ways to design depending on whether the client want a beautiful and comfortable house that cost all his budget or that he want a house with a good quality-price ratio.

During the first play, only half of the design teams decided to organise the way of working, but it was often to late du to the time. For the second play, each team was aware that an organisational information flows was as important as the technical information flow.

The experience of the first session and the analysis given in the beginning of the second session helped them to make up strategies before the second play. All the strategies used can be classified in three types.

For the first play, almost all the design teams had the same strategy. They designed by repair of an initial solution that was often not good at all. To modify the product, they used the "trial-error" strategy. But this tactic was applied almost unawarely. They did not control the process.

The second play allows them to elaborate some strategies more efficient. The main two strategies were the concession method and the aggregation criteria method. The first one consists in analyse the design product for a given state. The designers can see what are the criteria fulfilled and the one that have to be improved. Then the players can allow a degradation of the good criteria to enhance the bad criteria.

The criteria aggregation strategy requires a good knowledge of the environment. This is a more advanced process already used in value analysis. The team has to gather all the criteria and classify them from the less to the more important. By this way, the group knows what are the requirements on which they have to focus first. In this method the criteria classification chosen influence a lot the final result.

In the following table are the strategies used during the two sessions:

Table 2 Strategies used by the design teams

	Session 1	Session 2
Trial-error	7	0
Concession	0	3
Criteria aggregation	1	5

Some students have applied what they learned in functional analysis to classify the criteria (figure 6).

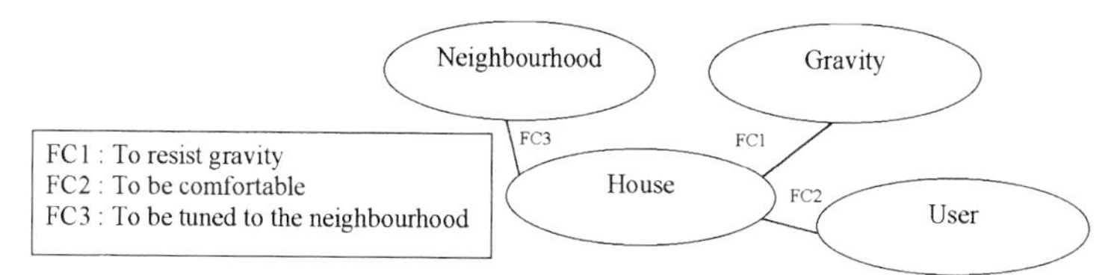

Fig. 6 Example of functional analysis

4. CONCLUSIONS

Most of the students understood the goal of the game. They were able at the end of these two sessions to explain how they see the design. They really understood that the design activity is more than an addition of technical skills. Some of them succeeded to find the limitation of the game in regard to a real life design. By thinking of limitation of the game, they can have an overview of the design activity. However, some students could not get beyond the game, and did not develop a personal idea of the design in general.

In a general way, the students are really interested in this game. It gives them a general overview of the design activity structure. These tutorial sessions are a ludic way to bring the students to think seriously about the design in a concurrent engineering context.

REFERENCES

(1) www.soce.org: Society of Concurrent Engineering.

(2) Chedmail, P., Bocquet, J-C., Dornfeld, J-C., "Integrated Design and Manufacturing in Mechanical Engineering", Kluwer Academic Publishers, 1997.

(3) Sohlenius, G., "Concurrent Engineering" Annals of the CIRP. Vol 41/2. pp. 645-655. 1992.

(4) Midler, C., "L'auto qui n'existait pas". Management des projets et transformation de l'entreprise, Interédition, Paris, 1993.

(5) Blanco, E,. "L'émergence du produit dans la conception distribuée. Vers de nouveaux modes de rationalisation dans la conception de systèmes mécaniques". Thèse de doctorat de l'INPG 1998.

(6) Charpentier, P., Garro, O., Toniolo, A.M., Brissaud, D., "Vers l'émergence d'un produit en conception distribué" Colloque PRIMECA, 3-4-5 avril 1995, La Plagne, France.

(7) Jacquet, L., Petiot, J.-F., Sommer, J.-L., Koyama, Y., "Analysis of the setting up of a concurrent engineering scenario", Kluwer Academic Publisher, sous la direction de : P. Chedmail, J.C. Bocquet, A. Dornfeld, pp. 3-12, ISBN 0-7923-4739-0, 1997.

(8) Ris, G., "Etude de cas : le jeu des CMAOistes". Conception de produits mécaniques – Méthodes, modèles et outils. Ss la direction de M.Tollenaere. Edition HERMES, Chap 22 pp 537, 550, Paris, France, 1998.

(9) Bucciarelli, L., "Designing Engineers". MIT Press. 1994.

Teaching time

D A HOLLAND
Faculty of Technology, Bolton Institute, UK

SYNOPSIS

This paper describes how, with the aid of a simple product, students are engrossed in the design process and experience the effects of time compression technologies at first hand. The chosen product was designed originally to demonstrate basic manufacturing techniques to undergraduate students. The exercise allows students to be totally involved in concurrent product development activities through the use of TeamSET® Concurrent Engineering software and in rapid product development through the use of JP System 5® educational Rapid Prototyping. The exercise demonstrates the effectiveness of systematic tools in the design process and the advantages of rapid availability of physical hardware.

1. INTRODUCTION

A challenge facing academic staff at Bolton during the mid to late 1990's was how to integrate modern design philosophies into courses whilst demonstrating to students the impact of Concurrent Engineering and Rapid Prototyping on the design process.

Design courses had sections on the design process as proposed by Pugh(1), concept generation, concept evaluation, design for manufacture using typical examples such as those given by Andreasen et al (2) concurrent engineering practices and rapid prototyping processes. These tended to be taught as independent topics within the curriculum. The formidable task was how to involve students at undergraduate and at postgraduate level in the total design process using the above techniques but in the relatively short period dictated by a modular course.

During 1996 a small team of academics involved in a consultancy investigated the methodologies of design for assembly and manufacture. This resulted in the purchase of the CSC TeamSET Concurrent Engineering Business Solution software. One reason for the choice of TeamSET was for its embodiment of systematic tools in the design process. The search for a suitable product which provided significant benefits from a Design for assembly analysis whilst being readily available and non-proprietary yielded the Model Cannon illustrated in Figure 1.

Fig. 1 Model Cannon assembled

2. PRODUCT DESCRIPTION

The Model Cannon is used as an Engineering Applications exercise in manufacturing with groups of undergraduate Product Design students. The original purpose of the Cannon was to involve the students in the manufacture of a simple product and uses hand, conventional and NC manufacturing processes. It is also used in an assembly exercise to demonstrate assembly problems caused by poor design. The Model Cannon has 23 parts as shown in Figure 2. Manufacturing drawings for the Model Cannon existed, as did a number of completed products manufactured from metal and polymer. The Model cannon has considerable scope for improvement as is illustrated in this paper and thus provided a suitable vehicle for use with students. Some students may not have had much experience in actual design when introduced to these topics.

Fig. 2 Model Cannon showing component parts (parts count = 23)

3. DESIGN METHODOLOGIES

TeamSET embodies the Methodologies of Quality Function Deployment (QFD), Concept Convergence (ConCon), Design for assembly (Dfa), Design for manufacture (Dfm) and Failure Mode and Effects Analysis (FMEA) as well as a Design to Target Cost (DTC) capability. The Dfa methodology is a result of collaborative work between the Lucas Organisation and the University of Hull. (3). The Model Cannon provides a simple medium to enable students to be involved with all aspects of the design process and to appreciate the benefits of the adoption of time compression technologies.

4. THE CONCURRENT PRODUCT DEVELOPMENT PROCESS.

Concurrent product development uses the concept of multiple interlocking teams. Payne 1996 (4). It is important that teams are selected a) to be able to work together and b) to be able to represent different areas expected within the team. To simulate the concurrent development process it is useful to have groups with a broad range of experience to represent the disciplines and roles needed within concurrent teams. This is often difficult since delivery tends to be to a specialist group of students, however with some guidance from tutors students have responded well.

The selected groups of Students are tasked to devise a design specification for a model cannon; some define a non-functioning ornament others a fully functioning scale replica of a particular era. They are then tasked to find out what other potential customers, their friends, family etc., would expect from such a product. This is “the voice of the customer” to be represented within the Product design specification. This leads to a QFD exercise using the TeamSET software. A typical output is shown in Figure 3. The Relationship Matrix enables the relationships between the customer requirement and the design requirements to be identified and hence used to inform the Product Design Specification (PDS). The Correlation Matrix enables the elements of the Design Requirements to be compared with one another and classify the strength of any correlation between them hence informing the PDS of the important issues. From the QFD exercise a PDS for the Model cannon is compiled and ideally a number of concepts generated.

Within TeamSET is the concept evaluation tool, Concept Convergence, which provides a systematic evaluation of initial concepts and enables the weak features of the best runners to be identified and strengthened. In an ideal world, the students would create concepts for the cannon specification that they have just devised via QFD. However, this is not very often possible due to teaching time constraints. Providing various concepts to the students and allowing them to evaluate them against their specification has simulated this exercise. A typical output is shown in Figure 4.

The identification of a strong runner, concept 3 in Figure 4, would focus design attention on the negative aspects, in this case polished material. This criterion is investigated to determine if it could be improved by convergence to one of the other designs.
The detailed concept design is then evaluated in a Design for assembly and manufacture (Dfa/m) exercise. In this case, a predefined PDS is used which should be comparable with some of the students own. Using the students PDS would be difficult to resource and provide

sufficient information unless they undertake it as a separate concept exercise and destroy some of the impetus created by the QFD exercise.

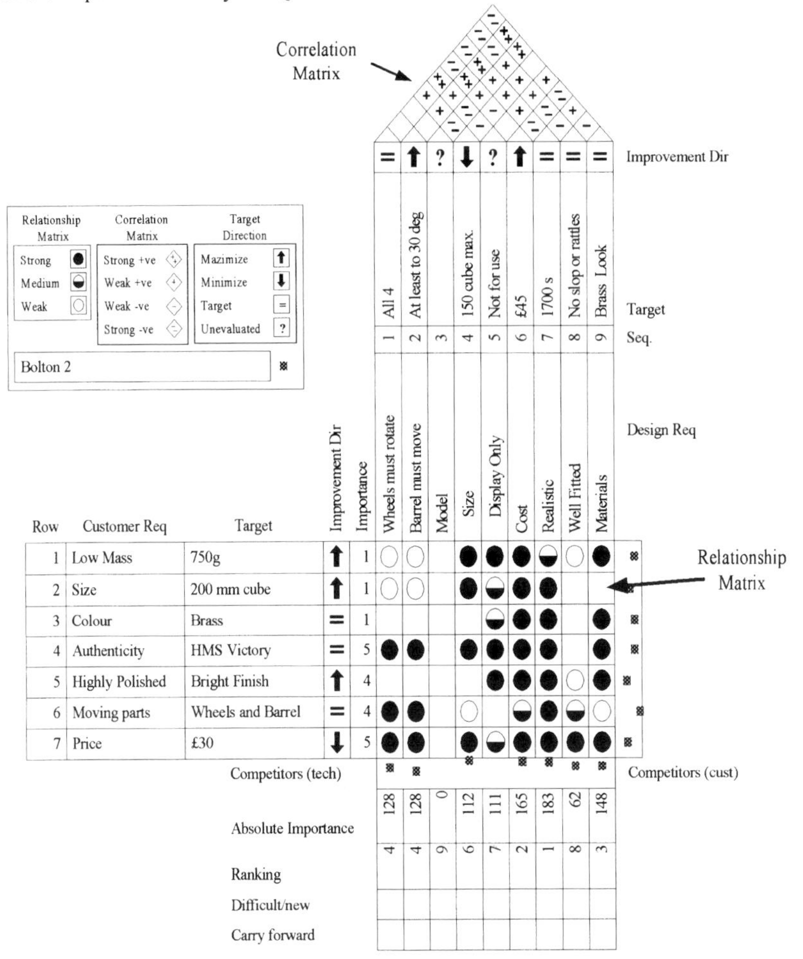

Fig. 3 Quality Function Deployment output – Model Cannon

The initial design evaluation yields output shown in Figure 5. A Functional Analysis (FA) indicates six "A" or essential parts. i.e. that the assembly may be designed in a minimum of six parts. The Handling Analysis (Hand.) indicates time penalties for difficulties in picking up and handling individual parts caused by design. The Assembly Flow chart indicates the assembly sequence and the times taken to complete these tasks. A Manufacturing Analysis (MA) indicates the cost of producing the components. These indicators are used to calculate Design Efficiency (A Parts/Total number of parts). Individual Handling and assembly indices are summed to give Assembly and Handling scores. These serve as indicators of the overall effectiveness of the design concentrating effort to appropriate areas.

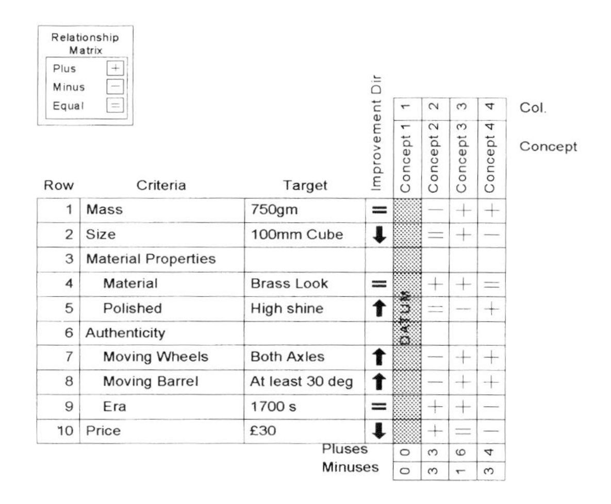

Row	Criteria	Target	Improvement Dir	Concept 1 (Col. 1)	Concept 2 (Col. 2)	Concept 3 (Col. 3)	Concept 4 (Col. 4)
1	Mass	750gm	=	DATUM	−	+	+
2	Size	100mm Cube	↓		=	+	−
3	Material Properties						
4	Material	Brass Look	=		+	+	=
5	Polished	High shine	↑		=	−	+
6	Authenticity						
7	Moving Wheels	Both Axles	↑		−	+	+
8	Moving Barrel	At least 30 deg	↑		−	+	+
9	Era	1700 s	=		+	+	−
10	Price	£30	↓		+	=	−
			Pluses	0	3	6	4
			Minuses	0	3	1	3

Fig. 4 Concept Convergence analysis – Model Cannon

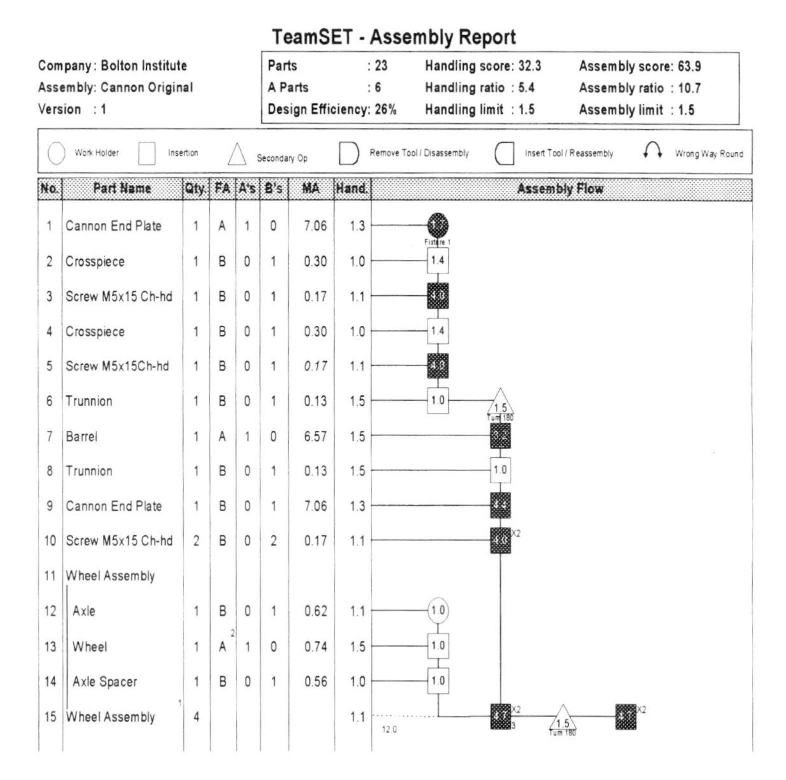

TeamSET - Assembly Report

Company: Bolton Institute
Assembly: Cannon Original
Version : 1

Parts : 23	Handling score: 32.3	Assembly score: 63.9
A Parts : 6	Handling ratio : 5.4	Assembly ratio : 10.7
Design Efficiency: 26%	Handling limit : 1.5	Assembly limit : 1.5

No.	Part Name	Qty.	FA	A's	B's	MA	Hand.
1	Cannon End Plate	1	A	1	0	7.06	1.3
2	Crosspiece	1	B	0	1	0.30	1.0
3	Screw M5x15 Ch-hd	1	B	0	1	0.17	1.1
4	Crosspiece	1	B	0	1	0.30	1.0
5	Screw M5x15Ch-hd	1	B	0	1	0.17	1.1
6	Trunnion	1	B	0	1	0.13	1.5
7	Barrel	1	A	1	0	6.57	1.5
8	Trunnion	1	B	0	1	0.13	1.5
9	Cannon End Plate	1	B	0	1	7.06	1.3
10	Screw M5x15 Ch-hd	2	B	0	2	0.17	1.1
11	Wheel Assembly						
12	Axle	1	B	0	1	0.62	1.1
13	Wheel	1	A	1	0	0.74	1.5
14	Axle Spacer	1	B	0	1	0.56	1.0
15	Wheel Assembly	4					1.1

Fig. 5 Design for assembly/manufacturing analysis-Original cannon

This is followed by a redesign exercise, in which the students try to minimise the parts count by combining "B" or unessential parts into "A" parts. This quite quickly yields design concepts such as those shown in Figures 6 and 7. Note that only 4 parts are used in this assembly due to the wheels not being independent of one another. A revisit to the Dfa analysis typically gives an output as shown in Figure 8. Note the 100% Design Efficiency which is not often achieved in more complex products.

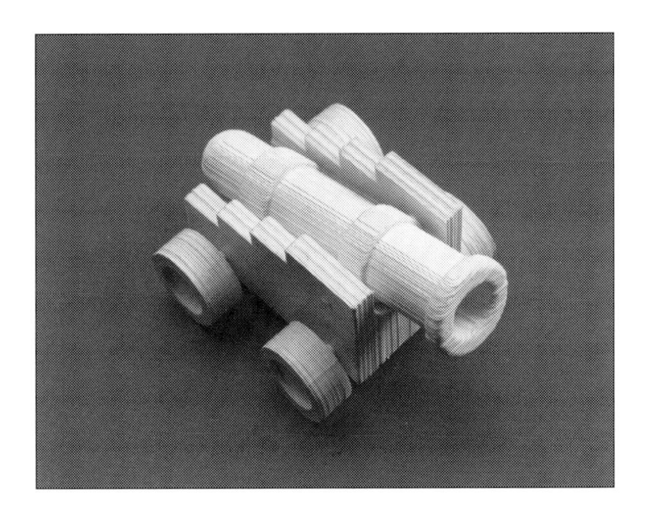

Fig. 6 JP5 Rapid Prototyped model of New Cannon

The revised product may be modeled as a 3D solid using CAD software and produced direct from the CAD data using the educational JP System 5 rapid prototyping system. This demonstrates the power of the technology to rapidly produce physical parts for concept evaluation, and serves to illustrate the concept of Rapid Prototyping. In this way, students are involved in the whole process.

A manufacturing analysis as part of the TeamSET software may show that the ultimate 100% design efficiency model shown in Figure 7 may be extremely difficult to manufacture because of the complexity of the machining or casting process needed. This stimulates debate in the merits of various manufacturing methods, and a further evaluation of the product using the Manufacturing Analysis software. In the examples shown, it is evident that although the redesigned components are quite expensive to produce (£13.59 total) this is much smaller than in the original design (£24.15). The handling and assembly scores have also been reduced. Handling from 32.3 to 4.8 and assembly from 63.9 to 8.6 indicating a significant reduction in handling and assembly times. The shaded areas of the Assembly reports in Figures 5 and 6 show where the handling and assembly score figures exceed the set limits thus highlighting where more work may be undertaken to improve the design and further reduce costs.

The last exercise is a design Failure Mode and Effects Analysis (FMEA) again conducted within TeamSET. This may yield further areas for design, manufacturing or other evaluation and highlight other potential problem areas.

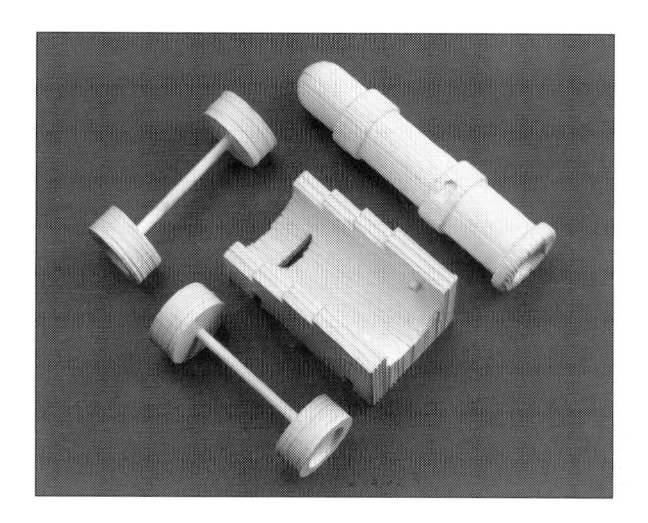

Fig. 7 JP5 Rapid Prototyped model of New Cannon (Parts count = 4)

TeamSET - Assembly Report

Company: Bolton Institute
Assembly: Cannon New
Version : 1

Parts : 4	Handling score: 4.8	Assembly score: 8.6
A Parts : 4	Handling ratio : 1.2	Assembly ratio : 2.2
Design Efficiency 100%	Handling limit : 1.6	Assembly limit : 1.6

Work Holder | Insertion | Secondary Op | Remove Tool / Disassembly | Insert Tool / Reassembly | Wrong Way Round

No.	Part Name	Qty.	FA	A's	B's	MA	Hand.	Assembly Flow
1	Chassis	1	A	1	0	4.93	1.3	1.0 Fixture
2	Barrel	1	A	1	0	6.53	1.5	2.5 1 Alignment; 1.5 Turn 180
3	Wheel & Axle	2	A	2	0	2.13	1.0	1.3 X2

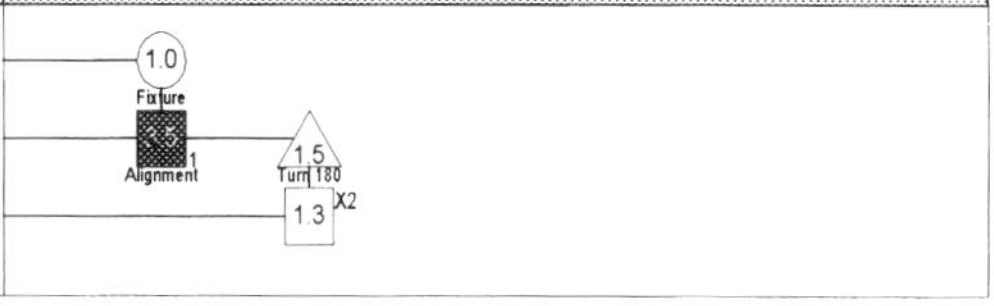

Fig. 8 Design for assembly/manufacturing analysis-New Cannon

5. EVALUATION

One problem with teaching this type of topic in a particular module of a course is that the students in the main are from one discipline. This often means that the students do not have the variety of backgrounds expected within a concurrent development team. The students tend to concentrate their initial efforts on the redesign rather than the systematic tools available and close control is needed to ensure that they see the benefits of the process. Many students find that the most challenging part of the exercise is in determining the assembly flow diagram. This aspect demonstrates to the students the effects of the design on the overall process and reinforces the message that the design activity is where costs are created. It is recognised that paper–based exercises may well yield the same outcomes. The advantages in using the techniques embedded in TeamSET are; that criteria may be pushed from one technique to another providing good continuity, that students may work independently of the group leaving a common database which may be later interrogated by others, and that I have found that students tend to respond much better to using the software based techniques. It is the author's view that the advantages of undertaking such an activity with the students, especially those who do have aspirations in the field of design, far outweigh the drawbacks. It provides them with a systematic team approach to the solution of design problems using defined tools. Students who have followed the activities do utilise the systematic tools available within TeamSET in their own project work and have appreciated the benefits of being able to produce representative models direct from 3D CAD solids via rapid prototyping.

6. CONCLUSION

Providing students with a knowledge base in the principles of rapid product development i.e. time compression technologies is enhanced by the use of a suitable product. Involving teams of students in the process of new product introduction with the use of systematic design tools within a concurrent engineering framework is refreshing and rewarding. This method of instructing students in concurrent product development practices within a systematic design environment and their use of time compression technologies is "Teaching Time"

REFERENCES

(1) Pugh S "Total Design Integrated Methods for Successful Product Engineering" 1997. Addison Wesley, Wokingham

(2) Andreasen M, Kahler S, Lund T, and Swift K, 1998 Design for Assembly", IFS/Springer-Verlag, Bedford

(3) Redford A, Chal J, "Design for Assembly Principles and Practice" 1994, McGraw-Hill, Maidenhead.

(4) Payne A, Chelsom J, Reavill L"Management for Engineers" 1996, Wiley, London.

Rapid Product Development

Integrated education and the rapid product development sector

L E J STYGER
STYLES Rapid Product Development, Teesside, UK

SYNOPSIS

Standard academic products such as the MSc and PhD are well suited for those destined to enter into research based careers, but not particularly appropriate for design engineers following a career in product development. Today's commercial environment demands that good product development practitioners will be well trained and proficient in a broad range of subjects, however, there are no standard "qualifications" for the "professional RPD engineer". This paper outlines the trends of the sector, provides a proposed framework for formal and professional recognition and finally, an over the horizon forecast on the future trends offered.

1. INTRODUCTION

In his opening remarks of the Mouldmaking 2K conference, D Barlow stated that he could be "presiding on the last meeting of a sector in terminal decline"[1]. The comment was made specifically with the UK toolmaking industry in mind and in this context was just and applicable. There is however a much larger question of total sector definition, change in market requirements and the longer-term prognosis of potential inherent opportunity.

Toolmaking is one of the sub-sections of the Rapid Product Development sector (RPD). The sector has massive potential when viewed in a holistic manner, yet the EU RPD sector currently lacks visibility and coverage, especially in the provision of applicable skills and education.

This paper covers three areas:

- A definition of the Rapid Product Development sector, industry trends and sector growth
- A proposed framework for a holistic educational and training system
- An over-the-horizon forecast on the future of the rapid product development sector without a holistic educational and training system

2. A DEFINITION OF THE RAPID PRODUCT DEVELOPMENT SECTOR, INDUSTRY TRENDS AND SECTOR GROWTH

2.1 A Brief History of the Rapid Product Development Sector

The origins of the RPD sector can be traced back to the late '80's. The RPD sector evolved primarily as a business process (supported by advanced technology) for designing and developing products rapidly. The key to the sector was the ability to "productionise" craft skills brought together from traditionally isolated specialist services such as engineering design, drafting, model and patternmaking and toolmaking etc.

Much time has been given to the technical aspects of RPD. It is however sufficient to state that the core driver for the RPD sector is CAE technology, but as with the traditional services, people remain the real issue for continued success.

There are some unique dynamics within the RPD sector. For example order books are typically short, debtor days (payment after delivery to the client) are long. Accurate business predictions are therefore impossible and the management of cashflow is primary. There is always a major conflict between workshop capacity and cyclicity of the marketplace. There is little understanding for the management of this type of business outside of the sector.

2.2 Definition of the Rapid Product Development Sector

The European Rapid Product Development sector is defined as those companies that are providing services involving the rapid design and development of products for a third party manufacturing organisation. The sector may include organisations practicing any and/or all of the following activities:

- Conceptual design
- Engineering design
- Test and analysis
- Modelmaking
- Prototyping
- Short-run or bridge production
- Value-added/niche manufacture, etc.

The distinguishing features of a "Rapid Product Development service provider" are:

- Rapid and responsive
- Customer focused
- Fast and efficient electronic communication

2.3 Industry Trends

Most EU RPD service supplier companies are SMEs that typically provide products and services to most of the OEMs and first tier suppliers throughout the EU. The client base is now demanding that RPD service companies offer a total service combining the major process and technical elements of product development (i.e. an "all under one roof service").

The new demands being placed upon the sector is forcing a change in paradigm that includes:

- New business processes
- New skills
- New technologies

Traditionally the RPD sector has placed great reliance on leading edge technologies. It is however the considered opinion of industry is that business process of RPD and most importantly people and skills are the key factors to future success of the sector[2].

If we make a comparison between the old and new paradigms of product development it is easy to understand why Barlow (2000) made his remarks concerning the "terminal decline" of toolmaking. A comparison between the old and the new paradigms of product development is shown in the table below.

Table 1 A Comparison Between the Old and New Paradigms of Product Development

A Comparison Between the Old and New Paradigms of Product Development	
Old Paradigm	*New Paradigm*
Highly fragmented industry	All under one 'roof'
No urgency or project management skills	'Rapid' culture
A craft industry	High project management
The 'Black Art' of product development	Continual communication
2D Specifications	Paperless - 3D Solid Model CAE
Outdated technology	Productionised manufacture
Little or no investment and development	Investment at a 'critical' scale

2.4 Sector Growth

In 1998, the EU RPD sector was valued at £30bn. During this period government estimates predicted that the market value of the sector would double by 2003. Current commercial indicators suggest that the predictions are on target[3,4,5].

Within the RPD sector, more revenue is usually generated from activities that are further along the process chain. Activities early in the process chain however allow client capture and feed the "productionised" functions that typically generate more revenue.

The diagram below illustrates the typical revenue differentials for the RPD sector.

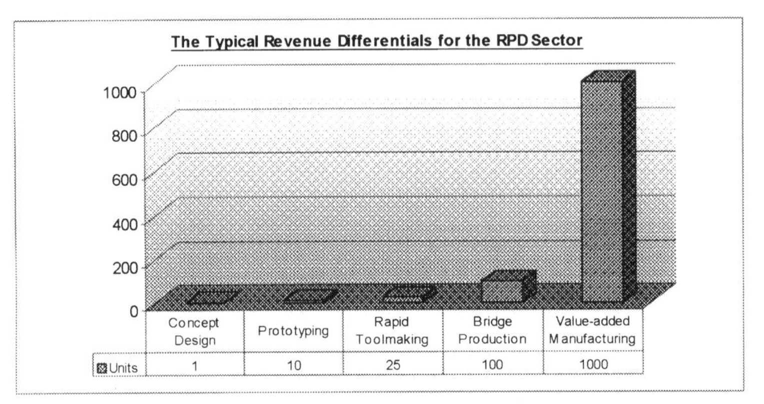

Fig.1 The Typical Revenue Differentials for the RPD Sector

The diagram below illustrates the typical monetary values the RPD sector.

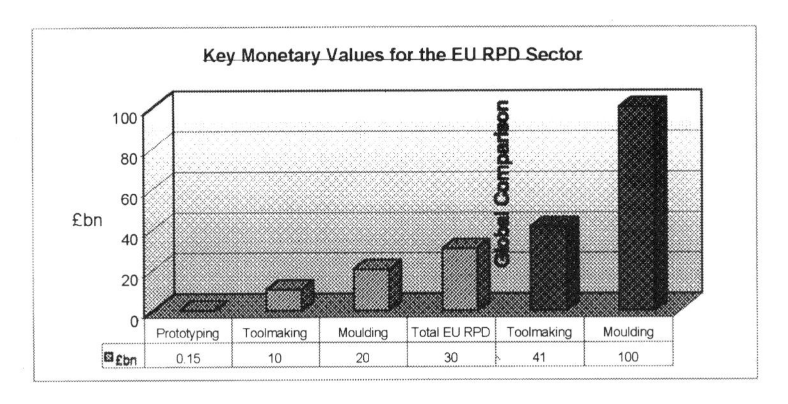

Fig.2 Key Monetary Values for the EU RPD Sector

Despite the high monetary value of the RPD sector, there is a general lack of sector visibility throughout the EU. The sector is therefore fragmented and undervalued by the customer base.

Importantly, the RPD sector currently employs 2.6million high-skilled professionals (i.e. people with formal higher-educational qualifications). It is expected that the RPD practitioner base will rise to 4million to coincide with the market expansion.

Although there is no definitive work relating to the number of people currently employed in the EU RPD sector, a reasonable estimate may be provided as follows:

Table 2 Formulation for the Estimate of the Number of Sustained Skilled, High-Value Jobs in the EU RPD Sector

Formulation for the Estimate of the Number of Sustained Skilled, High-Value Jobs in the EU RPD Sector
There are approximately 200k CAE seats in the UK [assume 1 operator per-seat][6]
There are approximately 1200 toolmaking companies in the UK. Each company has an average of 40 employees, therefore 1200 X 40 = 48k
Note this does not include any other portion of the RPD sector. However, let us assume that there are only 250k skilled people employed in the UK RPD sector (a probable gross underestimate).
The German market size is 2.5 times that of the UK[7]
Therefore we may assume that there are approximately 625k employed in the German RPD sector
We may assume that the rest of the EU is approximately 3 times the UK and German totals
(i.e. 3 X 875)
Therefore the current number of high-value jobs within the EU RPD sector is: **2.625m**
Note: this number is prior to any addition to cover the predicted doubling of the market size

There are concerns that there will be a chronic shortage of practitioners entering the sector with the right level of skills and qualifications. The lack of a skilled work-force will cause some companies to look further afield for expansion opportunities[8].

3. A PROPOSED FRAMEWORK FOR A HOLISTIC EDUCATIONAL AND TRAINING SYSTEM

Research is often wrongly confused with development[9]. Research activities generate knowledge as its final product, whereas development activities use knowledge to generate product.

The RPD sector does not typically become involved in contract research but rather contract development. However, whereas research training is well catered for in the standard academic products such as the masters and doctorate programmes, that typically demand a focused understanding on a single detailed subject, established educational providers do not typically cover development training in a logical or holistic manner.

The development engineer or scientist can require a broader base of understanding and learning; for example, design engineering, materials science and manufacturing systems engineering. As careers progress management and commercial training will also be necessary.

Companies involved in the Rapid Product Development Sector match the EU and UK government strategies of competitive high-value-added service organisations. Yet there are no standard "qualifications" and no recognised career path for the "professional RPD engineer".

There is a need for a multi-band educational and training standard for the RPD sector. Areas of technical speciality must be catered for alongside core general subjects. Importantly, there must be a "cradle to grave" approach to the whole process of learning and personal professional development that is transparent, recognised universally and transportable between institutions and geographic locations. A model framework is illustrated below.

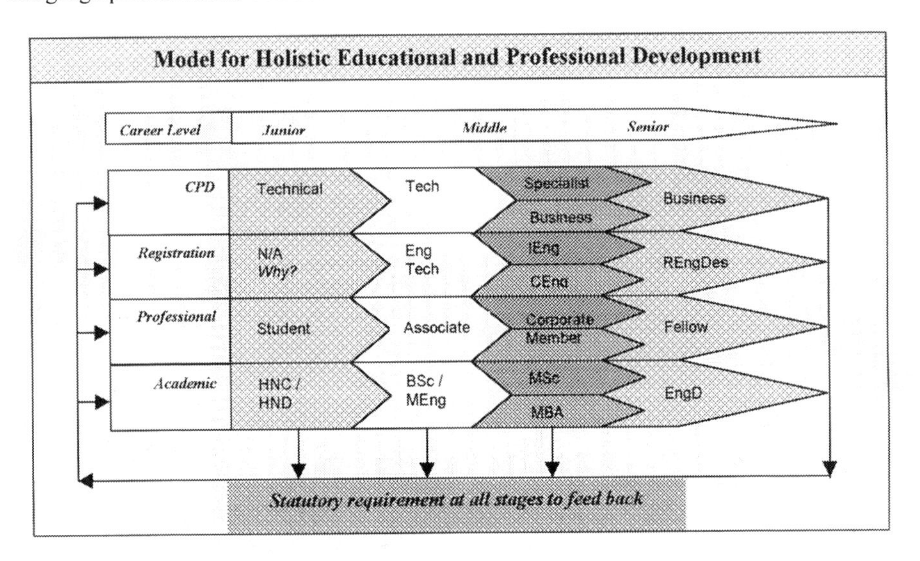

Fig.3 Model for Holistic Educational and Professional Development

The model will naturally become quite complex because individuals will enter the framework at different points. For example one would not expect an individual with three A levels to embark upon a four-year apprentice scheme but rather enter into the relevant degree and then into an industrial post. However, technical competence will still be a paramount requirement.

It may be argued that this model already exists in standard mechanical engineering and design engineering courses. However, the fact remains that traditional courses do not cover the core aspects of RPD, also a logical progression of learning and development is not always possible. Typically little thought and provision is given to those who progress "up the ranks" (a key element in this sector) and professional recognition does not always have parity with that of the academic world.

Time and flexibility of timing are important issues for the practising professionals that want to continue to attain further qualifications. As such there is a great need for further part-time, distance and "own speed learning" schemes. Perhaps following the Open University model or similar. However, these types of courses will still need to be accredited and recognised both within academia and indeed industry.

There is a need for accredited courses that have been designed by the sector and/or companies with core elements of in company assignment and industrial relevance.

There appear to be two camps within the engineering educational establishment:

- Pure Engineering
- Applied Engineering

Whereas the academic focus would appear to be fixed on the pure engineering subjects, the needs of industry are such that preference is usually given to people who have a track record of applied subjects. The problem is that there are not as yet any formal applied RPD engineering or RPD design engineering qualifications. Many may claim to be doing it, many may think that they are indeed doing it but there is no constant and no formal holistic route for RPD that links visibly (cradle to grave) the academic, professional and CPD elements critical to the individual and company.

4. AN OVER-THE-HORIZON FORECAST ON THE FUTURE OF THE RAPID PRODUCT DEVELOPMENT SECTOR WITHOUT A HOLISTIC EDUCATIONAL AND TRAINING SYSTEM.

There is a major opportunity in development, or more correctly the Rapid Product Development sector, for both the educational providers and the industrial practitioners. Government trends indicate a doubling of the market value from £30bn to £60bn within a five to ten year time scale. The employed based (those with formal higher-educational qualifications) will rise from 2.6million to 4million. There will be a chronic shortage of practitioners entering the sector with the right level of skills and qualifications. High-skilled and qualified people will stand a 100% chance of employment within the sector. Educational establishments offering the right training and qualifications will stand a 100% chance of fully funded courses applicable to industrial requirements and fully subscribed by high-caliber students.

There is little that can be done to stop the movement of low-value-added manufacturing to low-cost geographical locations. We must therefore expect to see a greater exodus eastward as new site selection is based on decisions of low-labour cost, infrastructure and high government incentives packages. We should indeed be pragmatic and question if we truly want to keep and artificially sustain low-value work in the EU.

However, the EU Rapid Product Development sector faces different growth and expansion problems. Questions of infrastructure, incentives and proximity to major clients will become incidental to successful growth. As a direct result of the lack of suitably skilled people, expansion and new site selection decisions will be based on the ability of an organisation to strategically acquire existing competitors in order to guarantee a critical mass of skilled people in one location.

5. CONCLUSIONS

There is considerable opportunity for both commercial and academic organisations within the European Rapid Product Development sector. There is however a tendency to focus on technical matters rather than the key driving issue of the sector, i.e. that of skilled people. The sector is not well catered for in terms of holistic or indeed focused qualifications. The lack of focus from the training providers will continue to compound the major concern of the RPD sector, that is the scarce number of good professionals and lack of new talent. Overall the lack of skilled professionals will cause companies to limit growth and/or acquire companies in order to grow. The process of acquisition will introduce further non-standard dynamics into the sector and add even further to the list specialist knowledge and skills needed for successful operation.

REFERENCES

(1) Barlow D. Chairmans Opening Remarks, Mouldmaking 2K, Institute of Materials, Coventry, 17th February 2000.

(2) Styger L E J. Adding Value through Product Development, Engineering Designer, November/December 1999, Institution of Engineering Designers, ISSN 0013-7858

(3) Anon. SME News March/April 1998

(4) Anon. DTI Technology Foresight First Interim Report, March 1998

(5) Confidential Privately Commissioned Industrial Report April 1998

(6) Telephone interview with N Ballard, Rambashi, 12th May 1999.

(7) Park J D, Mass Consulting Group, Carrington Business Park, Manchester, UK, February 1998.

(8) Styger L E J. Final Report Feasibility Study into a Collaborative Support Network for the UK Rapid Product Development Sector, Institution of Engineering Designers, May 1999. ISBN 0 9535796 0 3

(9) Vernon R. International Investment And International Trade In The Product Cycle, Quarterly Journal of Economics, 1966. Pages 190-207

Rapid design and manufacture – its role in the acquisition of professional artistry and judgement

G GREEN
Department of Mechanical Engineering, University of Glasgow, UK

SYNOPSIS

Young engineers and designers are increasingly expected to make an immediate contribution to the profitability of their employers. This suggests that students need to develop an appropriate level of professional know-how and judgement during their undergraduate study. It is argued that the above can, in part, be economically achieved via exposure to Rapid Design and Manufacture (RD&M) technologies offering rapid feedback on design ideas and opportunity for reflection, evaluation and judgement.

1. INTRODUCTION

Industry increasingly expects young engineering and design graduates to make an immediate contribution to the productivity and profitability of their business. Equally, the need to develop what Donald Schon refers to as 'reflective practice' (1) is gradually being accepted in engineering education. Both these issues require the adoption of 'learning by doing' and 'joint experimentation' modes of teaching integrated with more traditional teaching methods.
This approach to education supports the development of professional know-how and judgement within young engineers, a necessity in the education of future engineers and designers if they are to make the immediate contribution demanded by industry.

The issues preventing or delaying implementation of the above teaching modes are:

- Lack of specific knowledge of company technology or product
- Lack of confidence in acquired knowledge and skill
- Lack of practice in open-ended problem solving and development of know-how

The first of these issues is of course difficult to address in an educational environment. However some benefit can be achieved by linking projects, particularly final year projects, to specific industries. In this way a student can acquire some knowledge of the technology relating to a particular company or industrial activity.

The remaining two issues are increasingly being tackled by higher educational institutions throughout Europe (2). This paper describes how the provision of Rapid Design and Manufacture technologies can particularly assist in helping to increase confidence and develop know-how.

2. REFLECTIVE PRACTICE & EVALUATION

Schon recognised that the prevalent view of engineering education ignored the ability of experienced professional practitioners to deal with uncertain, conflicting and often unique problem situations where the required knowledge is revealed only in the action of doing. Schon termed this professional activity '*Reflection-in-Action*' and argued that students must learn '*a kind of reflection-in-action that goes beyond statable rules*'. To achieve the above he suggested employing two modes of operation within a teaching environment: telling and listening, demonstrating and imitating. The first mode of operation is familiar to all having followed an engineering education. The second mode is perhaps less familiar, requiring the following two main learning modes: follow me, joint experimentation.

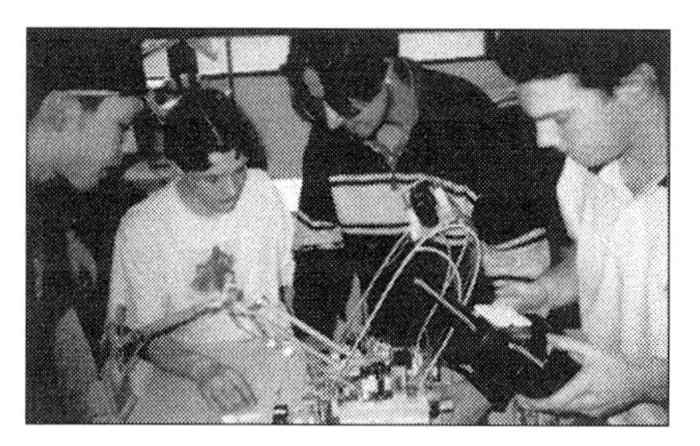

Fig. 1 Design, make and test – mobile robot

The first of these learning modes, *Follow me*, requires the lecturer to demonstrate how to design something followed by imitation by the student. The student doesn't have to understand exactly what is going on at the time, they have to accept that with practice will come appreciation and understanding.

The second learning mode, *joint experimentation*, requires lecturer and student to work together on solving an unfamiliar, open-ended problem. The student must feel confident enough to make suggestions and to criticise those made by the lecturer. In this environment the lecturer must be robust and willing to admit that they may not know the answer to the problem.

Fig. 2 Design, make and test – mobile robot

Both learning modes benefit from the provision of design, make and test resources. This enables the student to experiment through trial and error and to gain tangible feedback on the result of their efforts. It also enhances the ability of students to critically examine, or evaluate, their emerging design ideas and to link theory and knowledge with practical implementation. Clearly the use of physical models to aid evaluation is not new, they have been used extensively over many years to illustrate the link between design and manufacture (Figure 1,2). Additionally, they provide an opportunity for physical test and redesign to determine usability and aesthetic quality (Figure 2).

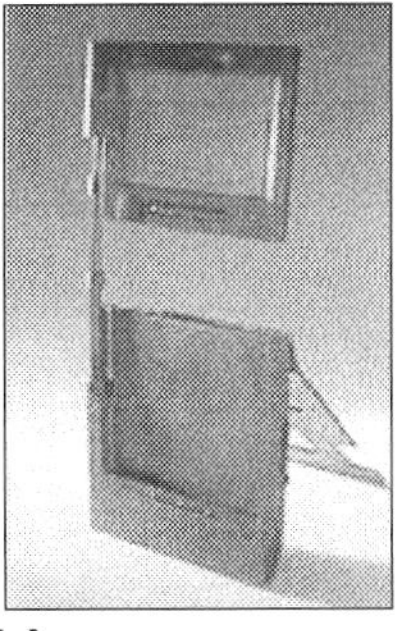

Fig. 3 Traditional appearance models

Physical models are an essential tool in developing judgement via critical evaluation and reflection. They are but one of the many forms of 'Design Representation', ranging from sketches (3) through to working prototypes. Each provides a mechanism enabling comparison between the options being proposed, and what is required of them. It is important that critical evaluation and reflection be maintained throughout the design process and that the appropriate methods and techniques are applied at the correct points within the process.

The following figure (Fig.4) illustrates one possible model of Integrated Design Evaluation that recognises the diversity of approach that must be learned and applied to allow effective evaluation and judgement of proposals to flourish (4). The range of evaluative judgements that can be made via comparison of the set of familiar design representations is evident within the model. It is through the skill of critical evaluation that the professional artistry of the engineer and designer develops. The acquisition of know-how and confidence in knowledge gained enables elegant solutions to complex problems to emerge.

Through the opportunity of rapid evaluation of options a student quickly begins to recognise elements of 'good' design, to see what works and to understand why it works. The key issue is of course the lack of time and resource to provide and sustain this 'rich' learning experience. It is not a new approach, crafts have employed this learning approach for centuries. The challenge is to merge learning while doing with other learning methods to provide an appropriate educational experience accessible by significant numbers of students. If such a learning experience can be made available in a cost effective and rapid fashion then the possibility exists for a well balanced learning method to emerge, one that both extends and challenges the intellect whilst aiding maturity of judgement. It is important to recognise that the above approach is limited in terms of both the time and cost required. Inevitably the number of iterations is limited, the opportunity for converging on optimum solutions is denied in many cases.

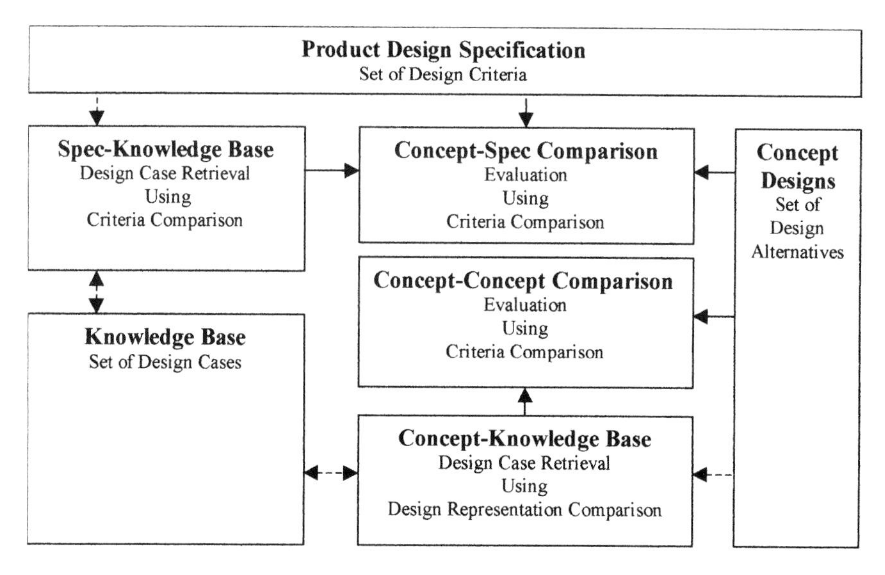

Fig. 4 Integrated Design Evaluation (IDE)

It is necessary to develop the capacity for students to generate ideas, then manufacture and test them quickly and often. Emerging rapid design and manufacture technologies offer a contribution to the achievement of this goal.

3. RAPID DESIGN AND MANUFACTURE

In recognition of the educational challenges/benefits and the associated costs involved, staff from the Universities of Glasgow, Strathclyde, Glasgow Caledonian and the Glasgow School of Art joined forces to submit a successful application to the Scottish Higher Education Funding Council (SHEFC). The resulting grant being used to establish a Rapid Design and Manufacture Centre (Fig.5).

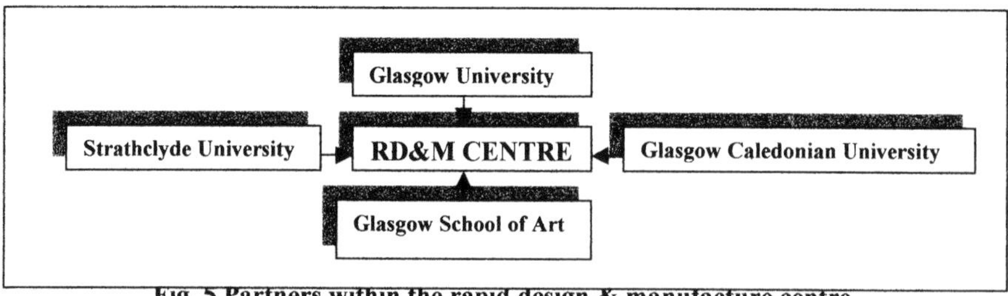

Fig. 5 Partners within the rapid design & manufacture centre

The RD&M Centre aims to undertake research, provide education and training to local industry and to support undergraduate education. It is the latter aim that is addressed here and is achieved via providing the opportunity for students to have physical models made of their design ideas in a time scale that permits reflection within the educational time limits.

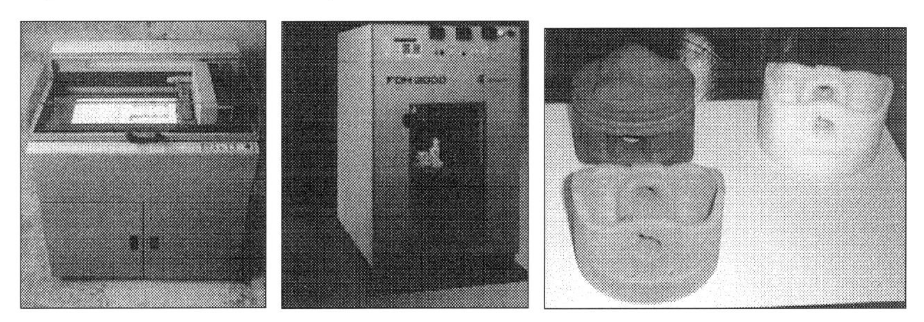

Fig. 6 Rapid prototyping technologies

At present the Centre is equipped with two Rapid Prototyping technologies (Fig 6), chosen to meet the needs of education and research whilst complimenting other technologies commercially available within Central Scotland. The technologies are:

- 3D Printing
- Fused Deposition Modelling

Each educational partner has a satellite site that allows students to generate CAD files for direct transmission to the Centre. Models can then be provided within a matter of hours making them available to the student for evaluation at there next project, lab, or tutorial session. Prior to the establishment of the RD&M Centre this 'rich' feedback would have been extremely expensive and would not have been achievable within the constraints of educational timetables.

3.1 Virtual design institute (VDI)

The RD&M Centre forms one third of the Virtual Design Institute (Fig. 7), once again the result of collaboration between all the Higher Educational Institutions in Glasgow.

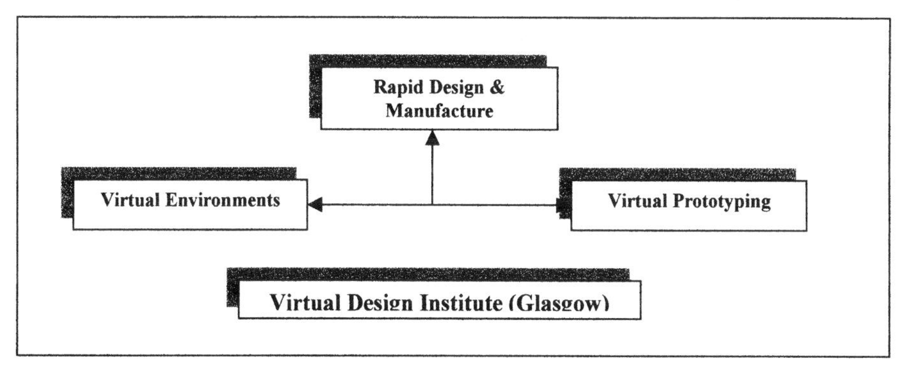

Fig. 7 Structure of the virtual design institute

This, perhaps unique, combination of facilities offers not only great opportunities to pursue research into the development of advanced design tools but also provides the capacity for students to experience the benefits of enhanced virtual prototyping of products and environments (Fig.8)

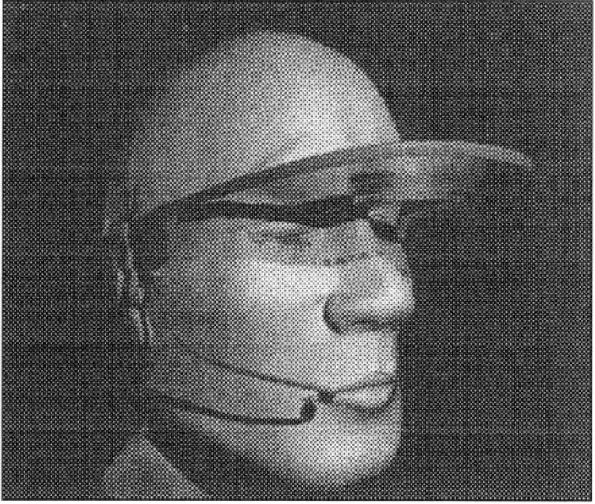

Fig. 8 Virtual environment and virtual prototyping

Clearly, the future holds the promise of the emergence of hybrid prototypes that will allow students exposure to an ever more 'rich' learning environment

4. EXAMPLE

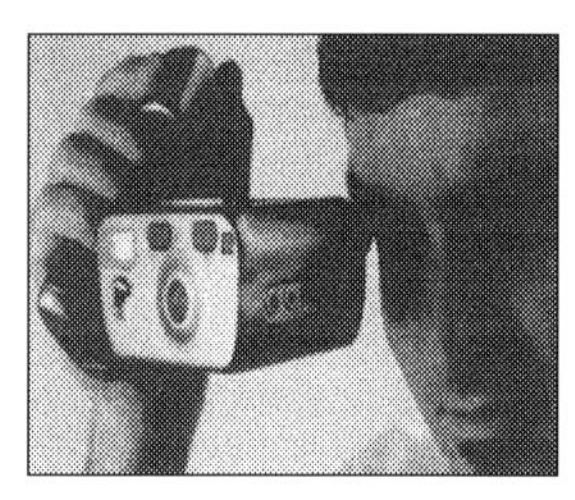

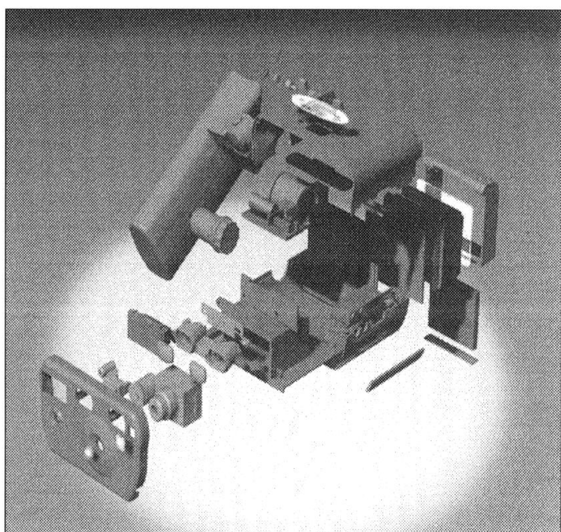

Fig. 9 Student project – design of camera

The following example provides illustration of the know-how and judgement that students are able to acquire when exposed to a learning environment that incorporates access to Rapid Design and Manufacture technologies.

The design of a camera (Fig.9) provides a good example of how RD&M technologies combine with Virtual Prototyping to permit a student to develop design ideas to a more complete level than was economically possible in the past. The images in Figure 9 show the development of the camera design from a CAD model through to a rapid prototype part and onto a presentation model. This particular project was undertaken by a 4th year Master of Engineering student of the Product Design Engineering degree programme, taught jointly by the University of Glasgow and the Glasgow School of Art. In this case the student was able to use the RP model to confirm manufacturing decisions and to validate product assembly and tolerance issues. Although issues could have been considered via 2D drawings the opportunity for the student to modify and test his design ideas would have been reduced. Equally the time available to the student to engage the issues of optimisation would have been restricted.

The unique result of this process is that the student is able to take their ideas as far as possible which has the effect of boasting student morale and enthusiasm for design and engineering because they get rapid feedback on the results of their efforts. Their confidence in their ability is increased; they see the reason for learning; they taste the 'fear' of getting it wrong. In short, they get an insight into the artistry and judgement that characterise a professional.

5. CONCLUSION

This paper has argued the need for engineering and design education to engage the professional know-how and judgement ability of students in parallel with the extension of knowledge and intellectual development. Expose to open-ended, design, make and test activities is identified as a very effective educational tool in support of the development of professional judgement. It is also suggested that the above can be achieved, within the economic and time constraints of the curriculum, with the provision of rapid design and manufacture technologies.

The current status and future development of a Rapid Design and Manufacture (RD&M) facility that, in part, supports engineering and design education within four higher education institutions within the city of Glasgow is described. The place of this RD&M facility within the Virtual Design Institute (VDI) is also addressed. It is argued that the ability to provide rapid and tangible/physical feedback to students on the success, or otherwise, of their design process is an essential ingredient in the acquisition of professional artistry and judgement within students. An example, taken from recent and current student work, was used to illustrate the effectiveness of the above educational approach.

REFERENCES

(1) Schon, D.A., *Educating the Reflective Practitioner*, Jossey-Bass Publishers, 1987, ISBN 155542 220 9

(2) Green, G., Gerson, P.M., Open Dynamic Design – towards a European model for an engineering design curriculum, in: N.P. Juster (Ed), Proceedings of 21st SEED Annual Design Conference and 6th National Conference on Product Design Education, 7-8 September, Glasgow, UK, 1999, ISBN 1 86058 208 7

(3) McGown, A., Green. G., Rodgers, P.A., Using concept sketches to track design progress, 4th Design Thinking Research Symposium: Design Representation, Massachusetts Institute of Technology, Cambridge, MA, USA, April 23-25, 1999, pp. I.89-I.108

(4) Green, G., Towards an Integrated Design Evaluation (IDE) Tool, *Acta Polytechnica-Journal of Advanced Engineering Design*, No. 4/2000, ISSN 1210 2709

Merging Design Cultures

The need by industry for practical skills in design education

K L EDWARDS
Elmac Group Limited, Newark, and School of Engineering, University of Derby, UK

SYNOPSIS

Manufacturing industry, in particular small to medium sized companies, demands a workforce comprising of people with a broad range of skills. In these types of enterprise, it is normal for employees to have to be resourceful and flexible in their approach, often working outside their immediate area of expertise and responsibility. A key aspect of success is a practical 'hands-on' approach to problem solving. The design function is especially affected because of its integrating nature and because of its positioning between art and engineering. Designing is also a team based activity, involving compromise, with the product being manufactured by others, whereas a craftsperson works alone, avoids compromise, and makes the end product. These fundamental differences in philosophy and working method present a dichotomy between the cultures of craft and design. The problem is compounded by a growing trend in education towards teaching more theory rather at the expense of practice. This paper discusses the value, to industry, of an appropriate mix of practical skills, and makes suggestions for improving the role of education in providing engineering graduates better equipped for a career in design.

1. SETTING THE OVERALL PROBLEM IN CONTEXT

Industry and industrial practices, mostly as a result of improvements in technology and the need to compete in a global market place, are changing at an incredibly fast rate. To cope with the culture of continual change, employers require a flexible work force, able to easily conform and adapt with the business (1). The need to continually review, plan and update the skills of employees as part of a life long learning process has never been so important. This not only preserves the performance of the business but also helps to retain key staff, avoiding the need to recruit unless it is absolutely essential. Therefore when it is necessary to recruit new staff, for whatever the reason, it is very useful to hire staff with up to date and appropriate experience. The situation described is ideal and a certain amount of training is normally required. However, most companies have been forced through the necessity to improve efficiency, to operate as 'lean' enterprises, leaving little or no resource available for intensive training of new staff. This situation restricts the introduction of new graduates, who have limited experience, making them difficult to integrate quickly into a business. The problem is particularly acute in small to medium sized companies (SME's), the majority employer in the UK. This is a sector not a company with a large and diverse range of problems that are difficult to tackle directly other than indirectly through networks.

Traditionally, the higher education institutions distinguished between education and training, concentrating on education and leaving training as a postgraduate activity in employment. The colleges of further education tended to have the bulk of their activity in part-time vocational courses and the more localised recruitment prompted greater concern with techniques of immediate application for local industry. The professional institutions took on the role of qualifying associations to define minimum standards for corporate membership. Typically these standards had three elements: education, training, and a minimum period in a position of responsibility as a practising professional engineer. However, the patterns of employment and to a lesser extent the engineering education system has been changing significantly over the last couple of decades. The trend is now clearly moving towards universities providing almost 'ready-made' engineers equipped with a skills-base that with a small amount of initial professional development allow them to become productive employees in the shortest possible time. The paper that follows discusses, through personal experience, the employment challenges facing industry, with reference to design engineers working in small companies, and what education in particular higher education can do to more readily help the situation.

2. RECONCILING THE DIFFERENCES BETWEEN INDUSTRY AND EDUCATION

In general, industry tends to operate to shorter time scales than universities, although it does work to long term business plans. Also, unexpected problems, when they occur, demand resource to be brought to bear to affect rapid solutions and to ensure minimal adverse effect on business performance. This is often carried out using 'standard engineering practice' and may be criticised for not having sufficient academic rigour. A specific training requirement normally follows. Universities on the other hand operate to longer time scales and are not set presently set up to be reactive enough to help solve industries' specific immediate problems and short term training needs. They can and do however collaborate quite effectively with industry on longer-term ventures that are more of a strategic nature.

The differences between industry and universities can be analysed from the perspective of their motivating factors or 'drivers'. Industry is driven by changes in technology and consumer demand (2). In simple terms, industry is motivated by profit and satisfying its customers, although this may not always be the ultimate end user of their product or service. Once legislative requirements are met they can selfishly focus on delivery (to time scale and budgetary constraints). Universities though have many different types of customer: students, government, industry and the community. They have limited resource and have to prioritise where to best place their effort while at the same time maximising income generation. Industry, while very important and essential to universities, is therefore only one of its customers. Universities, like companies, now tend to operate as lean enterprises with most of their staff fully utilised. If there is to be more effective relationships between industry and universities, partnership arrangements may need to be established, with finance and resource contributed from both sides.

Universities, as well as other types of educational establishment, are providers of people for use by industry (and others). It is therefore important to educate students with the key skills industry wants. The demands on industry today though are constantly changing with flexible working patterns the norm. The working environment is multidiscipline, either as individuals with broad ranging technical skills, or within multidisciplinary teams. Industry can no longer afford, in particular SME's, to spend considerable time and money training its new employees. There is in fact positive discrimination being practised in favour of older more experienced people over new graduates. Industry requires people with a broad technical education (such as engineering principles, design applications, information technology and business awareness) and in transferable skills (such as communication, problem solving and team working), specific product knowledge is best provided by companies themselves. To facilitate this, industry needs to be more pro-active and directly involved in curriculum development and if necessary assist in its delivery. This means universities will have to become involved more in directed training and development, complementing traditional study.

3. THE CURRENT CLIMATE FOR HIGHER EDUCATION PROVISION

The following extract from a report by the Engineering Employers' Federation, A New Millennium of Learning for Engineering (3), conveniently summarises the important future requirements for higher education provision:

- The higher education system as a whole must become more flexible to meet the changing needs of industry, society and individuals.
- 'Time-out' routes involving several years of intermittent work before graduating.
- An increase in local provision of higher education in response to grant withdrawals.
- Make more use of the information revolution and of learning methods pioneered by the Open University.
- Improved lines of communication between Universities and engineering employers, especially SME's.
- The UK's learning system must supply the right number of appropriately qualified people.
- The real engineering skill shortage currently lies at the intermediate or Incorporated Engineer (IEng) level rather than the Chartered Engineer (CEng) or Engineering Technician (EngTech) levels, and is likely to do so for some years. The levels are defined by the Engineering Council in their policy document 'Standards and Routes to Registration' (SARTOR 97), available in two parts (4).
- Those universities with a history of excellence in research will not necessarily be those which develop a future excellence in teaching courses to meet these new requirements.

Further, the Quality Assurance Agency (QAA) benchmark standards (5) refer to four sets of skills, summarised as:

- Knowledge and understanding
- Intellectual abilities
- Practical skills
- General transferable skills.

These are each mapped against the following headings:

- Mathematics.
- Science
- Information technology
- Design

- Business context
- Engineering practice.

Clearly, there is some agreement with the needs of industry, as already outlined and discussed. Interpreting the skill requirements, with the involvement of industry, is an important prerequisite to matching the education supply to the industrial demand.

There is also synergy with initiatives at a regional level. The 'Skills Action Plan' of the East Midlands Development Agency (EMDA), one of the government's Regional Development Agencies (RDA's) builds on the issues of a regional economic strategy to propose six priorities for action (6):

- Raise basic skill levels
- Promote the benefits of learning
- Make learning accessible
- Identify and tackle skills shortages
- Improve the skills of people at work
- Raise pre-16 attainment levels.

All of these priorities are not directly relevant but reinforces the scope of the problem, setting the needs of engineering and engineering design into the broader economic context.

4. THE NEED FOR MORE PRACTICAL SKILLS TRAINING

In higher education, a graduate gains not only a factual body of knowledge but also the methodology of thought associated with it. This is extremely valuable to an employer but because of the theoretical basis, its practical application is initially limited. In fact, the lack of practical skills, especially in the early years of employment, frustrates the development of the employee, extending the time taken to become a fully productive worker. The relevant dictionary definition of practical states: 'relating to action or real existence' or 'given to action rather than theory'. If the emphasis of the learning process in education can be changed to accommodate more practical skills acquisition, this will reduce the amount of time required to become productive and make the graduate more attractive to industry. A solution to this problem, that does not necessarily mean teaching different subjects, is to teach in a way that is more applicable to the industrial situation. This approach will mean a move towards more problem-based, team-working project oriented activities. Those universities already teaching design courses (engineering, product, industrial and similar types) will be familiar with this approach and the problems associated with its delivery such as increased resource levels for project work and a greater emphasis on continual assessment, and methods of assessing group work. The earlier this is begun in undergraduate engineering courses the better because it will provide the vital underpinning for the subsequent acquisition and application of more detailed knowledge to follow. Design in simple terms is about solving problems and communicating their solution. In reality, design is about creating the definition of a new or improved product to achieve a given function, performance, reliability and cost. In undergraduate engineering courses, design can be used throughout, instead of a separate subject, to provide that vital integrating focus, unifying the other technical subjects. Most students would find the experience of solving design problems initially difficult to assimilate but in time the relevance would become apparent. The outcome if applied successfully would obviate the age-old problem of irrelevance and dullness associated with teaching subjects in isolation, thereby stimulating interest. The experience would also be invaluable in synthesising solutions, a highly desirable skill to acquire.

5. WHAT SORT OF PEOPLE DOES INDUSTRY WANT IN THE FUTURE?

Looking into the future, five years ahead say, industry will require a work force that has a broad mix of technical knowledge, enhanced interpersonal skills, and commercial awareness (7). An even greater emphasis will be placed on the ability to articulate effectively and work as part of a team. To cope with a culture of continuous change, individuals will have to be extremely good problem solvers, resourceful, practical and above all flexible in their approach. In truth, industry desperately wants these people now, for the future these skills will need to be honed further.

The following list of personal attributes typifies the key skill requirements of today's successful engineer and it is not exhaustive:

- Practical with a degree of craft skills and ability to design experiments.
- Flexible/adaptable in approach.
- Resourceful and able to compromise intelligently.
- Self-motivated and independent thinker.
- Interdisciplinary team player and ability to lead when appropriate.
- Broad based technical knowledge – engineering and scientific principles, materials, manufacturing processes, etc.
- Numerate – basic calculations and estimating.
- Creative problem solver.
- Commercially aware – financial, contractual, and project management.
- Effective and articulate communicator – written and oral.
- Information Technology (IT) literate – word processing, spreadsheets, databases, computer-aided design, modelling and engineering analysis.

In addition, with the influence of the global economy and membership of the European Union, linguistic proficiency may also form a key part of the modern engineer's portfolio of skills. Specific product knowledge is often best acquired on the job by the employer, building on engineering principles acquired in education. It is acknowledged that the skill set described above will take time to acquire, improving with experience. Adapting the education system to deliver 'industry-friendly' young engineers is essential and immediate.

The personal attributes described are relevant to most manufacturing organisations but especially to SME's. Additionally for SME's, basic craft skills (workshop processes in mechanical and electrical technology) are desirable for fabricating prototypes and building test equipment. The education system is already beginning to adapt to the new demands placed on it but is slow to change. Over the next five years, new graduates will be expected to be, with the minimal of product training, effective employees contributing to the business almost from the day they start. The demands on industry are now such that it cannot afford to extensively train its workforce before it becomes productive. Also the need to continuously train, to keep up to date with new technologies and processes, through continual professional development (CPD) will increase (8). This training may need to be done via distance learning and greater access to modern university learning resource centres remotely, via IT, from the workplace and/or home. Universities need to follow the lead of industry by becoming more flexible in their approach and more pro-active in identifying future demands. This market driven customer led approach will be no different for this or any other university in the UK. The financing of education, with less government funding going to higher education and withdrawal of grants to students, is already affecting the situation. The need is greater than it has ever been for universities to collaborate with industry, to work together in developing business and training its future work force. A consequence of these financial changes is a move towards more local provision of higher education in a similar way to the further education sector.

6. THE CASE FOR ENGINEERING DESIGN

Design is now recognised by the government as one of the key priority sectors (9). Many companies, large and small, now see design as an important part of the business process. Design crosses many disciplines and for a long time proved difficult to categorise as a subject in university engineering degree courses. Today it is a conventional academic subject in its own right and degree courses totally dedicated to design are now common and offered by many universities. There is also considerable research effort going into developing the design process itself in order to design products more efficiently and effectively. A tangible output from the research is design support software to help designers develop new product more quickly and easily. The importance of working with industry cannot be overstated because it imparts practicality to the research. The other aspect of design and engineering research in general is the importance of disseminating the results where appropriate into teaching programmes, keeping the content of courses up to date. The research and/or the teaching outcomes however tend to be tailored towards the needs of the larger engineering companies and may not be appropriate to the needs of the SME. Today, industry and small companies in particular, as has been stated earlier, require a practical multi-skilled workforce. Engineering design, essentially a problem-solving process, is seen as pivotal in all of this, providing the integrating framework for engineering course subjects and with industry's assistance, the key skills for employment.

The problem of not producing the people industry wants is compounded by a growing trend in education towards teaching more and more theory at the expense of less and less practice. Although influenced by economics, this tends to produce design graduates with a deficiency in practical skills. This is compounded by an increasing reliance on the use of computers. In design, computers have an obvious and significant benefit for communicating manufacturing information, e.g. computer-aided design (CAD), engineering analysis, e.g. finite element analysis (FEA), and for simulation and modelling purposes. The use of modelling in design teaching typifies the situation where the use of computers is dominating the learning process. There is a place for computer modelling, but physical modelling in parallel is essential to the success of the design process. Typically, designers make physical models to validate design decisions, manufacture prototypes prior to full-scale production, and build and operate test equipment to check for correct functioning and performance of products. A culture needs to be developed of marrying real (physical) and virtual (computer) modelling, rather than using one at the expense of the other. For the future a balance needs to be struck between the theoretical and practical content of engineering design courses. This is best achieved with the collaboration of industry who have a vested interest in receiving appropriately qualified people with a blend of immediately useful and up to date technical knowledge and practical experience.

7. THE WAY FORWARD

There is a clear need, throughout the economy, for articulate team-oriented problem solvers with a broad grasp of business processes. Industry is changing rapidly and is dominated in the UK by the diversity of SME's. The current engineering courses are slow at producing engineers with the necessary skill base to enable them to integrate quickly into employment and equip them for life long learning. A mixture of work based learning and theory, the former setting the latter in context, would appear to be fundamental to the way forward. The recommended approach for future engineering course provision is the adoption of a philosophy of problem-based learning, underpinned by design, in close collaboration with industry.

REFERENCES

(1) The Investors in People Standard, Investors in People UK, 1996.
(2) Freeman-Bell, G and Balkwill, J. Management in Engineering – Principles and Practice, 2nd Edition, Prentice Hall, 1996.
(3) A New Millenium of Learning for Engineering, Engineering Employers' Federation, 1997.
(4) SARTOR 3rd Edition, The Engineering Council, 1997.
(5) QAA Benchmarking Standards for Engineering, 1999.
(6) Skills Action Plan, East Midlands Development Agency, 2000.
(7) Boyce, T. The Commercial Engineer, Hawksmere, 1992.
(8) Reach New Heights, New CPD Framework, IMechE/IEE/IIE, 1999.
(9) Labour Market and Skills Trends, DfEE, 1998.

Design futures – future designers

P A RODGERS and **A I MILTON**
Department of Design, Napier University, Edinburgh, UK

ABSTRACT

The tasks designers will face in the near-to-medium term future will undoubtedly be ever more demanding and complex. The convergence of computing and communications technologies will enable designers to collaborate and share tasks in design and development without ever having to meet physically. Moreover, recent social, cultural and economic shifts throughout the world will demand designers to be ever more flexible in the work that they are asked to undertake. Future designers will, thus, need to be sensitive to the demands from many specific sectors of the global marketplace, be able to exploit fully developing technologies as they evolve, yet be increasingly speculative in synthesising their solutions whilst challenging conventional wisdom. This paper describes a new Interdisciplinary Design programme that aims to provide students with the necessary skills to achieve success in the design world of the future.

1. INTRODUCTION

What will the design process of the future be like? What tasks will designers of the future have to undertake? With the confluence of computing and communications technologies developing ever faster, most notably observed in the widespread adoption of the World Wide Web, it is becoming increasingly difficult to predict accurately the future roles designers will have to face [1]. Developments in Internet technologies have enabled designers to communicate and collaborate in product development activities without ever having to meet physically, accelerating the transformation of design practice into a fully distributed process undertaken by "virtual" design teams [2]. Coupled with this, the significant social, cultural and technological shifts that have been witnessed during the late 1980s continuing through the 1990s have resulted in demands for designers to be more flexible in the tasks that they are asked to fulfil. Future designers will, thus, have a range of developments to take advantage of.

However, designers of the future should recognise that to fully exploit these developments they will have to be more creative and innovative, be increasingly speculative in synthesising solutions, challenge conventional approaches to new design, and have a sound critical understanding of visual, historical and cultural design totems. To this end, this paper details an Interdisciplinary Design programme that requires students to be accomplished in methods of their precedent analyses and to adopt a comprehensive and critical interpretation in relation to contemporary design practice, such as in the field of interactive games design, consumer desire, and speculative design.

2. FUTURE DESIGN WORKING PRACTICES

Presently, and for the foreseeable future, dispersed working practices, where design teams are located in different parts of the world will become commonplace. This is already a reality in design sectors such as the aerospace and the automotive industry [3, 4]. Dispersed styles of design working obviously make it difficult for designers to meet (physically) face-to-face due to the limitations of time and the costs involved. Thus, designers have to rely on communication and collaboration support tools to assist in their day-to-day design and decision making activities. An increasingly large variety of communication and collaboration tools are commonly used in modern design work, including video conferencing, electronic mail, the WWW and so on (Figure 1).

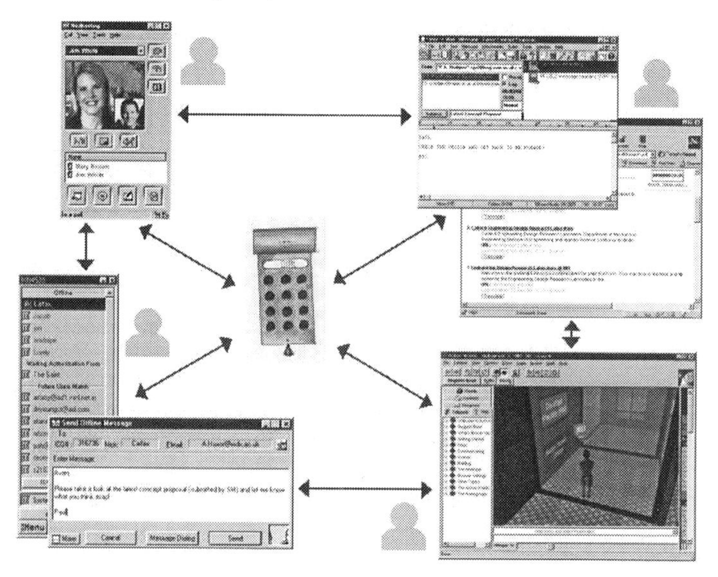

Fig. 1 Collaboration and Communication Trends in Modern Design Tasks

These tools include, clockwise from top left, NetMeeting, the Internet (*i.e.* email, web), 3-D virtual environments (*e.g.* Active Worlds) and ICQ. NetMeeting is a video conferencing package which provides a number of features including audio and video conferencing, text-chat, shared 'whiteboard' and the ability to share other Windows applications. NetMeeting is free, it provides good functionality and the audio and video are of a suitable quality for designing over the Internet. Specific applications for which NetMeeting is highly appropriate are one-to-one communications and where the collaboration goes beyond discussion alone (*i.e.* sketching, working on documents together, and so on). The main disadvantages of NetMeeting are that it is only available for Windows 95, 98, and NT and that only two participants can communicate using full multimedia at any one time.

The Internet, including email and the world wide web (WWW), is used widely by designers as a mechanism for collaborating and communicating with others, and for sending and receiving design information [5]. Specific methods used to communicate with others include electronic mailing lists[1] and news groups[2] and Internet-based communication and collaboration tools.

Active Worlds is a 3-D virtual environment that allows users to inhabit and communicate with other users who are logged in to the same Active World at the same time. In short, an Active World is a shared experience using the Active Worlds browser as an intermediary and common language. Users communicate with each other via their own personal avatar (*i.e.* a computer representation of the user). Active Worlds provides an infrastructure for electronic "social networks" or "communities" that enhance many of the "social activities" [6] that are prevalent within design work.

ICQ ('I seek you') is what is known as a "buddy list". This means that a user can be alerted when a colleague or friend is available on-line. Users add other nominated users to their contacts' list, and when any of these go on-line, the interface will make this clear by highlighting the name of that person meaning that they can be requested to have a real-time chat. If they are off-line, there are other, asynchronous, messaging systems to allow for communication. The ICQ software is free, "lightweight" and is available for a range of different hard/software platforms.

3. BDES INTERDISCIPLINARY DESIGN COURSE

3.1 Interdisciplinary Design Background

Interdisciplinary Design is a radical, intellectually stimulating programme which aims to equip students with the skills and knowledge required to exploit global opportunities in a rapidly changing profession. During the 1990s the blurring of design specialisms, the merging of digital media, and fluid patterns of employment opened new opportunities for interdisciplinary designers capable of communicating with a wide range of specialists, adaptable to new technologies, and above all able to conceptualise and creatively respond to commercial and cultural change. The BDes Interdisciplinary Design course draws on two key areas:

- ***Object*** – exploration and engagement with society through the conception and design of new objects informed by both industrial and product design;
- ***Spatial*** – through a thorough analysis of architectural and spatial design frames of reference including exhibition, retail and environment design.

3.2 Interdisciplinary Design Definitions

Interdisciplinary Design is defined, within the context of the design department of Napier University, as:

- a broad field of 3-D design including lighting, packaging, furniture, transportation and environment design that fall within the established programmes of Industrial Design and

[1] *e.g.* design-research@mailbase.ac.uk; idforum@yorku.ca
[2] *e.g.* alt.cad; comp.cad.pro-engineer

Interior Architecture and associated postgraduate studies in Material Culture at Napier University's design department.

- an engagement with branches of knowledge concerning the wider design environment such as consumer lifestyles, social and ecological issues, technological change and cultural context that together define a significant part of modern design practice.
- a convergence of digital technologies in design such as the Internet and electronic media, digitised text and imagery, computer-aided design and simulation which allows designers to move freely between once discrete design specialisms.

The expectation on every BDes student is to interrogate the world through design, to research, to reflect both analytically and critically [7], to re-conceptualise design problems, and to communicate solutions. The aim is not to pursue these as ends in themselves, but to act as the process for creative practice in interdisciplinary design.

3.3 Interdisciplinary Design Philosophy

The philosophy of the BDes programme is rooted in the belief that 21st Century designers should be capable of crossing traditional disciplinary barriers and be able to respond creatively to a volatile commercial and cultural environment. Accordingly, the BDes course is process based and instils capabilities in abstract conceptualisation through symbolic, virtual, physical and scenario modelling. This paradigm offers greater opportunities for more broadly based design projects than is typical in design discipline-specific courses such as Industrial Design, Product Design and Interior Architecture, for example.

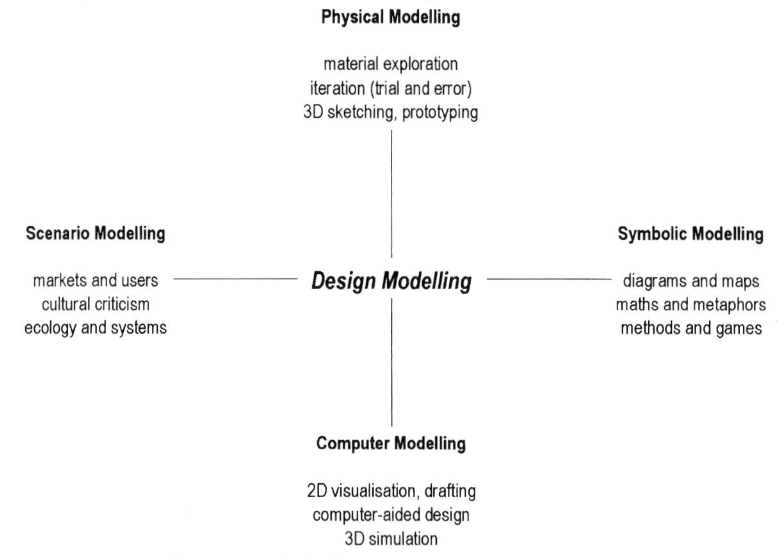

Fig. 2 Interdisciplinary Design Modelling Types

The key distinction between the BDes Interdisciplinary Design course and other three dimensional design programmes is that its aim is less to *realise* a solution through the established practice of a professional design specialism, but rather to *model* solutions as well as the brief, from the standpoint of a generalist discipline. The phrase *design realisation* is used here to cover a specialist design course (*e.g.* Furniture/Industrial Design, Interior Architecture) wherein the programme of study combines creative education with professional knowledge and specialist technological studies. *Design realisation* projects typically lead the student from brief to practical outcome, realised through professional specification and detailing, and an understanding of specialist production technologies.

By contrast, *design modelling* offers a useful and practical way to describe design as a process of re-conceptualisation that, for example, may transpose physical modelling in clay to diagrammatic analysis, computer simulation to environmental critique, marketing scenarios to graphic imagery. Thus, the outcomes of *design modelling* may be both concrete and abstract (Figure 2).

Thus, the aims of the course are to produce imaginative and creative graduates with critical and analytical abilities gained through reflective practice in three dimensional design. The major aim being to provide graduates with the skills for interrogating, modelling and envisaging material culture (*i.e.* both spatial and formal) through an understanding of contemporary design theory and practice.

4. DESIGN FUTURES: INTERDISCIPLINARY DESIGN CASE STUDY

This section provides a recent project example, "Design Futures", from year three of the Interdisciplinary Design programme. This project comprises two investigations into the "future of design". The first part is a brief set by British Telecom as part of the British Design and Art Direction (D+AD) student awards around the topic of "communication on the move" and the second is the creation of a conceptual forecasting report and design proposal which outlines a new spatial or formal design led market opportunity.

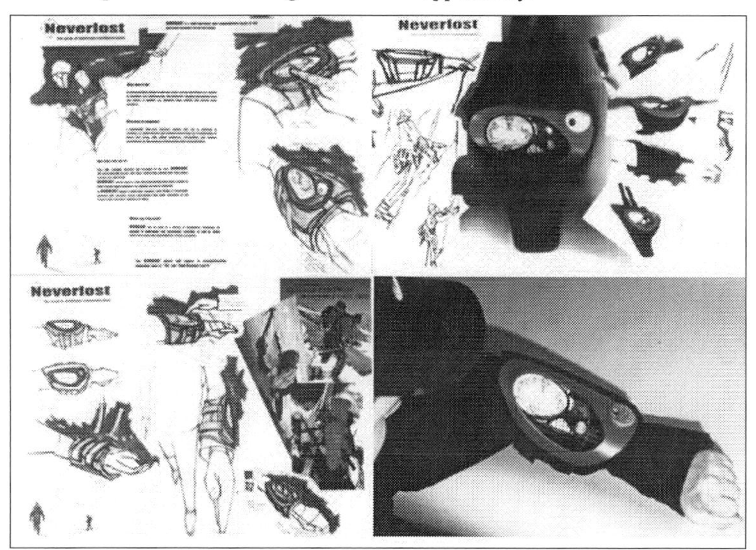

Fig. 3 Interdisciplinary Design Student Example

The example shown in Figure 3 illustrates a communication device aimed at mountain rescue personnel. The intention is that this communication device would be utilised by a range of users including the emergency services (*e.g.* police, paramedics) and both amateur and professional rescue helpers. The strap on device would afford users greater overall movement and flexibility during their rescue activities whilst offering a range of important information and communication facilities such as full audio/video communication, Internet browsing, up-to-the-minute weather reports, a global positioning system and a "homing" signal. Other proposals within this part of the project included an "interactive city map" for tourists, an "emotion" sensor/transmitter for couples and a recyclable cellular phone based around the "phonecard" idea.

5. CONCLUSIONS AND FUTURE WORK

Although the BDes Interdisciplinary Design programme has been running for a mere three years, the overall standard of the work produced is very high. Students from the more established Industrial Design and Interior Architecture programmes within the department regularly win national and International design competitions. However, recent results from cross-course projects (*i.e.* where Interior Architecture and Interdisciplinary Design students work on the same "spatial" problem) show that BDes students perform as well as, if not better than, their subject-specialist peers. Future plans are currently underway to create a MDes award which will be based on the philosophy of the BDes programme.

REFERENCES

(1) KING, P., "Design Process: The Future", *in* J. Peto (ed.), *Design Process Progress Practice*, Design Museum, London, 1999, pp. 76-87.
(2) TUIKKA, T. and SALMELA, M., "Facilitating Designer-Customer Communication in the World Wide Web", *Internet Research-Electronic Networking Applications and Policy*, Vol. 8, No. 5, 1998, pp. 442-453.
(3) ULRICH, K.T. and EPPINGER, S.D., *Product Design and Development*, McGraw-Hill, Inc., New York, USA, 1995.
(4) LINCOLN, J.R., AHMADJIAN, C.L. and MASON, E., "Organizational Learning and Purchase-Supply Relationships in Japan: Hitachi, Matsushita, and Toyota Compared", *California Management Review*, Vol. 40, No. 3, Spring 1998, pp. 241-264.
(5) RODGERS, P.A. and HUXOR, A.P., "Designing over Networks: A Review and Example of Using Internet Collaboration and Communication Tools in Design", *in* B. Jerrard (ed.), *Managing New Product Innovation*, Taylor and Francis, London, 1998, pp. 131-142.
(6) BUCCIARELLI, L.L., *Designing Engineers*, The MIT Press, Cambridge, MA, 1994.
(7) BONSIEPE, G., *Interface: An Approach to Design*, Jan van Eyck Akademie, Maastricht, The Netherlands, 1999.

Design: education and practice – closing the divide

M A C EVATT
School of Engineering, Coventry University, UK

SYNOPSIS

This paper further develops ideas in relation to the "two sides of design" i.e. "design as it is observed and design as it is practised" in terms of the "management of the design activity (the linear process model) and the creative act itself (the reflective model)" and suggests a graphical model of this holistic process. It also discusses the benefits of students being taught by academics and practitioners from both cultures i.e. engineering designers (the processors) and industrial designers (the reflectors) and the implications that this may have for course design.

1. INTRODUCTION

In (1) Dorst discussed two paradigms for design processing - namely the 'Reflective Practitioner' versus the 'Design Processor'. The former being usually thought of as the sort of activity undertaken by 'Industrial Designers' i.e. designers more involved with aesthetics and ergonomics, whilst the latter is associated with 'Engineering Designers' i.e. those more involved with performance and function.

The author in (2) further explored and developed these ideas in an attempt to clarify some of the apparent confusion between design as it is observed and design as it is practised - i.e. at the company level on one hand and by the practitioner at the other. It was suggested that the two sides of design are the management of the activity (the linear process model) and the creative act itself; the reflective model that is made up of alternate divergent and convergent operations. The former is characterised by fluent, flexible and original thinking and the latter by analytical, selective and evaluative thinking. These two different operations are the two kinds of design.

Three key points from this paper gave rise to a dilemma and are worth reiterating here: -

1] The author's observations that creativity is inhibited by the rigorous application of a design methodology. This is more typical in the teaching of engineering design rather than in the teaching of industrial or product design.
It may be contrasted with the more intuitive and reflective approach adopted by those engaged in the teaching and study of industrial design. Here the quality of the design outcome is more important than the actual journey. This approach is perfectly acceptable provided that all aspects of the design outcomes are fully justified.

2] Observations similar to these have been made at Delft Technical University. Keulen (3) characterised the two forms of design teaching and study as "learning by application" and "learning by doing". It might also noted that at Delft design in engineering courses is taught in the latter stages of the courses when students have developed the analytical engineering science skills perceived to make them ready for design activity. For the product and industrial design courses at Delft on the other hand design is taught and practised from the outset.

3] This finding is not dissimilar to that of French et al (4). They have made two interesting observations:

i) Engineering designers do not on the whole use the systematic approaches to design advocated by academics

ii) Engineering designers use the opportunistic approach to design rather than the methodological approach.

They further suggest that the former approach is more suited to incremental design than to radical design.

THE DILEMMA! It would appear that neither the opportunistic engineering designer nor the "trained by doing" product designer (or industrial designer, fashion designer or architect etc) actually follows logical design procedures when carrying on what would appear to be successful design careers.

It is often said that designers only become competent when they have internalised the design methodology, become intuitive, and moreover when they deny that a design process even exists. However, there is no doubt that the traditional academic paradigm is a good holistic description of the design process and can therefore be used to manage and control the design function.

This paper (2) also suggested that it is important that all students of design, whether they are being taught by doing or application, should experience the methods and processes traditionally adopted by each group in order to become better designers. Students need to be aware of, and to understand, the cultural differences and practices associated with the two approaches. This, it was suggested, works best when student projects are set by industrial clients and when design is accorded sufficient time within programmes of study.

2. DISCUSSION

McMahon (5) postulates that traditionally academic institutions concentrate their teaching effort toward "original" design as opposed to "incremental, variant or adaptive" design and that as, according to Pahl and Beitz (6), approximately 75% of all designed artefacts fall into this latter category design teaching should reflect this in some way. He goes on to suggest that as more and more of the analytical side of engineering may be conducted using computers more time might be devoted within design courses to the exploration of both original and variant design.

The author agrees with this in some measure but also sees some dangers in this approach.

Some recently validated courses appear to be increasing the amount of time spent on the analytical side in order to remedy a perceived lack of mathematical skills amongst students. This appears to be at the expense of design activity. Furthermore design may be seen to be subsumed into core module activities which are umbrellas where disparate activities are gathered under the guise of integration. It is hoped that enlightened institutions will realise that such core activities need to be design-led rather than discipline-led. Only then will more time be found for design.

Once again this year the author has taught groups of students studying engineering courses in the fields of aerospace, automotive and mechanical engineering. It has been observed that, more than ever, these cohorts of students do not understand the nature of what is expected as far as design is concerned. They mostly wish to be taught (or told the "right" answer) rather than go through a learning experience.

When questioned a number of students seemed to believe that engineering design was simply a matter of applying an engineering formula i.e. the next stage on from the standard type of engineering science question which may ask: -

"Calculate the maximum bending stress in a 2000mm long rectangular aluminium beam 50mm wide x 100mm deep when loaded at its centre span with a mass of 20kg. The beam is simply supported at its extremities."

This question is clearly designed to have only one solution and is about analysing a situation rather than synthesising a design solution.

The students felt comfortable with a question phrased along the lines of the following: -

"Specify the depth of a 2000mm long rectangular aluminium beam 50mm wide when loaded with a mass of 20kg at its centre span so that the deflection at mid span does not exceed 20mm. The beam is simply supported at its extremities."

Clearly there is a simple design outcome here and this is the sort of thing the students were expecting to do in my lectures i.e. more of what they are doing in engineering science modules now that science teaching is more application based.

However if the type of question moved on to require lateral thinking and the generation of many candidate solutions then dismay set in. It was not until the final part of the module where machine element selection and the procedural guide approach was introduced that they regained their accustomed cheerfulness.

This seems to support the argument that students are quite happy doing the sorts of activity associated with incremental, variant or adaptive design and find it reasonably easy. However if it is the lateral thought, the creative leap, or the innovative connection that is called for they find this difficult and need to be rehearsed as often as possible. This observation should not be a surprise as synthesis is a higher level activity than analysis and students must not be led to believe that such design synthesis is easy.

3. MERGING CULTURES IN PRACTICE

The MDes/BSc Industrial Product Design Course run at Coventry University has given the author an excellent opportunity to observe the operation of modules of design activity that are taught jointly by staff from the School of Engineering and the School of Art & Design.

In the second year of the course the group project module (three-eighths of the year's work) operates through this joint teaching format. The approach is taken further in the final year where both the individual project (0.5 of the year's work) and the final year group project are supervised jointly by staff from both the School of Art & Design and the School of Engineering. Furthermore group projects in both years are conducted with input from industrial clients or similar external sources.

These students are therefore exposed to two different cultures. The staff from the School of Art and Design tend to encourage and coach the students to draw, sketch and model many solutions The staff from the School of Engineering tend to encourage engineering rigour (both in terms of science and use of materials), project planning and adherence to specifications.

It has been found that this blend of the two cultures at the point of teaching is highly beneficial. That is not to say that it is without its problems as it represents a continuous learning experience for both the staff and the students. However the students quickly learn to distil the essence from both approaches and, it is believed, produce better design outcomes as a result.

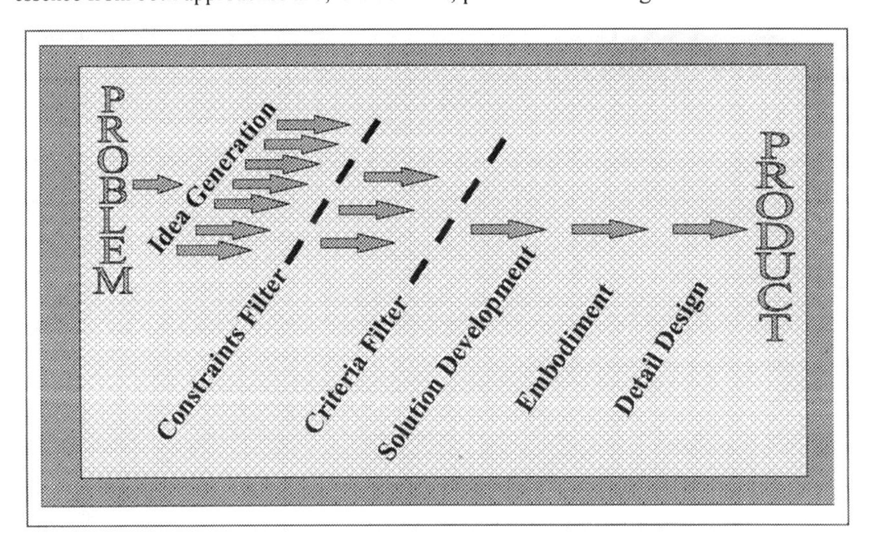

Fig. 1 A Linear Design Process Model

4. A GRAPHICAL MODEL

The author's "Problem to Product" diagram of the logical design process (figure 1) represents a starting point for the combined model. Indeed it is part way there in that it has in some measure the combined divergent/convergent 'funnel' leading to an optimum candidate solution for development at the embodiment stage. However it does not show the iterative loops or the fact that the same process happens at each stage in the execution of 'top down' design i.e. at the holistic product level, at the solution development level, at the embodiment level and at the component detail level.

Fig. 2 The Creative Loop

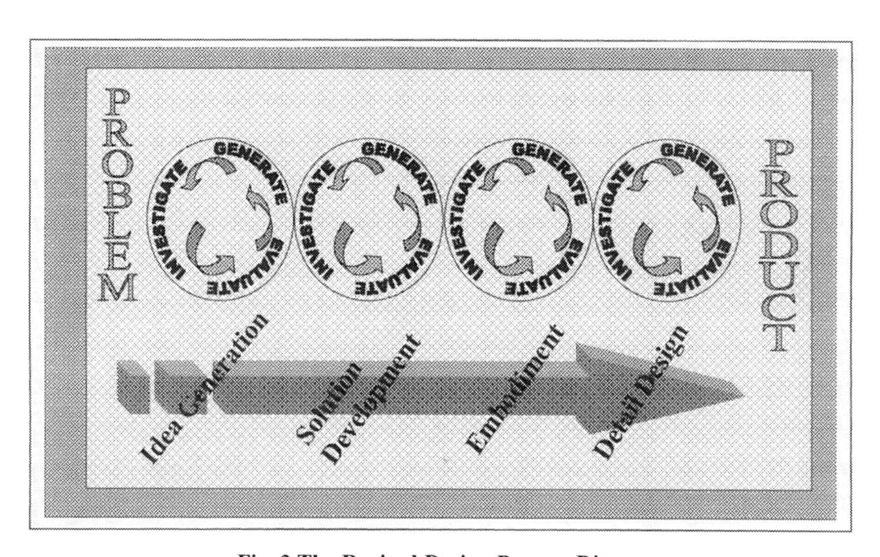

Fig. 3 The Revised Design Process Diagram

It is suggested that the 'Creative loop' (figure 2) is really the model building block as it occurs at every stage and it is what both kinds of designers practice whether they be processors or reflectors. The anticlockwise arrows indicate that the process is reflective and the deliberate randomness of their orientation suggests that a radical, tangential thought might appear. The words themselves suggest the divergent/convergent nature of the activity.

Figure 3 shows the linear design process diagram with the 'creative loops' attached at each stage. This diagram, with its iterative loops, keeps the mind focussed on the creative aspects of design whilst the large arrow emphasises the management of the process with its inevitable deadlines.

5. CONCLUDING REMARKS

From the above it may be seen that not only are there two kinds or sides of design but also two kinds of designer. Furthermore in (7) the author found that there are two types of student studying design. The students are probably best categorised as either predominately divergent thinkers or convergent thinkers. It is important all parties concerned in design education recognise and celebrate these differences. At Coventry University it has been successfully proved that on a Product Design Course staff from the two cultures working together can successfully and significantly improve students' design outcomes. It may be some time before the same can be done with the teaching of Engineering Design, however the author believes the message is clear.

With regard to the revised Design Process Diagram Figure 3 – this has been shown to some of the engineering design students who were having problems with the design modules and it appears that it is helping them understand the enormously complex process that is DESIGN.

REFERENCES

(1) Dorst K. "Two kinds of design, and two sides of design" 'ICED 99' MUNICH, Heurista, Zurich 1999 p. 1547-1552

(2) Evatt M.A.C. "Design across the boundaries" 'Design 2000' DUBROVNIK, 2000

(3) McMahon "Is a generic course for incremental design in any engineering discipline possible?" EDE 99 Glasgow, Professional Engineering Publishing Bury St Edmunds and London1999 p. 119-128

(4) Pahl & Beitz "Engineering Design", London, Springer-Verlag 2nd Edition 1996

(5) French M J, Chaplin R V, Langdon P M "A creativity aid for designers" 'ICED 93' Den Haag, Heurista, Zurich 1993 p. 53-59

(6) Van Keulen H "Design Teaching views and practices at Delft, the Netherlands" 'EDE 99' Glasgow, Professional Engineering Publishing, Bury St Edmunds and London1999 p. 51-58

(7) Evatt M.A.C. "Observation - The key to designing success" Proceedings of the 12th International Conference on Engineering Design in Munich 1999

Integrating the teaching of engineering science and design

D COLE
School of Mechanical, Materials, Manufacturing Engineering and Management, University of Nottingham, UK

SYNOPSIS

The objective of this paper is to describe recent experience of setting and assessing second year design projects that require students to apply engineering science. The motivation for setting projects of this type was to further integrate teaching of engineering science with teaching of design process. There was also a desire to adopt a more student-centred approach to learning. An example design project is described and the responses of the students to it are examined. The assessment method was found to play a role in ensuring a successful learning outcome.

1. INTRODUCTION

The University of Nottingham runs a highly regarded BEng/MEng degree course in mechanical engineering. Teaching of design features strongly in all years and is the core of the course. In the first year, design related teaching includes manufacturing processes, material properties and selection, 2D drafting, machine elements, conceptual design, design analysis, workshop practice and a design and make exercise. An objective of the first year course is to develop a base level of design awareness upon which the rest of the course can build. The second year course places emphasis on project-based learning and develops design ability to a level where students can work confidently on open-ended projects using a structured approach. Third year design activity is centred on a group design project, in which groups of four students undertake an extensive design task. In the fourth year, groups of four students work on a major project involving design, analysis, manufacture, test, marketing and financial appraisal. Complementing the design modules throughout the course are compulsory and optional modules in engineering science and related studies.

Human resources for design teaching include a Royal Academy of Engineering Visiting Professor, academic staff with design/materials/manufacturing expertise, and part-time tutors with experience of design in industry. Physical resources include extensive library and computing facilities, and a large and well-equipped workshop dedicated to undergraduate use. Class size in the second year is about 160.

The objective of this paper is to describe recent experience of setting and assessing second year design projects that require students to apply engineering science. The application of engineering science to design is an important skill that has to be learnt, and its successful use can result in a high degree of personal satisfaction for the students. The benefits of learning and applying the skill are improved decisions on the feasibility and optimisation of a design.

Teaching the skill provides opportunities for adopting a more student-centred approach to learning and for improving the integration of engineering science and design process teaching.

The next section of the paper discusses the design process and the reasons why difficulty can arise in applying engineering science. In section 3 it is argued that some principles from design optimisation theory can be used as the basis for a structured approach. An example of how this approach can be used in a conceptual design project is given in section 4. Assessment of students' work is discussed in section 5 and conclusions are given in the final section of the paper.

2. DESIGN PROCESS

The product design process can be broken down into a number of activities [1]: task clarification, conceptual design, embodiment design, detail design. The second year course exposes students to all of these activities, but this paper focuses on conceptual design. A systematic approach to conceptual design is taught, involving a list of requirements, function structure, combination (morphological) chart, and evaluation matrix. With appropriate choice of project, non-analytical skills such as creativity and synthesis are developed quite quickly. However it is necessary to discourage the use of engineering science too early in the project, otherwise too little time is devoted to these important non-analytical skills.

Once a concept has been selected and a sketch showing the working principles has been created, the work is assessed and feedback is given. The emphasis then moves towards engineering science. The students are given the task of confirming the feasibility of their concept and determining some of the key parameter values and material choices. It is here that difficulties can arise. Although the relevant engineering science and analytical techniques are taught in other parts of the course, application to the design task is not always found to be straightforward.

Ion and McCracken [2] classify design tasks into five types, A to E, see figure 1. They note that for a significant part of a degree course, problems are of type A, because they are suited to giving practice in a particular analytical technique. The single solution to the problem also makes it easy for students and staff to assess progress.

TYPE	A	B	C	D	E
INSTRUCTION	Find maximum stress	Select cross-section dimensions	Design a support for load W	Satisfy a list of requirements	Satisfy a market need
GIVEN	L b d W L, b, d, W	L b d W L, W, design stress	L W L, W	List of requirements	Market need

Fig. 1 Types of design problem. Based on a figure in [2].

Problems set in design classes need to be of type D or E, to introduce the many non-analytical skills that the design task requires in addition to engineering analysis. But it is not straightforward for students to apply engineering science to design problems effectively if they only have experience of type A or B problems. Difficulty arises because the question is not fully defined and there is not a unique answer. Without guidance, some students attempt to reduce the problem to one of type A and in doing so make restrictive assumptions or neglect important criteria. Erroneous conclusions about the feasibility or optimum performance of the design are the result. Formal methods of design optimisation provide a basis for a structured approach.

3. DESIGN OPTIMISATION

Having chosen a promising design concept the students are given the task of confirming its feasibility and determining some of the key parameter values and, if appropriate, material choices. The main learning objective is to select and use relevant principles from engineering science in the design situation. In essence the task to be performed is a design optimisation. Although classical methods for design optimisation exist, for example [3], learning their use for a general case requires significant time and would distract from achieving the main learning objective. However, it is proposed that the underlying ideas and principles of design optimisation can provide a structured approach to applying engineering science effectively to design problems. The main steps in a classical design optimisation procedure are described below, and simplifications adopted for second year conceptual design projects are explained.

1. *Function*. The function of the component or device should be clarified. The function is the primary purpose, for example, to withstand a force or to store energy.

2. *Idealise*. Sketch an idealisation of the design, showing the geometric parameters and other items of relevance, such as loads and deflections. The idealisation should be as simple as possible, including only the parameters of definite significance to the problem.

3. *Objective*. Decide on the objective of the design. The objective is that which the designer is seeking to maximise or minimise. This might for example be mass or cost. Often there are compound objectives, and techniques such as weighting factors or utility functions can be used to resolve the conflict between them [3,4]. To simplify the optimisation task a single objective should be chosen. An equation for the objective in terms of the design parameters and functional requirements should be written.

4. *Constraints*. Identify the important constraints on the design. These might include specified values, such as the stiffness of a cantilever beam, or variables that have maximum or minimum values, for example, stress in a beam. Write equations for the constrained variables in terms of the design parameters and functional requirements.

5. *Free variables*. Identify the free variables. These are the variables that the designer has freedom to vary in order to achieve the objective and satisfy the constraints. The free variables should not include any material properties (such as density or Young's modulus) because these have discrete values dependent on the material choice. There may be practical limits to the values of the free variables, but they should be checked at the end of the analysis rather than included as constraints in step 4.

6. *Number of free variables and constraints*. Determine the nature of the optimisation problem. If the number of free variables N_f is less than the number of constraints N_c, the

selection of optimum parameter values is possible, but not straightforward [3,4]. To simplify the optimisation task ensure that $N_f \geq N_c$, but note that if $N_f = N_c$ there is no freedom to choose the values of the free variables independently of the other parameters and if $N_f - N_c > 2$ there can be too much freedom to deal with easily.

7. *Eliminate free variables.* Substitute the constraint equations into the objective function, eliminating one free variable with each constraint equation. Organise the resulting equation into three groups of parameters:

objective=f_1(functional requirements) f_2(geometric parameters) f_3(material parameters)

8. *Select material.* Generally the three groups f_1 to f_3 will be uncoupled. If material parameters appear in the equation, these should be investigated first. A short list of materials that minimise (or maximize) the objective should be selected, by using either tabulated data or selection charts [4].

9. *Select values of free variables.* Examine the values of the free variables. If $N_f = N_c$ the values of the free variables will depend solely on the material choice. If $N_f > N_c$, there will be some freedom, in which case graphical relationships can be drawn to aid decisions on the values.

The simplified procedure provides a structured approach to applying engineering science to open-ended design problems. The features of the procedure are that: (i) the function, objective, constraints and free variables are considered before calculations are begun; and (ii) the numbers of objectives, free variables and constraints are chosen to ensure that the problem is straightforward to solve. The goal of the procedure is to guide students whilst requiring them to do the analytical thinking.

One of the benefits of teaching the procedure is that no other analysis should need to be taught, maximising the time available for teaching the important non-analytical skills such as creativity. There is no need at this stage to teach mathematical optimisation techniques such as steepest descent or constrained optimisation methods. Whilst these algorithms undoubtedly have an important role in design, their use can discourage obtaining a fundamental understanding of design trade-offs using simple engineering analysis.

4. EXAMPLE

Conceptual design projects are chosen to be interesting and fun, with an easily understood context and the possibility of several different solutions [1]. Care is taken to ensure that the design problems are new to the design tutors as well as the students. Small groups of students (3-5) follow a systematic design procedure and present their work on an A1-size poster that must include:

- list of requirements (including demands and wishes)
- function structure diagram
- combination (morphological) chart
- evaluation matrix (either weighting and rating, or controlled convergence)
- sketch showing the working principles of the selected design

Each group gives a ten minute verbal presentation. The posters and presentations are assessed and marked by the design tutors. Marks are normally awarded equally to each member of the group. Significant engineering calculation is not encouraged, to focus attention on the non-analytical aspects of the design process. The application of engineering science is practised in the second part of the exercise, which is performed individually. An example design project involves a human-powered device that will get a rope to a person in difficulty in water up to 50m from land. A variety of solutions are generated, but most involve firing a projectile (some sort of buoyancy aid attached to a rope) through the air.

The initial calculations are straightforward: determine the initial velocity of the projectile, perhaps making simple assumptions for the drag of the air and the rope; then determine the energy required by estimating the mass of the projectile and the rope. Difficulties can arise if accurate answers are sought; students need to be encouraged to make suitable approximations, or to calculate upper and lower bound values. Proposed methods of energy storage include compressed air, coil springs and cantilever springs (crossbow). To demonstrate the application of the simplified optimisation procedure, the feasibility of a cantilever spring will be investigated.

1. *Function*. The function of the component is to store energy.

2. *Idealise*. Each half of the crossbow can be idealised as an elastic cantilever beam, figure 2. It will be assumed that the cross-section is square, dimension d, and uniform along the length L. The energy is stored by applying a force W to the end of the beam.

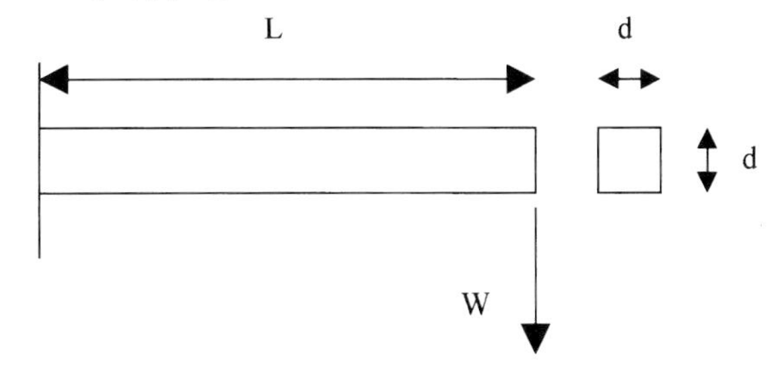

Fig. 2 Idealisation of one half of the crossbow spring.

3. *Objective*. The objective is to minimise mass m. Cost could also be chosen. The mass is given by:

$$m = Ld^2\rho \tag{1}$$

where ρ is the density of the material.

4. *Constraints*. The important constraints are stress and energy. The maximum stress must be less than the failure strength of the material. The maximum stress σ is given by

$$\sigma = \frac{6WL}{d^3} \tag{2}$$

The energy required to launch the projectile, calculated from the desired launch velocity and the estimated mass, is typically 400J. The energy e stored in the cantilever is given by:

$$e = \frac{2W^2L^3}{Ed^4} \tag{3}$$

where E is Young's modulus.

5. *Free variables.* The free variables are W, L and d.

6. *Number of free variables and constraints.* $N_f=3$, $N_c=2$, hence $N_f \geq N_c$ as required, and $N_f-N_c=1$, to ensure some freedom in selecting values for the free variables.

7. *Eliminate free variables.* Substituting (2) and (3) into (1) eliminates all three free variables to give the following equation for the objective:

$$m = (18e)\left(\frac{\rho E}{\sigma^2}\right) \tag{4}$$

8. *Select material.* Looking at the group of material parameters in equation (4) it can be seen that the mass is minimised if the stress is set to the maximum allowable, that is, the failure strength S_f of the material. Materials that give a low value of $(\rho E/S_f^2)$ include spring steel, glass or carbon fibre reinforced polymer and wood. Best of all is rubber. The high performance of rubber suggests that a concept in which energy is stored in rubber under tension might give the lightest device of all. However, supposing that the feasibility of the crossbow concept is to be pursued, select glass fibre reinforced polymer (GFRP) as the first choice of material, for which approximate property values are $E=30GN/m^2$ and $S_f=200MN/m^2$.

9. *Select values of free variables.* Using equations (2) and (3), d can be eliminated and a relationship between L and W derived, shown in figure 3. Practical constraints on L and W can be drawn on the graph to determine the combinations of possible values. It may be considered that no combinations of L and W are practical, in which case the feasibility of the concept can be questioned.

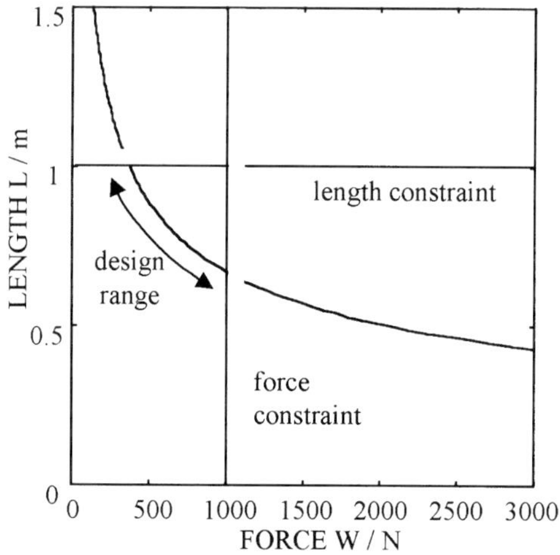

Fig. 3 Relationship between force W and length L for GFRP material.

5. ASSESSMENT METHOD

Before being introduced to the simple optimisation procedure, many students had difficulty in applying engineering science to the design situation, despite having a good grasp of the science. Their difficulties arose because they had not identified all of the important design criteria, or because they had over-constrained the problem. Adopting the procedure allowed the students to apply the science effectively and arrive at appropriate conclusions about their design concept. However, success did still depend on a good understanding of the science.

Assessment and feedback are an important part of the learning process, particularly in design, where emphasis is on methodology. An example of an assessment sheet given to students at the start of the individual part of the conceptual design exercise is shown in figure 4. Similar sheets are used for the other design exercises in the course. The sheet performs several functions:

- Informs the students of the assessment criteria.
- Allows students to assess their own work before submission.
- Helps ensure consistent marking standard across design tutors.
- Provides a structure for feedback from design tutors to students.

The criteria do not depend on any particular design solution, but focus more on the methodology and the quality of the analysis and presentation. The criteria allow credit to be given for each aspect of the procedure, including identification of appropriate functions, objectives, constraints and free variables. The marking scheme is based on the criteria, although precise details of the scheme are not given to the students. A moderation exercise across markers is held before final marks are awarded. Students are generally satisfied with the feedback that they receive and with the marks awarded. Their work on third and fourth year projects demonstrates effective learning in the second year.

In future it is planned to further integrate the assessment and teaching activities, which should assist students in maintaining progress during the design exercises and encourage full participation of all students during group work.

6. CONCLUSION

- Application of engineering science to conceptual design problems is an important skill for which a structured approach is needed.
- A procedure based on simplified optimisation methodology has been used successfully in second year conceptual design projects.
- An assessment method based on marking criteria circulated to students is thought to have assisted the learning process and will be further developed in the future.

REFERENCES

(1) Pahl G and Beitz W 'Engineering design: a systematic approach' 2nd edition, Springer, 1995.

(2) Ion W and McCracken W 'Can we successfully teach design to engineering students?' Proceedings of the 1990 SEFI Conference, Dublin, p173-178, September 1990.

(3) Johnson RC 'Optimum design of mechanical elements' 2nd edition, Wiley, 1980.

(4) Ashby MF 'Material selection in mechanical design' 2nd edition, Butterworth-Heinemann, 1999.

Conceptual Design Exercise – Individual Report
Assessment Sheet

NAME: MARKER:

- Your work will be marked with the guidance of the criteria given below.
- You may find it useful to assess your work against these criteria before submitting your work.
- Put a tick in each row according to how well you think you have done.
- If you do this, the marker will enter his/her own assessment in a contrasting colour. The marker may comment on significant differences, and offer advice on how to improve.

	E	D	C	B	A	Comments
Excellent, near professional standard					A	
Good, some fine tuning possible				B		
Competent, but room for improvement			C			
Poor, more work needed		D				
Very poor, needs a lot more work	E					
ENGINEERING:						
Have only components critical to the feasibility of the concept been studied?						
Has the function of the component been identified and an idealisation sketched?						
Are the choice and number of objectives, constraints and free variables appropriate?						
Material selection: Appropriate criteria? Wide range considered?						
Is the decision on feasibility sensible? Have appropriate values of free variables been chosen?						
Accuracy of calculations?						
PRESENTATION:						
Definition of equations & algebraic terms?						
Correct units included with numbers?						
Use of explanatory sentences and comments on calculated values?						
Final design values clearly identified?						
Are the written conclusions supported by the calculation results?						
Overall neatness and clarity? Contained within six sides?						

Fig. 4 Typical assessment sheet for the conceptual design exercise.

Profiling the prospective engineering design student

J B WOOD and **T CHAMBERS**
Department of Architecture and Building Science, University of Strathclyde, Glasgow, UK

ABSTRACT

To create the ideal prospective engineering design student is the goal of a project being undertaken by the CADET Unit at the University of Strathclyde in conjunction with Govan High School. It takes the philosophy of the University of Strathclyde's highly successful Building Design Engineering course that promotes the use of IT as a common language in interdisciplinary team working and cascades this down to the secondary school sector. It is attempting to bring together the various strands of the 5 – 14 curriculum to create scientifically rigorous individuals with both an appreciation of and an eye for good design. To create individuals who can think with both sides of their brains

1. BACKGROUND

"Organised thought is the basis of organised action. Organisation is the adjustment of diverse elements so that mutual relations may exhibit some predetermined quality."[1] *A.N. Whitehead.*

CADET, of the Department of Architecture and Building Science in the Faculty of Engineering, is researching the teaching of Design through IT in both Year One and Year Two of the Building Design Engineering Course and in the Secondary Sector of education. The latter involves a pilot project, which is participatory in nature and focuses on teaching design in the built environment through IT within Art and Technology courses within the school. This paper describes CADET's contribution to the BDE Design Course and its interface with Govan High School, the participating secondary school in Glasgow.

In reviewing the Design Course within BDE over the past 3 years, CADET has identified particular elements that contribute to good design practice through the use of Information Technology. A characteristic of the course is the integration of history, theory and the principles of visual communication with the Design Studio achieving a balance between the practical, the theoretical, the aesthetics and cultural dynamics of the social context. Design precedents are chosen, which encourage the student to reflect the rich and varied sources of knowledge and practices from arts and science alike.

In a series of introductory lectures and tutorials, beginning with the 18th century, a semiotic approach to the analysis and synthesis of the visual information presented by a selection of Design Precedents are considered in their particular historical context, and are supported by theoretical references (indicated in this paper). But in any historical review, the past must become *"the subject of a construction whose locus is not empty time, but a particular epoch, the particular life, the particular work."* [2] The outcomes of this analysis should be informed by principles of the theory and practice of visual communication and represented in visual presentations throughout the course.

In proposing *"that the age of science has developed into an age of organisation,"* [3] Whitehead (1917) could have been referring to engineering and architecture, and perhaps more so to the education of engineers and architects in the 21st century. Between 1912 and 1922, Whitehead, with regard to the pre-selection and organisation of the effective learning experience, warned against teaching too many subjects in a disconnected way, identifying education's central task a *"delicate adjustment of many variable factors."* [4] CADET accommodates such an *adjustment* by teaching design through IT and seeks to place these variables firmly in a social and environmental context. CADET believes that this should include a proactive approach to the social applications of design – which feature in a number of live projects in partnership with the City of Glasgow, and in collaboration with an arts organisation and other social agencies. In particular, the Govan High School Environment Project demonstrates the potential for a meaningful integration of art, technology, as well as the principles of mathematics and science, which together will achieve a better understanding of the applications of design in the built environment.

If the built environment is to represent our cultural aspirations, our sense of a common culture and shared values, we must embark on a continuing dialogue that enhances our understanding and appreciation of the built environment's semantic potential. Information Technology, with its special capacity for manipulating discursive and non-discursive symbols, is pivotal in such a creative dialogue.

2. INTEGRATED DESIGN COURSE

In the review of the design studio credits of the Building Design Engineering course, bridging architecture and engineering, adjustments have been made to meet the various specialist norms, as much as the demands of the change brought about by IT. What we believe is required of an effective designer is, as with the teaching of mathematics and science, the *art of thought* 'namely, the art of forming clear conceptions'.

Where it concerns the quality of the built space, design is part of the 'art of deductive reasoning', while simultaneously remaining open to the intuitive response to the play of ideas. In 3D Studio, the design course explores both the rational and intuitive responses in exploring the structuring of space - an exploration through the analysis and synthesis of the urban environment in considering the relationships of *number, quantity and space*. A synthesis arrived at in this way leads to an understanding and appreciation of architectural concepts, rather than simply an identification of individual 'types' or styles. The design course therefore demands an inquiring mind – through which all of the cognitive faculties come alive, " *thoughts gain vividness by an immediate translation (of ideas) into acts."* [5] The student, therefore, should be visually aware and susceptible to working within a rational framework while open to the poetic structure of the world – one that is confident and articulate in the performance of a symbolic language. In facilitating an integrated approach, the course material includes specific core elements. These are; the interpretation of visual information, the principles of visual communication, an understanding of the process of abstraction, exploring concepts of form and meaning in structures, aesthetics in the built environment and a rational and critical method of inquiry that informs the design process.

2.1 The semiotic

A semiotic approach to design is useful in two ways. Firstly, it provides a method of interrogating and interpreting design while, secondly, it underlines the importance of design in the built environment as representations of cultural value. The latter underpins the potential aesthetic value of those products that are too often seen as being simply utilitarian. Designers work within a network of professional and social relationships, operating with *"a social storage of world knowledge"*[6] which reveals that *"the whole of culture should be studied as a communicative phenomenon based on signification systems"* [7]

Particular 'presuppositions' concerning art, design and the aesthetic experience have historically existed that isolate the aesthetic from everyday life – and foster processes that exclude a wider audience. A semiotic approach, to both the production and reception of design, that values the concept of the artistic product as 'open work' has the potential to engage a wider audience installing *"a new relationship between the contemplation and the utilization of a work of art."*[8] Where these presuppositions have created a split between the functional and the aesthetic, the aesthetic is too often considered less relevant – contributing to an impoverishment of the visual impact of the environment, which itself adds to the sense of alienation. This sense of alienation, so characteristic of the early modern period (Romantic) was in fact a failed reception of technology, which produced skewed perception of the struggle between *nature and industrialisation.* What was, and is still required, is a focus on the triadic relationship between society, nature and technology – so inextricably linked. Design theory, therefore, must inform design practice by demystifying *canonical works* by locating them firmly in the wider cultural context. The latter will inevitably include the contributions that both science and art have made to our cultural heritage.

2.2 Form

It is essential that in any research activity the student should appreciate that *Form* is an idea that we *have*, that has been appreciated over time with experience of other artistic productions. It is not natural or god-given, it is something we produce in the process of interacting and interpreting objects and ideas – an interaction that is inextricably bound to the process of reflection and abstraction. To *know* is to have clear ideas. For our thoughts to have *form*, and to engage in a process of reflection, is to apply a method of thinking - to *unite reflection, ideas, knowledge and method.*

2.3 The aesthetic

The aesthetic is dependent on the percept, and percepts *'are always bound by relations'*. In the analysis of the design precedents, as well as their own design solutions, the student is required to identify and appreciate the necessary fusion of function and aesthetic. In the case of objects or events in the built environment, the aesthetic is seen as informed by both the historical and cultural context as well as the individual designer's own subjective response.

2.4 The rational

Whereas the curriculum of the 19th and 20th century has tended to separate the rational from the aesthetic, the logical from the intuitive, 'science' itself has never been complacent in this regard. In his study of the brain, J.Z. Young draws upon research and literature on the functioning of the brain that ranges over the last hundred years and more. Young certainly is not inhibited in acknowledging the importance of creativity. *"Proper study of the organisation of the brain shows that belief and creative art are essential and universal features of all human life. They are not mere peripheral luxury activities. They are literally the most important of all the functional features that ensure human homeostasis."*[9] Although very special in nature, these human activities are after all concerned with beliefs, judgements and sensations as with other life experiences.

As distinguished from mere sensations or unconnected thoughts, beliefs and judgements have form, a sense of coherence and clarity. In searching for this clarity and coherence in the design process, the student should be encouraged to apply the rigour of mathematics and science together with the intuition and imagination associated with the arts. While mathematics deals with concept and form, and science with the operation of these in the real world, the teaching of science also deals with the *"art of thought ... namely, the art of forming clear conceptions."* We *"choose just those determinate ideas and definitions of ordering that will enable us to construct the system of experience in the least complicated way."*[10] In visual communication this might very well be defined as *poetic form*.

2.5 Poetics in the rational

Leibniz provides that quintessential poetry in describing that in which we exist. *"They (minute perceptions) it is that constitute that indefinable something, those tastes, those images of the qualities of the senses, clear in the mass but confused in the parts, those impressions which surrounding bodies make on us."*[11] In mathematics, again, we find (Poncelet) proposing a geometry that is no longer based on 'size and measure' *"but on a concept and study of pure relationships of position."*[12] But if that form is to be realised, such as in Meis van der Rohe's Concrete House, the designer must visualise those *structuralist* principles by which *"starting from a certain point, the different figures can be progressively generated. The whole system must be built up by the successive production of higher grades of figures from basic ones; in the plane – the straight line ...."*[13] to the construction of a manifold of forms structuring space. As Hartoonian observes, it is in the use and play of actual material that Van der Rohe *"would find the proper design device to achieve his objectives."*[14] By a close analysis of the design precedent the students' understanding of the principles of construction and their impact on the 'semantics' of built form become embedded in their own design practice.

In the context of Design in BDE, experience of the built environment is concerned with a 'multiplicity' *(the manifold of experience)*, that has to be perceived as order and form rather than the mere number of particulars or elements. As with mathematics, the design student must progress from quantitative relations *"to one that treats quite generally all relations – expressions that have to do with order, similarity, in short, a subordination of the science of quantity to the science of quality."*[15] This qualitative relation embraces, not only abstract scientific thought but also aesthetics. *"Now mathematics is nothing else than the more complicated parts of the art of deductive reasoning, especially where it concerns number, quantity and space."*[16] Building on the student's preliminary experience of mathematics, the design course aims to illustrate such relations of 'number, quantity and space' through the analysis and synthesis of elements of the urban space through the use of CAD.

3. OUTREACH (meeting future needs)

In the Information Age, an understanding of design practices that shape and attempts to sustain the built environment, must appreciate the social and ethical as well as the practical and the aesthetic. The resulting environmental ethos should inform all sectors of education concerned with the teaching of design. With this in mind, CADET has initiated a pilot project in Govan High School, Glasgow, which focuses on the teaching of design in the built environment through IT within both Art and Technology courses. This participatory project involves a partnership with arts, education, health and social service organisations, recognising that design has a role in the regeneration of the urban environment. Within the secondary curriculum, design in the environment must surely call for the integration of mathematics, science, visual communication, design and information technology. The challenge therefore is to recognise, with Whitehead, that education "*is the acquisition of the art of the utilisation of knowledge*."[17] A utilisation, the aim of which is to reach an outcome that represents common interests, beliefs and values. Such an ethos that eluded early modern practice, emerged in the utopian aspirations of the early 20th century but was sidelined by the explosive energy of the city as an economic machine. The language of design now embraces ideas of sustainability, partnership and in general, an environmental ethos. In order to explore and articulate these ideas in the design of built spaces, the designer must be skilled in communication – not exclusively within a narrow specialist practice, but also in the public domain. IT has a unique potential for engaging the wider community through interactive media – a necessary aspect of the design process if a sense of ownership, and participation in the production of the social space is to be achieved.

4. CONCLUSION

Since the design course should facilitate the student's ability to interpret the visual information and understand abstract concepts, the theoretical component seeks to achieve a balance between the discursive and non-discursive languages – between the logical and visual languages. It is through these *languages* that the complexity of the world and cultural phenomena are revealed to us. In recent reception theory it is claimed that in the act of interpretation the work of art comes to life (Walter Benjamin). This emphasis on the productive nature of interpretation is an invaluable tool in promoting an ethical approach to the built environment that, in turn, would create a sense of ownership in the built environment. Whereas, from the legacy of the Romantic world-view, the medium of representation itself, with its Fine Art bias, created a barrier between producer and spectator, IT has the potential to facilitate our expressive and communicative capabilities. Visualisation, with its direct reference to popular culture, can enable young designers to both express their own ideas and communicate the complex information required in the production of those most public of our visual arts – architecture and engineering that shape our built environment. An effective designer is one that has an understanding of the relationship between the form and meaning of the built space, and crucially, its impact on the social space of the modern urban environment.

A practice built on the integration of the theory and practice of the arts and science, and informed by an environmental ethos, would overcome that historical rupture that marked the split between architecture and engineering in the 18th century. From such a creative praxis, which combined the analytic and poetic, would be produced a social space while enhancing the built space - the two are inextricably linked. The student would appreciate, and counter the separation that has been exposed, by Hartoonian, as the drift from the classic *techne* to the modern technique. By such analysis, Hartoonian provides the design precedent that can exemplify the change in paradigm made possible by innovations in technology as much as shifts in ideology.

In the public domain, the designer's role in a creative dialogue must be to produce the concept, which as with scientific theories, has *"the greater explanatory power: that explains more; that explains with greater precision; and that allows us to make better predictions".* [18] Popper characterises such theories as *"free creations of our minds, the result of an almost poetic intuition, of an attempt to understand intuitively the laws of nature."* [19] But, as with science, creations that *"can be controlled and tempered by self-criticism, and by the severest tests we can design. It is here, through our critical methods of testing, that scientific rigour and logic enter into empirical science .... From a logical point of view, all empirical tests are therefore attempted refutations."* [20] Where the Design Brief is a means of defining the 'problem', the design solution should be that which survives 'attempted refutations', and this the result of a creative and productive dialogue in the public domain that deals with the poetics of space as much as the functional.

With an ability to translate the visual material of the built environment into analytical elements as well as intellectual concepts, the student will develop the capacity to conceptualize the structure of the built space. Space as a cultural phenomena must be *considered "as objects or events with meaning and hence signs...*(that) *... are defined by a network of relations."* [21] Culler continues that in *"studying signs one must investigate the system of relations that enables meaning to be produced."* [22] Through a semiotic, or poetic, view of space the *"conditions are ripe to consider space as a form that can become a spatial language making it possible to 'speak' about something other than space"* [23] As 'object' it has that capacity to represent value, identity and a sense of culture - space, both social and built, 'the mutual relations' of which 'exhibit some predetermined quality' – the aesthetic of the urban form.

REFERENCES

(1) A.N. Whitehead, The Aims of Education, Ernest Benn 1970 ISBN 0-510-43501-7
(2) W. Benjamin, One Way Street, Verso 1992 ISBN 0-86091-836-X
(3) A.N. Whitehead, The Aims of Education, Ernest Benn 1970 ISBN 0-510-43501-7
(4) A.N. Whitehead, Ibid.
(5) A.N. Whitehead, Ibid.
(6) U. Eco, Semiotics and the Philosophy of Language, Macmillan 1984 ISBN 0 333 36355 8
(7) U. Eco, A Theory of Signs, Indiana University Press 1979 ISBN 0-253-35955-4
(8) U. Eco, The Open Work, Hutchinson Radius ISBN 0-09-175896-3
(9) J.Z. Young, Programs of the Brain, Oxford University Press 1981 ISBN 0-19-286019-4
(10) A.N. Whitehead, The Aims of Education, Ernest Benn, 1970 ISBN 0-510-43501-7
(11) G.W. Leibniz, Philosophical Writings, ed. G.H.R. Parkinson. Dent 1973, ISBN 0 460 11905 2
(12) E. Cassirer, The Problem of Knowledge, Yale University Press 1978,ISBN0-300-01098-2
(13) E. Cassirer, Ibid.

(14)G. Hartoonian, Ontology of Construction, Cambridge University Press 1997 ISBN 0-521-58645-3
(15)E. Cassirer, The Problem of Knowledge, Yale University Press 1978 ISBN0-300-01098-2
(16)E. Cassirer, Ibid.
(17)A.N. Whitehead, The Aims of Education, Ernest Benn 1970 ISBN 0-510-43501-7
(18)K. Popper, Conjectures and Refutations, Routledge 1991 ISBN 0-415-04318-2
(19)K. Popper, Ibid.
(20)K. Popper, Ibid.
(21)J. Culler, Ibid.
(22)J. Culler, Ibid.
(23)A.J. Greimas, Social Sciences, a Semiotic View, University of Minnesota Press 1990 ISBN 0-8166-1819-4

Multi-disciplinary projects produce more effective graduates

M LEWIS
School of Engineering, Coventry University, UK

ABSTRACT

We all tend to be thrown in at the deep end with respect to work. There is an expectation that our qualifications and experience will enable us to be effective and efficient quickly. This would be true if we had already been employed elsewhere but if we have just graduated then there is a serious problem. Working in a multi discipline team for the first time can be daunting, so how can we give our students this confidence building experience? The paper looks at the main options - their structure, benefits, assessment and typical examples.

1. INTRODUCTION

Most engineering degrees, traditionally, do not cross discipline frontiers and many are narrow and specific to the point of being regarded as introverted. The students do not get exposed to related disciplines, they are not aware of how their work fits into the grand plan. What is required is a mechanism to expose the students to a 'real' situation where they can practise their chosen discipline interacting with others to produce a viable product. There appears to be two main ways of doing this - industrial placements and 'live' projects with industry. We must get away from the attitude that calls industry the 'real' world; this does our work a great disservice. We all work in a real world and our job is to make the students' one relative and realistic to that in industry. The maxim - a team is a group of individuals working together for a common cause - is an useful one to remember when thinking about what we are training the students to be.

2. PLACEMENTS

In the past placements were used as the only way to expose the students to the work place. They allowed them to practice as a professional engineer but totally depended on the company to give them appropriate work, which would enhance their capability, as well as getting the job done. Too often it was seen as 'work experience' with the student carrying out very general tasks, not being given much responsibility and certainly not being stretched in application of knowledge. It is pleasing to note that this attitude has changed with companies valuing the work done by the placement student in the same way as its permanent staff. This is what placements should do, act as a catalyst for the student to develop confidence and competence by doing a valued piece of work in a professional manner in a professional situation, working in a multi disciplinary environment. This enables the student to put their contribution into perspective and really appreciate that you cannot achieve much on your

own. The placement is assessed and contributes towards the sandwich degree thereby recognising the important part that it plays in preparing the graduate for a career in engineering. We use the City & Guilds Licentiateship award thus giving the students a nationally recognised award. This benefit is also recognised by both industry and students :-
Rank Xerox - 'places a high value on good design.......this starts through ensuring that students receive a sound education in design disciplines and through placements in the real world of business'. (we must dispel the notion of the University not being in the real world !).
Leighton Rees - 'little did I know how important my placement year would be to me when I started the Industrial Product Design course. It proved most enjoyable with excellent tutorial support from all the staff, however it was the placement year that made the difference to my life. I was fortunate to gain a place with TNO Industry in Delft Holland. Staff helped me settle into their way of life very quickly, even providing evening classes to help me learn a new language. I was also fortunate to be approached by TNO in my final year for a position within the company allowing me to expand my knowledge and offer the total package as an Industrial Product Designer'.
Some students choose for various reasons not to do a placement year even though this is strongly advised. We have for several years been comparing their results with those who did go out. These results are from the Industrial Product Design course.

Table 1 Degree classification *v* Placement.

	Placement	Year 2 class	No placement	Year 2 class
1st class	6	3	-	-
2.1	8	8	4	5
2.2	-	3	11	10
3rd class	-	-	3	3

This pattern is fairly consistent over the years that the course has been operating with the final year tutors noting the difference in approach between the two sets of students. The ones who have been on placement show an increased confidence and competence that is lacking in those who have not had the undoubted benefit of a placement.
The School of Engineering believes that placements are so important that we have two permanent staff whose sole task is to run the placement operation. Even though the benefit is stressed to all students there is a marked difference in take up of positions. The following table shows take up by course as percentage of eligible students :-

Table 2 Placement take up *v* Course

	% placement
Industrial Product Design	77
Automotive Engineering Design	88
Electrical/Electronic Engineering	20
Mechanical Engineering	48
Manufacturing Engineering	25
Aerospace/Avionics Engineering	72
Engineering Management	32

The take up of placements on the more 'traditional' courses is very low so how do these students get the exposure to industry that gives them the extra competence necessary to make that instant impact. A way of increasing their capability is by giving them a 'live' project supported by industry.

3. INDUSTRIALLY SUPPORTED PROJECTS

We have for many years been running a final year Group Design Project. This has the following objectives :-

1. Develop a design problem from a project 'brief' through to a complete integrated design solution.
2. Develop professionalism and design abilities.
3. Practise organisational, engineering and team building skills.

Companies that have provided projects include British Timken, Clayton Heaters, Futurerail, Morris Handling, Autopia, APK, Lindstrand.

Teams of 4/5 are used and a mix of disciplines is the target. Even if there is not the required spread the project still has to be complete. The work is divided up in an equitable and hopefully amicable manner by the team members. The student will work with fellow students in his/her group to ensure that each individual part is co-ordinated into the design as a whole, and that all relevant aspects of the solution have been investigated and included in the specifications and the final design report. This ensures that they are aware of the global problem and cannot focus exclusively on their own particular part.

They are reminded that :-

'a team is a group of individuals working together for a common goal'.

The sponsoring company will introduce the project giving sufficient information for the team to understand the problem and begin the research into the processes necessary to solve it. A visit to the factory may be made if it would help the process but not if it may result in pre-conceived ideas of what the solution may be. The industrialist may make a further 4 visits to see progress and possibly assist in assessment. Each team will keep a file containing the following : -

- an up-to-date detailed project plan with target dates,
- minutes of all meetings,
- records of all decisions taken,
- copies of drawings/relevant information.

The aim is to make this project operate as close as possible to the way it would be carried out in industry including the commercial aspects. Monthly reports indicate how they are progressing and if they are considering all aspects. This enables the tutor to guide the team towards a cohesive, complete solution.

The assessment has 5 areas:-

- presentations,
- industrialist's assessment of progress and soundness of work,
- formal report,
- oral examination of group and individual contribution,
- peer assessment.

This gives a comprehensive view of quality and quantity of work. The breakdown is a mixture of group and individual marks:-

Table 3 Mark allocation v parameters.

	Group %	individual %
presentation 1	2	2
sponsor meeting 1	4	
sponsor meeting 2	8	
presentation 2	4	6
report/drawings/group file	8	36
Display	10	
oral examination		20
Total Mark %	36	64

The schedule is set so that from October to December is design specification/concept generation/ solution selection. This results in a sponsor meeting to review the work and approve the chosen concept, which all the team has worked on. From January to May the detail design is carried out with each team member doing their own section but still ensuring that the result is complete and fully justified. (Pugh's Plates provides the overall framework.) The sponsor still monitors progress bringing in their own 'local' knowledge to prevent any problems in implementing the design at their factory. A secret peer evaluation is applied at the end to be used with all the delivered documentation to ensure a fair distribution of marks.

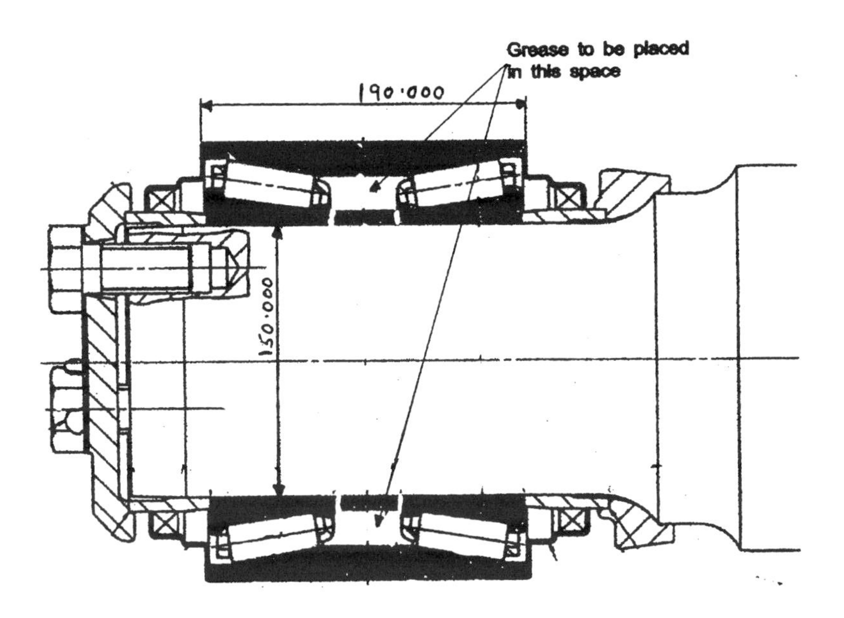

Fig. 1 Bearing installation.

One of the major projects carried out last year involved a 32kg double cup taper roller bearing. The brief was simple:-

'To place 200±1gms of grease accurately and consistently into the annulus between the two races.'

Initially the grease had been injected into two holes in the centre of the bearing. We made a Perspex bearing which showed that the grease basically stayed in two large blobs, not doing its job. The customer then asked that the holes be deleted but the internal annulus still be filled with grease. The equipment had to meet the following requirements:-

Operated manually and be capable of being integrated into the existing PLC controlled production line,

Run off 24volts d.c. or hydraulic(self contained) or factory 80psi air.

The disciplines that the team needed to come up with a viable solution appeared to be:-

Design methodology, mechanical design, CAD drawing, manufacturing, materials, power source and control methods, specification writing, system modelling, Health and Safety factors, component selection, electrical – the list goes on.

A Gantt chart controlled the work and there was one 2 hour timetabled slot per week and the team met in their own time as well as keeping a logbook of all activities. The team worked extremely well producing a sponsor viable solution of high quality and integrity.

The solution was an annular grease delivery system powered by the factory compressed air system. A fixed volume of grease was injected via a limit switch set cylinder so fine adjustments could be made as it was the weight that was important. Figure 2 shows the configuration.

The grease is injected through the centre channel, pushing the annular sealing ring up, allowing the grease to fill the space between the spacer ring and the double cup. The ring is held closed by a circular spring ensuring an even pressure to shut off the grease once the limit is reached. The design could easily be fitted to the existing system, which made the overall cost of implementing the solution less. The industrial sponsor was impressed by the solution having the following comments:-

'there could be a problem with the annular ring, making it slide evenly would demand some more calculations and maybe there should be a look at an alternative way to control the grease flow.'

material cost, at £240, was very acceptable.'

'the pneumatic control system was well thought out and would fit easily with existing equipment.'

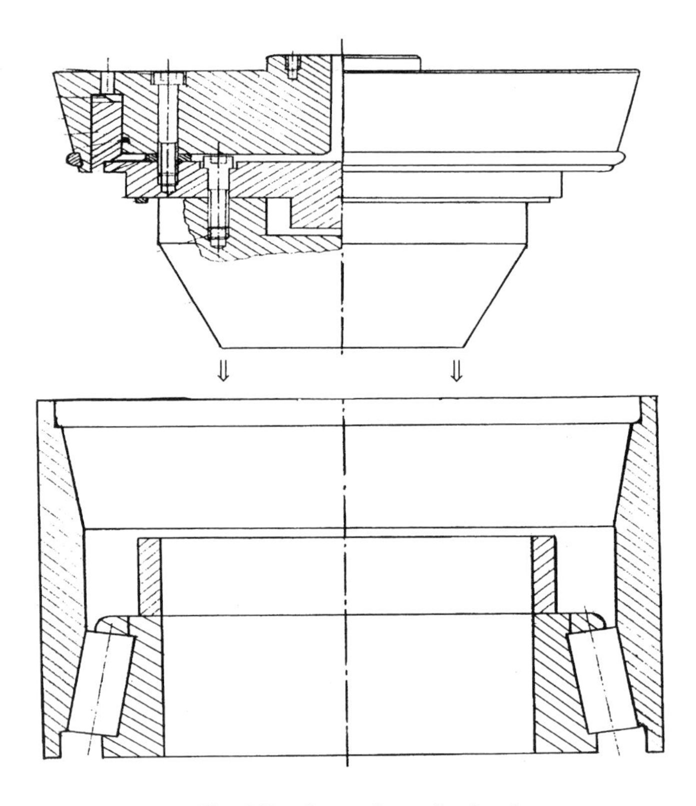

Fig. 2 Bearing and annular head.

4. CONCLUSIONS

The customer got a viable, comprehensive solution supported by a detailed report and the students got a taste of the outside world. None of the students had any industrial experience; none had even considered a placement. One comment was:-

'I did not realise how hard it was to work in a team, there was so much to understand about other team members' work to enable me to get mine done well.'

Three out of the team of four were also doing the double individual final year project and the overall feeling was that this team project had helped to structure and consider their individual problem in a better manner.

The company stated that they looked for graduates who were 'street wise', either having done a placement or been part of a team project, and that they would support us in the future with suitable projects. They also mentioned that they had just got rid of two young graduates, who had good degrees, but could not or would not appreciate that they were part of a team and had to work together.

It is not by accident that we have placements and team projects for industry, it is because we believe that they are an integral part of a student's education and that industry expects the graduate to have relevant experience when they start work.

The graduate must come through the door with their feet on the ground, running.

'The Karlsruhe Model' – a successful approach to an academic education in industrial product development

A ALBERS, N BURKARDT, S MATTHIESEN, and **D SCHWEINBERGER**
Institute for Machine Design and Automotive Engineering, University of Karlsruhe (TH), Germany

ABSTRACT

The increasing complexity of product requirements combined with a customer tendency to prefer more simple and profitable solutions needs the integrated use by the design engineers of targeted combinations of organisational, technical and systematic tools based on an holistic view of the entire product development process. The education of design engineers has to react to these new professional requirements and excellence in professional competence must be generated. The education of mechanical engineering students in machine design can make a decisive contribution in obtaining technique and social competence for later success in study and profession.

At the University of Karlsruhe the Institute of Machine Design and Automotive Engineering has constituted the new education concept called "The Karlsruher Model" in order to take up these challenges. This paper will present the basic course content and methodological approach and will describe our initial experiences with this new educational concept.

1. INTRODUCTION AND MOTIVATION

Successful product development leads to competitive products and safeguards both the existing and future market position of an industrial concern. The present market situation can be described as having

- Short product life cycles
- Lower price expectation for even higher standards of quality
- Growing requirement for more specialised products
- Short product launch cycles

The conditions for a successful product under these demands is a complete holistic view of the total development process from market demand to product launch described with the term **"Integrated Product Development"**. This development philosophy is based fundamentally on the creation of multi-disciplined project teams with the ability to quickly adapt to the step changes in a dynamic marketplace and possible fundamental changes in base technologies. This requires all the members of the development team to personally identify with the development process and to be organised so that all relevant data is to hand. Therefore new requirements are growing up for qualification, particularly to the combination and the weight of competence by design engineers (Fig.1). Required is an outstandingly creative and flexible engineer, able to think about the task "in the round" and fitted out with management skills – a specialist with the ability to act as a successful generalist. Competence in the fundamental technologies, social interaction, and design techniques are the essential core qualifications of a design engineer which up to now have not been taught in a traditional university curriculum in the broadness required. The university education of design engineers has to react to these new professional requirements and an adequate professional competence must be generated [99/14].

2. EDUCATION OBJECTIVES

The education objectives can be structured as follows according to the core competencies required for development engineers:

2.1 Competence in Fundamentals

The traditional university education provides a broad fundamental engineering knowledge and offers discipline specialisations. The limits of these specialisations must be reconsidered and varied, where necessary, to achieve an integrated process of development. For the development engineer it is necessary to consider technical, economic and organisational systems in terms of the complexity of the product, heterogeneity of product components (mechanical, electronic, hydraulic, data processing etc.) and their combination into a superior marketable product.

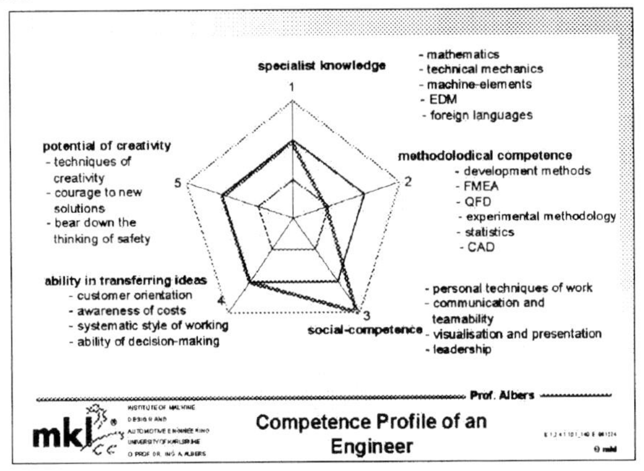

Fig. 1 Competence Profile

The ability to analyse problems, develop solutions, operate work stations and processes is an essential part of the competence in fundamentals. The continuous updating of information about development, relevant materials and components from trends in market and research requires an efficient strategy for information procurement, data processing and the readiness for a Life-Long-Learning even beyond all "Comfort Zones" of individual specialisations. The split between a product specific specialisation on the one hand and the integrated development process on the other hand requires an efficient management. Internal processes regarding information, planning, decision making and execution are to be co-ordinated in order to avoid loss of time, misunderstandings and errors which can appear across specialisation interfaces.

This management task is not the job of the project manager by himself but part of the working process of the whole team [99/9].

The following educational objectives have been derived from competence demands outlined above at the Institute of Machine Design and Automotive Engineering. They are developed to teach the following ways of thinking and their correlation:

- *Process thinking* - Product development as a process chain
- *System thinking* - Product development as a systematic process
- *Innovative thinking* - Product development as an innovative process
- *Problem thinking* - Product development as a problem solving process
- *Integration thinking* - Product development as an integrated process
- *Organisation thinking* - Product development as a management process
- *Cost thinking* - Product development as a cost optimised process
- *Time thinking* - Product development as a time optimised process
- *Customer-/Quality thinking* - Product development as a customer-oriented process
- *Market Thinking* - Product development as a market-oriented process

Social Competence

A successful integrated product design is based on a goal-oriented and innovative culture of dialogue in enterprises with the following kinds:

- *Problem-solving culture*: Seeing problems as a chance and challenge to think of possibilities instead of difficulties
- *Constructive error culture*: Solving conflicts co-operatively, analysing causes, initiating perspective variation
- *Creative culture*: Promoting flexibility in thinking, creating bases for cross-functional thinking, imagination, creativity and inventive chaos
- *Fractal culture*: Employees as responsible, self-controlling, closed-loop control systems in the product development process
- *Courage-of-conviction culture*: Promoting constructive obstinacy and courage of conviction, breaking moral cowardliness and hasty uncritical acceptance.
- *Comfort Zone culture*: Application of the employees in accordance with their talents and interests, promoting fun

A distinct communication behaviour of employees is necessary for a dialogue culture described above. Decisive here is the outwardly directed behaviour of the participants in co-operation with other colleagues involved in the development process. The educational objectives which cover all these requirements are:

- Communication ability
- Co-operation ability
- Ability to resolve conflicts

2.3 Methodological competence

The Institute of Machine Design and Automotive Engineering defines "method" in this context as tools required for the technically and socially competent development engineer to convert steps of the product development process into real concrete progress in the generation of a target product. The support of these tools to translate the product idea from the product concept and -design to product manufacturing and recycling is an important requirement for an efficient treatment of the development processes [99/5].

Therefore the following education goals are defined by the Institute:

- Teaching approved techniques compatible with each process step of the product development process
- Teaching criteria to select efficient techniques
- Teaching application experience and safety

3. CONVERSION OF EDUCATIONAL GOALS

The education model is based on the following structures (Fig.2):

Basic courses **MKL** (Maschinenkonstruktionslehre I-III) and the graduated course Integrated Product Development **IP** ("Integrierte Produktentwicklung).

4. MKL

The lecture is fundamentally understood as a superordinate element of the teaching model and offers the basics for the other two modules of the doctrine. It explains the product development process by using examples from the industrial development practice. Two guiding examples are important here: the drive-line of a car and an extruder. Both ones accompany the entire lecture and the discussion of special machine-elements.

The **exercise** deepens the theoretical knowledge and shows how to use it in practice. Special elements of the exercise have to be processed within the framework of the third module of this teaching model: the workshop. In the **workshop** student teams work on concrete practical tasks. At the beginning of this event, the emphasis is the practical understanding of machine-elements by the students. The workshop shows special construction-elements and -systems which guide through the whole event and which are already completely known in function and design by the students.

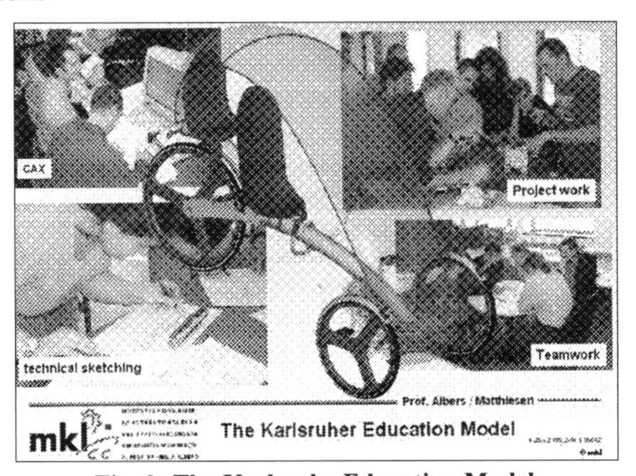

Fig. 2 The Karlsruhe Education Model

4.1 Lecture

Traditional lecture-concepts offer a sequential presentation of different machine-elements. Those lectures are an attempt to handle all machine-elements completely. The function, the design and the layout of each machine-element is exactly described. Nowadays a complete treatment of all existing machine-elements is not possible and efficient any more. The main reason for this is the increasing number of new machine-elements while having a constant capacity of time to impart the lecture. The new Karlsruhe-lecture-concept is based on a completely different view. Its main aim is to communicate an overall and complete science of machine-elements in spite of the increasing number of machine-elements.

Machine-elements are now considered to be on a higher level of abstraction. By this they can be handled with the modern tools of design-methodology. Most machine elements can be understood as a system of several components. In this case, every element of a system fulfils a function with the help of one or several contacts to another system component. The actual function and therefore the desired effect is implemented by the contact of one surface with another. Consequently these surfaces become functional surfaces.

To fulfil the function of the machine-element every functional surface is in contact with another at some time. Those two functional surfaces form an **working surface pair** (wsp). Strain, construction and design of these functional surfaces depend on the function of the machine element, the marginal demands and the contacted functional surfaces. Wsp's of completely different machine elements are often designed in the same way because the same elemental functions are realised by them. An example for such an elementary function is the lubricated contact under Hertzian stress. A task of the further structure of the machine-element is to keep the functional surface in its defined position. Therefore it is called **channel and support structure** and has to be designed in accordance to its functional performance. Similar to the functional surface the layout and design of the supporting structure depends on the function of the machine-element and the surrounding demands.

At the beginning of this new lecture-system the students are confronted with the theory of the **working surface pairs** and **channel and support structures**. By this machine-elements are put onto a high level of abstraction already at the beginning of the studies. At the same time the ability to think in an abstract way is taught and deepened. This skill is very important for the future mechanical engineer. During the further process the theory will be explained on the basis of selected machine-elements examples. The aim is to teach the ability to apply the abstract theory on concrete examples. At first the machine-elements are regarded from the viewpoint of the guiding system. The next step is the discussion of their elemental characteristics and their interactions within the entire system. A special example in this context is the disc spring as an element of an automobile-clutch. The parallel discussion of aspects of design, manufacturing, cost and dimensioning and the explanation on practical examples leads to an entire view and comprehension. Corresponding to this the emphasis of the lecture differs from the topics of conventional textbooks on the science of machine-elements. They are the basis for further learning of factual knowledge during private studies. While discussing further important machine-elements the lecturer can refer to the machine-elements used as examples. In a next step the student is able to transform the knowledge of a higher degree of abstraction to the problems of this special machine-element. **Herztian stress** for example occurs between the roll barrel and the outer ring of a bearing. These two system components represent the isc roll barrel surface - bearing running part. The wsp of two interacting cog-wheels and their pairing of tooth profiles is another example for occuring Hertzian stress (Fig. 3). As a result in a higher level of abstraction the interpretation and design of tooth profiles and roll barrel surfaces is similar. The lecture does not have the ambition of treating all machine-elements completely. Its main aim is to impart the ability to understand and analyse unknown and complex machine systems. Wsp`s and supporting structures are tools which help to sort new elements into known basic knowledge. In this way the ability of independent synthesis is promoted. About 15% of the contents of the lectures treat non-mechanical mechatronic elements and systems. This helps to show the expansion of modern machine construction. One example system is the automatic driving train system again.

The lecture is done completely with the support of digital techniques of presentation. Therefore it is possible to include digital video sequences into the lecture and to impart knowledge whilst saving a lot of time.

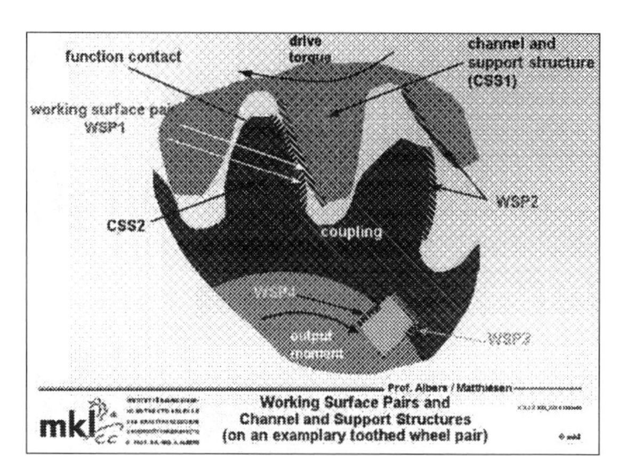

Fig.3 Working Surface Pairs and Channel and Support Structures

4.2 Exercise

According to the new teaching model the exercise is a special event. An exercise guiding assistant imparts special knowledge while acting and explaining in front of the students. This „guiding style“ is similar to the lecture, tutor-orientated and with little interactivity. Simplified, this guiding style can be called authoritarian and patriarchal. During the exercise questions and suggestions are welcomed, but the extent of the discussions is limited by the amount of students (275 students in the summer semester'99).

During the **exercise** the theory from the lecture is picked up and becomes more intense. Special exercise problems refer to the guiding elements and guiding systems which have been discussed in the lecture and in the workshops. The student learns to convert and transform the internalised knowledge to concrete problems.

4.3 Workshop

Main aim of the workshop is to impart the already mentioned soft skills. From the beginning consistent teamwork is expected and practised. Construction work is divided up by the team members independently, different experiences have to be communicated. In most cases students are not used to teamwork. Therefore this aspect has to be imparted under guidance of six assistants and eighteen student scientific aides during the weekly workshop. At the beginning of the workshop the attendants still intervene in the teamwork in an authoritarian manner. During the second and third semester the attendants withdraw more and more from the solution finding process. They only intervene to fulfil advisory or participatory tasks in the process. During these semesters this attendance is understood as a special form of coaching. This process leads to a continuously increasing independence of the students. During the first two semesters the students deal with an easy guiding system and the guiding elements. 18 compulsory and 9 optional workshops offer the possibility to take apart a mid-sized gearbox and to analyse the various system components. The workshop task during the first semester is the imparting of technical freehand drawing, the analysis of building parts in design and function, surface-analysis and measurement under consideration of various fabrication techniques and the analysis of fits and first synthesis considerations.

During the second and third semester machine systems with an increasing degree of complexity have to be designed. These events also take place in student teams. The teams have to define interfaces of the construction for themselves. Single constructions are coordinated, put together by the team and finally graded by the attendants. The final exercise is an industry-oriented construction-problem . Its solution is also unknown for the attendants [99/8].

5. IP

1. Lectures (4 hrs / semester week)
2. Workshop (3 hrs / semester week)
3. Project work (120 hrs total)

They are offered parallel in the winter semester as a main subject.

5.1 Lectures

The Students are introduced to product development of enterprises in lectures with particular reference to the requirements of small and medium sized companies. Based on practical experiences and examples from industry, the theory of systematic planning, design, cost control and management of the development and innovation process is introduced and discussed as a team-oriented adoption of effective techniques viewed a problem solving process. Strategies of development and innovation management, system analyses and team leading are presented and discussed. The lecture is designed for a limited number of participants (est. 20 students) and is a break from traditional lecturing arrangements. This offers the possibility of teaching in discursive form with the use of multimedia tools to aid presentation of the subject. If necessary the official time allotted to lecture can be relaxed to enable open-end discussions.

5.2 Workshops

In the workshops knowledge is actively built up and developed with the first real application experiences. This is achieved through:

- A direct and practical translation of the learnt methods directed to the development process.
- A simulation of group-dynamic processes by means of exercises. This requires a flexible timetable for the "Methods" and their application for each workshop.

A total of 13 Workshops covers the following topics:

- Team processes
- Hosting- and communication techniques

- Product profiling, list of requirements, project design
- Application oriented creativity techniques
- Online-research
- 3D – hand drafts
- TRIZ, ARIS, Invention Machine
- Introduction to patent law

Single workshops are accompanied by guests from industry as required (e.g. SAP, CAD-Manager, STN and other). A 3D-CAD education is given supplemental to the workshop in a 5-day crash-course.

These workshops are set up to deepen and extend knowledge and are not used for the direct concurrent support of the student project work.

5.3 Project work

The content in this project is the development of a product from the idea up to the virtual prototype (3D-CADModel) with an independent student development team. It is shown here as an example of an mid-sized enterprise at the Product-Development-Centre of the Institute with the attendance of the head of Inst. and his assistants as a simulated management and respective development teams.

Hardware and software equipment (MS-Project, Pro/Engineer, Invention Machine, IM-Phenomenon, Access to the Internet/WWW and Databases) is set up in closed working areas for each the project team. At the end of the project a presentation is given in front of the management. In certain cases the management awards prizes for the best solutions. Each team makes an evaluation of their group performance and their individual team members in a feed-back briefing. The results are handed to the management for a assessment.

5.4 Organisation

The number of participants is limited to 20 students. The selection of students is made in an assessment with the head of Institute and the responsible staff members participation in this selection process is a duty on all teaching staff.

This education module closes with an oral examination which is evaluated as a main subject within University course.

5.5 Experiences

A tremendous interest on this educational model is present within the students, in spite of the very high work-load required so that a candidate selection must always be made. A high motivation and keenness is shown by the students. The project work produced patentable product developments, the product presentations were to a professional standard including a number of functional prototypes. The most innovative and unconventional solutions have surprisingly translated into concrete product ideas (Fig. 4).

It shows that graduates of this subject, who carry out their Diploma work in industrial companies, are able to translate their knowledge directly and successfully as the relevant feedback shows. Also the initial evaluation discussions with Graduates of this module showed a tremendous acceptance by all kinds of industrial companies.

Therefore it can be stated that this Karlsruhe originated education model promotes a professional competence for graduates [99/5].

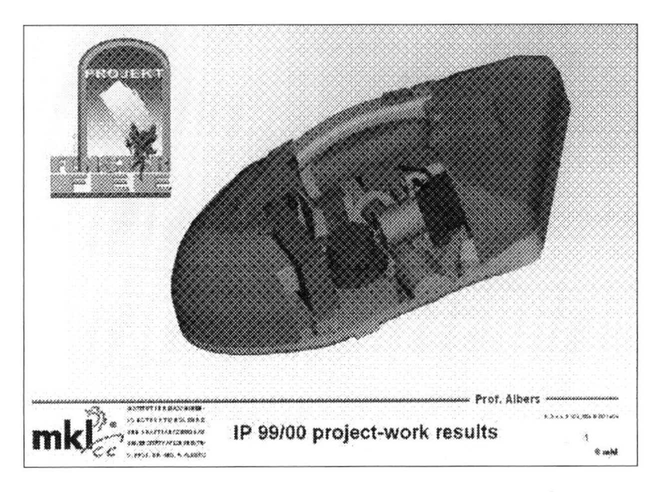

Fig. 4 Example of a result of project-work (Window Cleaner)

REFERENCES

(1) [99/14] Albers, A., Matthiesen, S.: Lichtentaler Protokolle - Modelle für die Maschinenelemente Lehre. February 24, 1999 Baden-Baden. Universität Karlsruhe (TH), Darmstadt 1999

(2) [99/9] Albers, A., Matthiesen, M.: Maschinenbau im Informationszeitalter - Das Karlsruher Lehrmodell. 44. Internationales Wissenschaftliches Kolloquium, Maschinenbau im Informationszeitalter 20.-23.09.1999, Technische Universität Ilmenau, 1999

(3) [99/8] Albers, A., Birkhofer, H., Matthiesen, S.: Neue Ansätze in der Maschinenkonstruktionslehre. Beitz Kolloquium 09.07.99; Universität Berlin, 1999

(4) [99/5] Albers, A., Burkardt, N.: Proficient Designers - A Challenge to Academical Education. International Conference on Engineering Design, ICED 99, Munich, August 24-26, 1999

Environment in engineering education – new approaches to curricula content

I P SOLOMONIDES
Department of Design, The Nottingham Trent University, UK
A R CRISP
Department of Mechanical and Manufacturing Engineering, The Nottingham Trent University, UK

SYNOPSIS

The teaching of environmental themes in design and other degree programmes is currently popular and appropriate. Integration comes at a time when courses are being written from a 'leaning outcomes' position. Provision in the department was based on traditional precedents. This is now changing as new courses begin and students from different backgrounds enter university. As a consequence, the successful integration of environmental themes will demand new approaches to curriculum development and a new conception of what teaching such issues involves. This paper describes new course designs and discusses some of the factors associated with teaching that may affect a successful outcome, notably the conception tutors have of what teaching is.

1. INTRODUCTION

The authors have previously described (1, 2) the development from its embryonic stage through to its present order the BEng Honours Degree course in Occupational Health and Safety Management, which has since developed into the BSc Honours Degree course in Health, Safety and Environmental Management, currently residing in the Department of Mechanical and Manufacturing Engineering. It is therefore timely to detail the current situation relative to the development of all courses within the department's remit with particular reference to the influence being played by the original development methodology employed on the Health and Safety course. It is also necessary to describe the rationalisation of courses, which has led to the demise of the BEng Honours Degree course in Manufacturing Engineering and the BSc Honours Degree course in Industrial Management.

This development and rationalisation comes at a time when eclectic influences from many governing bodies (in particular the accrediting institutions, the HEFCE and QAA), begin to impinge upon the activities of programme authors and designers. This paper describes the methods employed when attempting to incorporate contemporary issues: particularly the concepts of *'sustainability', 'clean manufacture' and 'ethical and moral principles'* within the various programmes operated by the department. This activity coincides with the introduction by the professional bodies of new datum for 'entry qualifications' coupled with an increased awareness by the educational authorities of the need to evidence 'quality'.

This has inevitably led to a self-questioning process by academic staff relative to their learning and teaching practices of environmental issues, professional integrity, engineering management and engineering in general. A fresh recognition of the specific requirements now sought by programme documentation has also led to new developments in presentation making the entire process of programme design 'student outcome' led. This paper describes the attempts to integrate and focus these many influences and impingements whilst designing MEng, BEng and BSc programmes for Mechanical Engineering and specifically a new and exciting programme i.e. BA/BSc Product Design; the special features of which are:

- Student experience of cross faculty teaching i.e. multi-disciplinarianism;
- The sharing of art and engineering experiences;
- Team teaching to encourage cross fertilisation of intuitive and analytical approaches to design;
- Programme pathways to provide a diverse experience, enabling students the opportunity of choice and ultimately award;
- The sharing of resources and expertise distinctive to both faculties;
- Input from recognised industrial and commercial practitioners;
- The input from all sources of contemporary issues, particularly environmental concerns.

2. BACKGROUND

The last decade has witnessed significant changes within educational philosophy as to the relevance of environmental education. It has become a major component of most educators curricula, no longer peripheral or add-on but mainstream and integrating. The English and Welsh National Curriculum Council had by 1989 developed a scheme summarised as follows;

'*by the age of sixteen all pupils should have had environmental educational experiences, which range from local to global in scale .....* '(3)

The basis of this education would be developed from two themes, firstly, knowledge and understanding, and secondly, skills. The knowledge and understanding would operate at different levels i.e. local to global and incorporate different influences such as moral, political, ethical, economic etc. The second theme, skills, has become a dominant feature in all aspects of education in England and Wales during the last decade, essentially seen as the ability (taught if necessary) to research, communicate and disseminate ideas about a chosen subject, in this instance the environment.

The Toyne Report of 1993 addressed the issue of environmental education post 16. Titled '*Environmental Responsibility: An Agenda for Further and Higher Education*' it proposed that all institutions of further and higher education both adopt and make public their comprehensive environmental policy statement, a policy and strategy for the development of environmental education and an action plan for their implementation (4). Further evidence that environmental education was now on the agenda and taken seriously came in 1996 when the UK government announced its strategy for environmental education; '*Taking Environmental Education into the 21st Century*'. The strategy provided for both further and higher education as well as the 16-19 age group and a general framework for the National Curriculum, its main objective being;

' to instill in people of all ages, through formal and informal education, and training, the concepts of sustainable development and responsible global citizenship; and to develop, renew and reinforce their capacity to address environment and development issues through their lives, both at home and at work'.

However, this impetus inadvertently caused education and the professional bodies many problems, some of which are discussed in this paper, namely:

- What is environmental education?
- Is it defined as ecocentrism or technocentrism?
- What is meant by sustainability?
- Can a compromise be sought between educationalists, politicians, economists and a broad grouping of peoples described by their ethic, moral or religious stance?

The Department of Mechanical and Manufacturing Engineering has throughout this period sought to develop its portfolio of courses in line with such contemporary issues as described above yet mindful of the influence of the professional bodies and its historic teaching strategy.

3. COURSE DEVELOPMENT

The department's portfolio of courses has seen significant change throughout the last decade in many areas i.e. type, number and discipline of courses offered. For example, the provision of Higher National Diploma courses is now the province of the colleges of higher or further education. A national decline in the number of students wishing to study engineering, as indicated has prompted many Universities to rationalise their engineering provision. This has led at the Nottingham Trent University to the formulation of a common first year for engineering and the closure as a distinct course of the BEng Honours Degree in Manufacturing Engineering. Although offered at level 2 and 3 it is deemed to have little future even though industrial bodies claim a shortfall of skilled and graduate engineers.

The higher education sector has increasingly been asked to account for itself relative to the quality of educational provision overseen by the Higher Education Funding Council for England (HEFCE). This has led academics to question their teaching styles and practices and review their peers in both formal and informal manners[i]. The split, which existed between the Universities, Polytechnics and Colleges of Higher Education has been removed 'on paper' by the restructuring of higher education granting almost all institutions offering degree courses University status and title. This has led to the publication of league tables and a growing competitiveness between the universities for ranking, students and research monies. This has in turn moved the universities to develop a business culture, which has to integrate with their teaching and learning philosophies as well as their mission statements on the environment and peoples' welfare. At the same time the professional bodies such as The Engineering Council have sought to enhance the standing of engineering within the global community by raising the standard of graduate education. These eclectic influences have stimulated the department of Mechanical and Manufacturing Engineering to rationalise and develop a new grouping of programmes, which will not only produce in the technical disciplines quality graduates but will produce graduates educated broadly in contemporary studies of which environmental matters would be included. The programmes and groups are:

Group 1
- BEng Honours Degree in Mechanical Engineering
- BEng Honours Degree in Integrated Engineering
- BSc Honours Degree in Engineering (Mechanical and Integrated)

Group 2
- BA/BSc Product Design
- BSc Honours Safety, Health and Environmental Management
- Diploma in Safety, Health and Environmental Management

4. TEACHING AND LEARNING STYLES

The courses split naturally into two groups as described above, which due to the many influences described are delivered in different manners creating the need to take careful consideration of the content and approach required for environmental and contemporary issues. The Institution of Mechanical Engineers published their interpretation of the required educational base for the professional engineer (5) stating that graduates should *'know, understand and be aware of'* environmental factors. The problem envisaged by the authors is that due to the modular structure of engineering degree courses, environmental issues may be considered only in a single module, and more worryingly, be taught in the style of an engineering module as opposed to the action oriented, open ended inquiry required by environmental issues. It is as if the paradigm, the model by which tutors in these subjects conceive of and deliver teaching and learning is somehow at odds with the demands of environmentally based content.

Engineering education is in the main based on a positivistic and deterministic tradition. Perry (6) showed that the majority of engineering students tend to graduate with a dualistic view (that is a relatively dogmatic, subjective and fixed ideas of what is 'right' and 'wrong') rather than a relativistic view (realising that the issues under discussion might be interrelated). Later studies (7) have revealed a relationship between student conceptions of what learning is, and tutor conceptions of what learning is. Student conceptions of learning were first reported by Saljo (8) and can be described as being broadly reproducing or transforming; the simple acquisition of presented knowledge or the integration of knowledge into personal understanding. Similar, variable conceptions of what teaching is and should be, can be expected to be demonstrated by tutors.

The point being that both students *and* faculty can have profoundly differing views of what teaching and learning consist of, let alone what forms of teaching and learning are required within a multifaceted area such as environmental education. In discussion with colleagues we have found significant differences between the views and attitudes held by academics in the same department. Of course this perversion is expected in institutions such as universities, but could lead to a rift between those perceived to be either dualistic or relativistic in outlook. This difference can be clearly seen for example, in the comparison between research paradigms often associated with science and the humanities (figure 2). Anderson and Burns (9) call these 'contrasting dimensions of inquiry', a title that would seem to fit the tenet of this paper, whilst as far back as 1894 Wilhelm

Wilderband proposed a distinction between what he called 'nomothetic' and 'idographic' explanations. The former is concerned with generating laws about phenomena and the latter is concerned with focussing on the uniqueness of events and what those events mean[ii].

Verifying hypothetico-deductive theory

Discovery Phase				*Confirming Phase*			
Theory →	Deduce hypotheses →	Research design →	Collect data →	Analyse data →	Test hypotheses →	Interpret →	Modify theory

Discovery of grounded theory

Discovery Phase				*Revisional Phase*		
Collect Data →	Analyse data →	Induce theory →	Collect more data →	Analyse data →	Revise hypotheses to fit data →	Generate theory

Fig. 1 Two approaches to theory construction (9)

In attempting to deal with these potential conflicts the department has tried several initiatives, including during the last academic year, the formation of year teams. These teams met weekly to discuss issues relative to the day to day teaching of students in an attempt to harmonise the approach taken across various modules. Whilst successful in promoting dialogue between subject areas, the meetings could not and were not intended to alter the conception of members of staff in relation to how teaching should be conducted or the values and beliefs held by faculty.

There is therefore a significant problem that has yet to be resolved. Faculty will continue to examine the teaching and learning milieu largely based on the positivistic tradition. Unless staff can adopt a more interpretative or phenomenological approach to examining their practice, it is unlikely that they will promote the action-orientated research stance needed when examining environmental issues or even recognise that students will themselves have a range of beliefs and values concerning the environment.

In order to deal with this last issue, the forthcoming academic year will for the first time involve undergraduates in a skills audit, designed to reveal their knowledge and stance toward contemporary issues. Not only will this make transparent the inevitable gulf between student ability and staff expectations of that ability, it could be used further in staff-development activities to raise awareness of differing conceptions as described here. Indeed it could further seem that a skills audit is essential to gauge the contemporary issues knowledge of students enrolled on engineering courses. The Institution of Mechanical Engineers is clear that General Studies 'A' level is not to be included in the UCAS score (5). This against a general calling for the broadening of 'A' level studies moving toward the baccalaureate system. These factors have led to the creation of 'group one' type degree programmes for engineering in modular format where the environment could form only 1/3 of a 10-credit module in year 2, or as little as 1.2% of the course. 'Group two' courses, have developed in a less restricting framework. Indeed the BA/BSc Product Design is an innovative and idealistic course designed to cross the disciplines of two distinct faculties incorporating the best practices of both whilst creating a course suited to the aspirations of the wide ranging technologist. A skills audit is a fundamental part of the early induction programme and environmental issues are developed in all modules. This is comparable with the BSc honours degree in Environmental, Safety and Health Management where the integrating theme throughout the course is the environment. However, the base for teaching is the interpretation of 'environmental issues' by the course team, which may not reflect the holistic approach to environmental teaching envisaged by the schools of thought developing theories such as 'deep ecology' and 'ecocentrism'. Indeed the base for the individual modules concerned with environmental matters appears to be the rigid policies and legal transpositions of the EU.

Although environmental issues can be shown to be the integrating theme of both the BSc Honours Degree in Safety, Health and Environmental Management and the BA/BSc Honours Degree in Product Design, the taught material in many of the modules can be described as of a rigid and prescriptive nature. The curriculum needs to be broad enough to accept ideas from all quarters and structured around debate enabling both students and staff to educate themselves(3) 'about the environment, for the environment, in and from the environment'.

To exemplify the attempt to meet at least some of these requirements, two of the courses from 'group two' are described below. The arrows are indicative of an environmental theme; the heavy arrows indicating coherence through several modules, the broken arrows indicating reference to issues rather than a focus on issues.

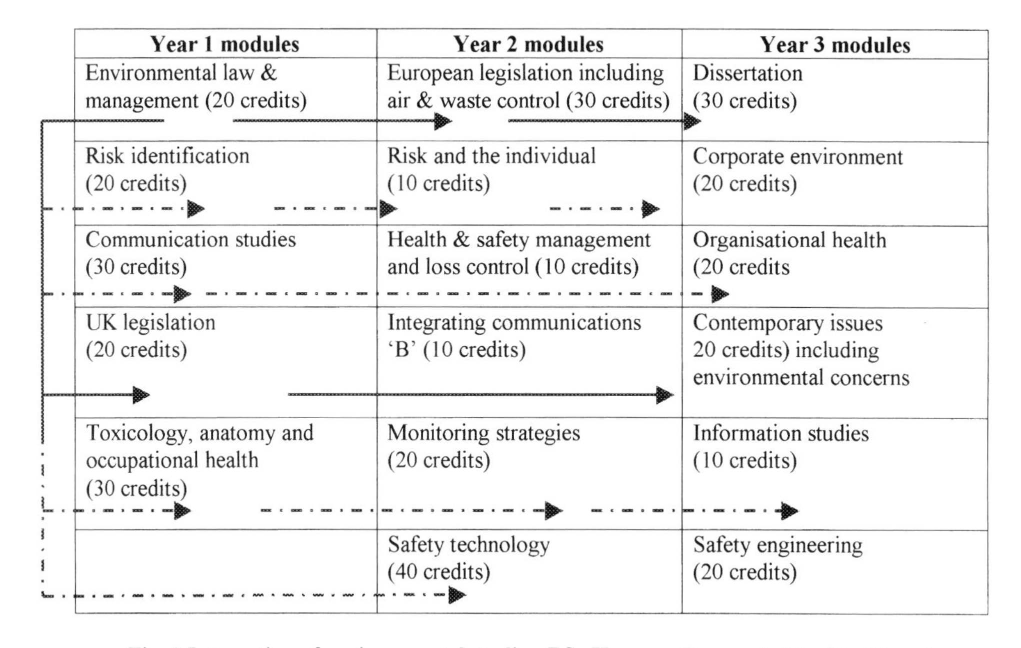

Year 1 modules	Year 2 modules	Year 3 modules
Environmental law & management (20 credits)	European legislation including air & waste control (30 credits)	Dissertation (30 credits)
Risk identification (20 credits)	Risk and the individual (10 credits)	Corporate environment (20 credits)
Communication studies (30 credits)	Health & safety management and loss control (10 credits)	Organisational health (20 credits
UK legislation (20 credits)	Integrating communications 'B' (10 credits)	Contemporary issues 20 credits) including environmental concerns
Toxicology, anatomy and occupational health (30 credits)	Monitoring strategies (20 credits)	Information studies (10 credits)
	Safety technology (40 credits)	Safety engineering (20 credits)

Fig. 2 Integration of environmental studies, BSc Honours Degree Safety, Health and Environmental Management

Year 1	Year 2	Year 3
Design studies 1, including ethics & environmental considerations (20 credits)	Design studies 2, (20 credits)	Design studies 3, (20 credits)
Design technology 1, including green design (30 credits)	Design technology 2, BSc/BA (30 credits)	Design technology 3, BSc (40 credits)
Graphical methods (30 credits)	Management & marketing toinclude 'green funding' (20 credits)	Minor projects BA, (40 credits)
Design projects including environmental & contemporary issues (40 credits)	Design project 2, BSc/BA (50 credits)	Major projects BSc/BA, (60 credits)

Fig. 3 Integration of environmental studies, BA/BSc Honours Degree Product Design

The challenge that faces the staff is the methods by which environmental issues will be taught and discussed across the course portfolio ensuring individual students receive comparable tuition. Individually the staff must ask themselves what they know and understand by the term *'environmental issues'* and accept that personal development is required.

5. CONCLUSION

There is a tension to be resolved between the desired teaching in engineering at degree level of environmental issues and the way in which that teaching is undertaken. This tension is not only manifest in the edicts of professional and accrediting bodies, the curricula of degrees, but also in the underlying assumptions, conceptions and orientations both staff and students may hold when perceiving tasks in hand and interacting with environmental issues. Whilst the background skills and attitudes of students may be attributed to former experience, the main influence on their formation as graduates should rightly be associated with the context of their degree education. This context is heavily influence by factors such as teaching, workload and assessment that in turn are often formulated in accordance with the personal stance of the teacher. Within engineering and science based faculty it can be argued that this stance is predominately deterministic and resides with a positivistic tradition. Faculty therefore needs to examine their curricula and obtain feedback from their students in order that some indication and realisation of their perspectives might be achieved. The approach that students adopt and depth of understand they achieve may be more heavily influenced by their tutors approaches to teaching than is currently realised. This is not limited to just environmental issues.

REFERENCES

(1) Crisp A, Randall T and Solomonides I (1995) 'Reorganising Design Teaching: Design for the Whole Curriculum SEED '95 Proc. 17th Annual Design Conference Lancaster University UK
(2) Crisp A, Solomonides I, Laing F and Harrison T (1997) 'Teaching Environmental Education Through Integrated Themes' ENTRÉE '97 Proc. International Conference on Teaching Environmental Training in Engineering Education Sophia-Antipolis, France
(3) Palmer J (1998) Environmental Education in the 21st Century Routledge
(4) HMSO (1993) Environmental responsibility: An agenda for Further and Higher Education
(5) The Institution of Mechanical Engineers (IMechE)(1998) The Formation of Mechanical Engineers
(6) Perry W (1970) Forms of Intellectual and Ethical Development in the College Years: A Scheme Holt, Rinehart and Wilson, New York.
(7) Bruce C and Gerber R (1995) 'Toward's University Lecturer's Coneceptions of Student Learning' Higher Education Vol 29 pp.443-358
(8) Saljo R (1979) Learning in the Learner's Perspective: I – Some Common Sense Conceptions (Reports from the Department of Education, the University of Goteborg)
(9) Anderson L and Burns R (1989) Research in Classrooms Pergamon Press

---

[i] See previous discussions by Solomonides I and Swannell M (1996) 'Encouraging Students, Making the Passive Active', in Whisker G and Brown S (eds) Enabling Student Learning, Kogan Page pp 102-115. The department also participated in a national project aimed ant developing best practice. The 'Sharing Excellence' project is reported at http://www.celt.ntu.ac.uk/se/. Further discussion is found in Solomonides I (1998) 'Intervention and Motivation, What affects What?' in Brown S, Armstrong CS and Thompson G (eds) Motivating Students Kogan Page pp 25-35.

[ii] Various research undertaken concurrently in the UK (e.g. Noel Entwistle, Paul Ramsden, Graham Gibbs), Scandinavia (e.g. Ference Marton, Roger Saljo) and Australasia (e.g. John Biggs) over the last 30 years has established the notion of 'student approaches to study'. This describes the 'why' and 'how' students interact with learning tasks. Based on perception of the task in hand the student will adopt an approach to study (deep, surface, achieving or a mixture of these) and will interact in a transforming or reproducing manner. It is likely that these student approaches to study and allied concepts are significantly affected by tutor approaches, orientations and conceptions.

Inside out and outside in – taking product design education into the real world and bringing the real world into product design education

M WILKINSON and **V THOMAS**
School of Technology and Design, University College Northampton, UK
R SALE
Faculty of Design, Engineering and the Built Environment, University of East London, UK

SYNOPSIS

Product design education has a long established practice of involving part-time and visiting staff in undergraduate and postgraduate programmes and the benefits on both sides are clearly recognised. A natural extension of this is the live-client project. In this case there is the potential for conflict between academic requirements and commercial imperatives. The paper represented by the authors reviews live-client and commercially sponsored projects from both academic institutions and, in the light of the relative successes and failures of these projects, propose guidelines for creating the conditions where both academic and commercial objectives may be met.

1. BRIDGING MODES

Traditionally ways in which product design education and the 'real world' of practice are brought together cover a broad spectrum from total direct immersion of the student, as in work placement, to indirect exposure through tutorial contact with part-time staff.

A review of the opportunities in any institution offering a Product Design course will probably find some if not all of the following in operation:-

- Part time staff
- Internally briefed projects
- Externally briefed projects
- Competitions
- Visits to manufacturers, design offices, etc.
- Websites
- Work placements
- Teaching Company schemes
- Consultancy through HEI design centres
- Partnerships and exchange of staff

Successful outcomes in any of the bridging methods mentioned above is dependent on identifying clear objectives for the contact and establishing ground rules for interchange that all parties understand and have agreed from the outset.

In the situation of a consultancy based design project protocols are well established including the brief and a legally binding contract between commissioning agent and designer which governs all aspects of the transaction from ownership of IPR, timeframe, fee and level of expected output. A design project initiated by a commercial client to be worked on by students in an academic environment and within the structure of a syllabus will not be subject to legal or business imperatives of a consultancy transaction and so requires more consideration as to its purpose.

So called 'live design' projects, with commercial client bodies or other external organisations, can provide excellent vehicles for exposing students to commercial concerns in a controlled, sheltered environment. Manufacturing, marketing, technical, human factors and communication issues can be explored and developed by students and staff in a low risk/ high learning relationship.

The context of the live project made up by the design brief, briefing process, contact with staff from the company, site visits, extra resources, etc. can provide opportunities for students to measure their own ability and skills to respond to a perceived 'real world' challenge and through positive feedback grow in confidence and effectiveness as designers.

Experience has shown that tensions and conflicts can occur in the live project context when the objectives are not clearly understood by all participants or when perceptions and assumptions are not addressed from the outset. For example students may assume that a live project is an opportunity to produce designs for products which may go into production. Raising these kind of expectations may neither be fair to the student, relevant to the educational objectives nor useful to the product development strategy of the client company. A live project in a design course is in a real sense no more real than a project initiated by teaching staff. Students are operating in an education environment the product of which is increased knowledge, skill, and experience towards achieving improved effectiveness as designers.

Live projects are not about replicating the world or practice as this is patently unrealistic but provide an opportunity to explore issues addressed in the curriculum which relate to the 'fit' of design activity in business development and NPD process. It should not be about who in the year cohort can best produce product designs which match a particular market/manufacturing context but how the opportunity for contact with an outside "client" can be put to best use to inform and inspire design activity, and better prepare the student for future practice.

Quality of the contact will be judged as each of the participants (students, staff, client groups) perceives it.

The notion of the contact as a one way transfer of knowledge and wisdom is not a helpful model and condemns some if not all students to failure. The notion of the contact as a free exchange of ideas and information in an informal and positive environment will most probably result in benefits for all participants. The kind of exchange and nature of the exchange is important. This process of sharing information will usually start a long time before students are involved and should continue throughout the project.

An approach whose goals are demonstrable learning outputs and whose basis is information sharing and idea development will also answer the perception that educational live projects undercut work from the professional market place. It can be argued that exchange based relationships which are non-contractual and non-confrontational can actually stimulate the possibilities for new design commissions which will ultimately benefit professional design practice. Educational institutions are ideally suited at taking a more reflective, critical, longer term view of product design strategy and it can be argued that the live project if well structured provides an excellent forum for this interchange where end users, buyers and others involved in the whole-life of a product may be able to contribute. This approach is not necessarily characteristic in the traditional consultancy / client relationship but design practice may well be the net beneficiaries of "inside out-outside in" activities in product design education. In this sense the educational institution is acting as a catalyst, raising awareness of design in business and perhaps acting as an introduction agency to companies/businesses who may not be design based but who would like to know more about what design does and could do for them.

A high level of open interaction between students and client body may result in the tutor's role shifting temporarily from a direct teaching-learning one to a mentoring-learning one. Whereas the tutor is normally perceived as a focus for both guidance and assessment – a live project introduces a three way situation where direct conflicts or single flow of exchange became more conversational and a wider range of perspectives can be brought to bear.

The two case studies outlined below demonstrate two possible alternative approaches to live projects, the first is centred on raising awareness of recent developments in materials and processes, the second on raising awareness of end user requirements. Both involved a high level of exchange between external organisation and students.

2. CASE STUDIES

University College Northampton : BSc (Hons) Product Design
Design for Blowmoulding : 2nd Year Project with Krupp Plastics & Rubber Machinery

Initial contact with Krupp, a division of ThyssenKrupp, came as a result of a direct approach by UCN for technical assistance with an ongoing research project. The insight which Krupp gained into Product Design at UCN encouraged them to propose a joint project.

Krupp were entirely candid in stating their expectations of the collaboration: as process machinery manufacturers they are always keen to meet their customers' customers, the specifiers of material and process. In the normal commercial supply chain the specifier would not necessarily be aware of recent developments in processing and the process machinery manufacturer would not have direct access to the specifier. The impeding factor in this relationship is the converter, who has a vested interest in only appraising the specifier of processes which are currently in-house.

Hopefully the Product Design students of today will be amongst the plastics specifiers of tomorrow and Krupp's intention is to address them directly.

A second benefit which Krupp anticipated was access to concept designs which could be integrated into promotional material to demonstrate the kind of product which could be manufactured using leading edge blow moulding technology. There was no expectation of direct commercial exploitation of student design work.

The project was initiated with a multimedia presentation from Krupp introducing the company, outlining the basics of the blow moulding process and discussing recent developments. A second presentation was given part way through the project to respond to specific technical and moulding issues. There were also two informal studio visits for one to one tutorial sessions.

Thus the students were able to benefit from high level management and technical input from Krupp's UK and German teams and respond to real business imperatives and manufacturing considerations in their projects, whilst being encouraged to 'push the envelope' outside normal commercial constraints. Two examples of projects are illustrated, in Figure 1 (display portfolio) and Figure 2 (gardening watering system).

In their briefings Krupp were concerned that each student should be as free as possible to write their own brief to explore new and innovative applications for blow moulding. The student response to such freedom was generally very enthusiastic but some students did have difficulty in defining an appropriate area for investigation and, in these situations, staff involvement and intervention was higher than would normally be the case with an internally briefed project.

University of East London : BSc (Hons) Product Design
Interactive System Design : 2nd Year Project with U3A

The design brief required an investigation, analysis and solutions to problems associated with the usability of public information terminals. Previous commercial trials has indicated a lack of take-up by users of new ticketing machines and information kiosks. The focus of the project was on the full range of potential public end users, including those who may have been previously excluded i.e. :-

- Elderly users
- People with visual and hearing difficulties
- People with dyslexia or learning difficulties
- People who are not familiar with IT systems
- People with restricted mobility
- People with manual dexterity problems

The range of products chosen for the project were:-

- Integrated public transport travel planning and ticketing
- Tourist information point
- Public 'e' mail / internet service

Students were expected to work in teams of 3 with a leader responsible for preparing project schedules, client liaison, keeping project log books, recording team activities, attendance, etc. (figures 3 and 5)

The "live" nature of the project is by the inclusion of an expert end-user group representing a different age/experience/background to the student. This dimension was supplied by University of the Third Age (U3A) Design Group of Bromley, South London. The U3A

Design Group is made up of 20-30 retired or semi retired designers, engineers, architects whose ages range from 55 to over 80. The purpose of the link was to challenge assumptions about user requirements perceived to be resident in the student cohort.

In common with Product Design students from similar courses the cohort is young (20-24 years) and largely male.

Technical / marketing context was provided by kiosk manufacturers and service providers and from use of the internet. Each team was asked to construct a coherent business case for the terminal and identify commercial benefits to service provider, site owner and operator.

Teams worked up their concepts to full size wherever possible to enhance information sharing and feedback with the client group (figure 4). U3A members (6-8) visited on four occasions during the project and participated in informal discussions and active testing of design ideas for product interface and hardware solutions.

Contact with U3A proved to accelerate student's understanding of the needs of previously excluded user groups. Students were made aware of the full impact of their design decisions and were placed in a situation where there were different questions to ask. Having a knowledgeable reference group enabled a high level of design iteration at each presentation, where barriers to use were clearly identified, and solutions considered.

It was observed that teaching staff as well as students were equally able to draw upon U3A support and knowledge thus strengthening debate in the teaching /learning exchange. U3A members noted their input was most effective at an early stage of concept development and generally felt useful and wanted by the students.

Generally there was a feeling of raised motivation and sense of purpose which was not evident in projects of a similar nature which had not included a 'live' element.

3. GUIDELINES

- Early involvement in design process
- Clear understanding of commercial and educational requirements and expectations
- Framework of defined teaching and learning outcomes, even with 'blue sky' brief
- Agreement on terms of engagement (i.e. ownership of IPR, funding, etc) at inception.
- Agreed deliverables upon project completion

Figure 1. Display portfolio, Kevin Moore, UCN

Figure 2. Garden watering system, James Clarke, UCN

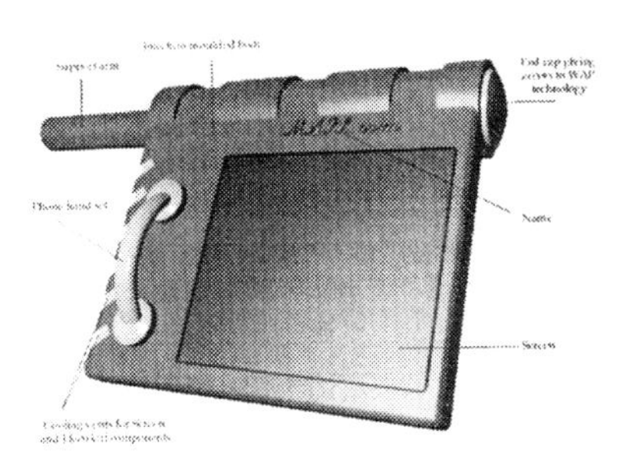

Figure 5. Interactive Communications Booth Stuart Wells, UEL

Figure 3. Public Transport Information Interactive Terminal Dana Mohammad, Patrick Owusu-Sarpong, Kiran Seunarine, UEL

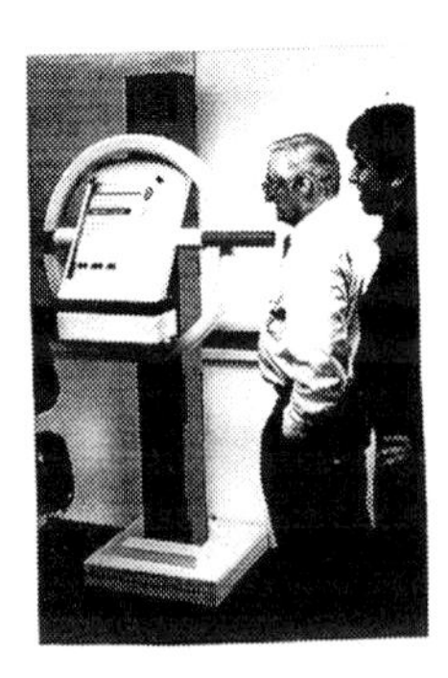

Figure 4. Journey Planner and Ticket Dispenser Najid Chaudhary, Ian Drew, Joseph Owusu-Sarpong, UEL

KBE research, practice, and education – meeting the design challenge

P J LOVETT, K OLDHAM, and **C N BANCROFT**
Knowledge Enigneering and Management Centre, Coventry University, UK

ABSTRACT

This paper contrasts the benefits to be gained from the use of knowledge-based engineering with its limited adoption to date. This incongruity presents a challenge to educators and practitioners, which has been taken up by the Knowledge Engineering and Management Centre at Coventry University, using the methods described herein.

1. INTRODUCTION

Knowledge-Based Engineering (KBE) offers substantial benefits to the engineering industry, particularly in the areas of product and tooling design. It makes tacit design knowledge explicit, so that it can be shared. It encapsulates expertise from a range of disciplines in an automated application, accelerating the design process dramatically, enabling more options to be investigated, and resulting in a better design. However, considering the potential rewards, the adoption of KBE has been less widespread than might have been expected.

This paper first describes KBE, the principles upon which it is based, and its application to design problems. There follows an account of the work of the Knowledge Engineering and Management (KEM) Centre at Coventry University. Several factors that contribute to the poor uptake of KBE have been identified, and work is in progress to reduce their influence. These are the lack of:

- methodological support for KBE system development
- evidence of the practical benefits of KBE
- KBE skills and understanding within engineering circles

The remaining sections describe the means employed by the KEM Centre to combat these obstacles, namely:

- The compilation of two methodologies for KBE system development – aimed at large and small organisations and projects
- The building of software to demonstrate the benefits that KBE can provide
- The transfer of KBE skills to industrial partners through close collaboration on KBE software projects
- Formal training by means of computerised provision and face-to-face instruction

2. TERMINOLOGY

A great deal of time is devoted to debating the merits of competing definitions of terms in this area. To obtain general acceptance it is often necessary to make definitions too broad to be of real use. The only claim made here is that the following meanings are those which apply within this paper:

- Knowledge management -
  The aggregated knowledge of a company's employees can be regarded as an asset. Knowledge management attempts to gain the maximum value from this asset by using it effectively (collecting it, sharing it, and using it).
- Knowledge engineering -
  The practice of capturing knowledge, representing it, and embedding it within computer programs.
- Knowledge-based systems -
  Systems consisting of a repository of knowledge, which is distinct from the inference engine that processes it. They are often used for diagnostic purposes.
- Knowledge-based engineering -
  A discipline that uses knowledge embedded within computerised applications to support engineering activities. It evolved from CAD, and close links remain. While it is suitable for a range of purposes, including product configuration, cost estimation, and training and presentation packages, it is CAD-based KBE that is the subject of this paper.
- Knowledge -
  This term is discussed last as it is used differently in different contexts (because knowledge management and KBE have independent origins). In knowledge management circles its meaning is fairly broad. It often refers to that which was previously known as "information", before the term "knowledge" became fashionable. Its use is more apposite when it is applied to that which is particularly difficult to extract and express, or which answers questions beginning with the words "how" and "why" rather than "what". In KBE (and the other terms discussed above) knowledge is usually expressed in the form of rules (e.g. if dimension_a > 50 then scrap = true) and relationships (e.g. A is a variant of B). Knowledge is applied to information to interpret or manipulate it.

3. KBE FOR DESIGN

One of the most profitable areas for the employment of KBE is that of product design. This is because of the many influences on design that call for knowledge outside of the designer's domain (1). The principles of concurrent engineering demand that consideration of factors such as manufacturing limitations and costs, properties of materials, industrial standards, legal constraints, etc. cannot be left until after the geometrical elements have been determined (2). KBE enables knowledge relating to such matters to be embedded in design tools, minimising or eliminating the need to refer to specialists at design time. This shortens the duration of the design process, and removes from specialists the burden of repeated mundane queries.

KBE can be regarded as an extension of parametric modelling, in which geometrical elements of a design may be described indirectly, in terms of their relationship to others that have been given absolute values. Both are implemented in the form of rules, although KBE is not confined to working with geometrical aspects. However, the use of parametric modelling within KBE to automate repetitive tasks related to geometry allows the designer to concentrate on more creative aspects.

Members of the KEM Centre have been working with Precision Disc Castings, of Poole, UK, on a KBE project to automate the production of a set of 13 engineering drawings (3). Software has been written which requires only that the user input certain essential dimensions and angles of a casting to be manufactured, using a graphical interface (i.e. there is no need to be able to use a CAD package). The drawings have previously been produced using AutoCAD 14 alone, which takes 6 to 8 weeks. Using the new application (which was written in Intent! and is linked to AutoCAD 14), they can now be produced in 6 to 8 minutes.

4. METHODOLOGIES

While investment in KBE can return substantial rewards, building KBE applications is not a trivial enterprise. A common approach to managing complex development activities is to follow a methodology. This guides the developer in the tasks to be executed and provides the structure necessary for project management. Moreover, there are considerable advantages to the standardisation that results, especially of documentation, not least the ability to maintain the application as the knowledge embedded within it changes and must be updated..

While methodologies exist for knowledge-based system development, there is currently no industrially accepted methodology specifically directed at KBE applications. The KEM Centre is attempting to rectify this situation, and is working on two complementary methodologies for KBE system development.

5. MOKA AND KOMPRESSA

5.1 The aims of the methodologies

It is hoped that MOKA (Methodology & tools Oriented to Knowledge based engineering Application (4)) will form the basis of an international standard. The Centre is contributing to its development as a member of a consortium comprising: Knowledge Technologies International, Decan E-company, Aerospatiale Matra, BAE Systems, Daimler-Chrysler, and PSA Peugeot Citroen. As can be deduced from the participants, it is aimed at large manufacturing concerns.

KOMPRESSA (Knowledge-Oriented Methodology for the Planning and Rapid Engineering of Small-Scale Applications (5)) is aimed at smaller companies and projects, and at organisations with little experience of information technology. Consequently the methodologies complement each other rather than compete, and together are suitable for projects of all sizes.

5.2 The effects of organisation and project size

There are many differences between small and large organisations that must be taken into account when devising a methodology to suit their needs. These include

- The existing level of IT (and perhaps KBE) skills
- The size of development budgets and vulnerability to risk
- The problems of knowledge validation in relation to the number of knowledge sources

In addition, the size of projects affects

- The amount of time and effort it is worth devoting to learning a methodology
- The amount of time and effort it is worth devoting to using a methodology
- The degree of formality of a methodology most appropriate for the quantities of knowledge and information involved (i.e. smaller volumes need less rigid mechanisms).

5.3 Differences between the methodologies

Table 1 compares the MOKA Phases with the Activity Groups of KOMPRESSA. Although both methodologies are concerned with the whole KBE system development life-cycle, MOKA concentrates on the Capture and Formalise phases, which deal with structuring and formalising the collected knowledge. The Design activity group is the largest and most detailed part of KOMPRESSA.

Table 1 The MOKA life-cycle and KOMPRESSA activity groups

MOKA Phase	KOMPRESSA Activity Group
Identify	Initial Investigation
Justify	Requirements Analysis
	Application Classification
	Tool Selection
Capture	Design
Formalise	
Package	Implementation
	Validation, Verification and Testing
Activate	System Realisation
	Maintenance

MOKA uses informal and formal modelling (of the product and design process). The informal model consists of a set of diagrams showing the elements and how they are related, and structured forms containing the details. These are easy to use and understand, and constitute the "knowledge handbook", which promotes knowledge sharing. The formal model is based upon the Unified Modelling Language, and is more suited to people developing information systems. All KOMPRESSA models are informal, and easy to use for newcomers to information systems and modelling. The reduced level of formality is possible because of the smaller volumes of knowledge.

There are some tasks to be carried out when using a methodology that need to be performed regardless of the size of the project. When the project is large these form only a small proportion of the effort needed for system development. KOMPRESSA attempts to keep these "overheads" to a minimum, as their effect on small projects would be considerable. It is neither possible nor desirable to do this with MOKA, which must include mechanisms to deal with large amounts of complex knowledge. In short, smaller projects require a smaller methodology, and vice-versa.

It is hoped to provide a fuller comparison of the two methodologies in a further publication.

6. REFIT

REFIT (Revitalisation of Expertise in Foundries using Information Technology), is a project funded by the ADAPT initiative of the European Social Fund. It aims to enhance the competitiveness of small foundries in the West Midlands by encouraging and supporting the use of KBE. It is attempting to achieve this in the following ways:

- By building demonstrator applications with the five project partners. This serves two purposes. Firstly, it presents evidence of its practical utility to those who doubt the worth of KBE. The software being developed can be seen to provide solutions to real problems within an industrial environment. Secondly, by involving the industrial partners in the development process, skills in IT and KBE are transferred to them, and will remain after the lifetime of the project.
- The project funds the KOMPRESSA methodology.
- A workbook is being written to document the experiences of the REFIT team and to provide case studies for future reference. It will also contain a formal description of KOMPRESSA, and a "User Guide", written in less technical language.

7. NO DEFECTS

The KEM Centre is also providing more formal training in KBE, in the form of NO DEFECTS ("New Opportunity for the Development and delivery of Education for Foundry Employees utilising Computer-based Training Systems"). This is a one-week introductory course in KBE for employees in the foundry industry, based on the results of the REFIT project and KOMPRESSA. It is funded by the European Social Fund, and while it is targeted at personnel within small foundries, it is also relevant to other sectors of industry.

Initial delivery will be via face-to-face tuition, and will include a practical element, to enable students to gain first-hand experience with KBE software. Topics include the advantages of using a methodology, KOMPRESSA, software selection, setting the scope of a project, knowledge capture, programming, and validation, verification and testing. The course material is to undergo rigorous industry-based evaluation, and will also be made available in multi-media format on CD-ROM later in the year.

8. OTHER ACTIVITIES OF THE KEM CENTRE

The Centre provides consultancy and business support in applications of KBE not directly related to design (especially product configuration). It is also active in several areas of research related to KBE. Furthermore, members of the Centre deliver a module on KBE that forms part of a University MSc course, and consideration is being given to the presentation of a full MSc course on KBE.

9. OUTLOOK

There are still problems to be overcome if KBE is to fulfil the potential that it undoubtedly offers for knowledge-based design. The lack of IT and KBE skills, of confidence in its ability to deliver real benefits, and of methodological support, are all serious hindrances to its uptake, and must be tackled if it is to achieve its promise. The KEM Centre has taken up this challenge: REFIT has attracted a great deal of interest, both within and outside the foundry sector, and is creating a skill base that will outlive the project. The first release of MOKA is now imminent, and that of KOMPRESSA will take place by the end of year. NO DEFECTS will provide training for large numbers of individuals. In addition, the cost of software and hardware needed to create KBE applications has fallen dramatically in recent years. Finally, there is currently great interest in knowledge-related technologies in general, and it is to be hoped that this will lead to a greater appreciation of KBE.

It would be misleading to claim that there were no questions of a technical nature remaining in this field. Issues of particular interest to members of the Centre include:

- The validation of knowledge when there is a limited number of knowledge sources (perhaps only one).
- Maintaining knowledge and ensuring its consistency as it changes over time.
- Forms of knowledge representation that permit knowledge sharing and exchange.

Staff at the KEM Centre are contributing to research in these areas. However, the benefits that KBE can provide *now* can be spectacular, as the REFIT project has demonstrated.

There can be real dangers in ignoring the opportunities provided by KBE, as evidenced by the foundry industry. Many long-serving employees are now approaching retirement, but the apprenticeship schemes formerly used to pass on their skills have not been maintained. If action is not taken quickly, the knowledge of these experts will be lost forever.

An increasing number of factors have to be considered when designing products, and it is important that full advantage is taken of the help that is available. KBE offers tremendous scope for sharing knowledge – within and between disciplines – and designers should be encouraged to participate in the development and use of KBE applications.

REFERENCES

(1) P. A. Rodgers and P. J. Clarkson, “An Investigation and Review of the Knowledge Needs of Designer in SMEs”, The Design Journal, 1(3), pp 16-29, 1999.

(2) G. Coates, A. H. B. Duffy, W. Hills and R. I. Whitfield, “Enabling Concurrent Engineering through Design Coordination”, Proceedings of the 6th Annual Conference on Concurrent Engineering (CE ’99), Bath, UK, pp 189-198, 1999.

(3) P. J. Lovett, A. Ingram and C. N. Bancroft, Proceedings of the 15th International Conference on Computer-Aided Production Engineering (CAPE ’99), Durham, UK, pp 709-715, 1999.

(4) R. Brimble, K. Oldham, M. Callot and A. Murton, “MOKA: A Methodology for Developing KBE Applications”, Proceedings of the 8th European Conference on Product Data Technology, Stavanger, Norway, pp 361-366, 1999.

(5) C. N. Bancroft, S. J. Crump, P. J. Lovett, D. Bone and N. J. Kightley, “Taking KBE into the Foundry”, Proceedings of the 7th Annual Conference on Concurrent Engineering (CE 2000), Lyon, France, (to be published), 2000.

User-centred design research methods – the designer's perspective

A BRUSEBERG and **D McDONAGH-PHILP**
Department of Design and Technology, Loughborough University, UK

ABSTRACT

Direct contact with users can provide a valuable resource for designers, who often design products for use outside their own experience. Design research is currently being undertaken, which is concentrating on developing a training package and guide for the use of focus groups during the designing process – aimed at both undergraduate and practising designers. The initial stage of the project investigated the resources available to designers with particular emphasis on those resources that were regularly utilised. This paper reports the results of a series of interviews with practising industrial designers – to examine methods used during the concept generation stage, and highlight designers' perceptions toward focus group methods.

1. FOCUS GROUPS FOR USER CENTRED DESIGN

A focus group is a collection of individuals that have been brought together to discuss a particular topic, issue or concern. A moderator (a chair) provides a framework and structure to the meeting, integrating open-ended questions to promote discussion. The method relies upon the interaction between the individuals, encouraging *synergy* within the group (1). Focus groups assist in qualitative data collection, providing detailed insights into individuals' beliefs, experiences and perceptions, rather than statistically secured facts (2).

For the designer, user needs should be paramount. Products satisfy a number of needs, from the functional to the emotional. The user needs should be considered as early as possible – ideally, before and during the concept generation phase. Product developers, such as Industrial Designers, need to immerse themselves in tangible data (e.g. anthropometric, market research) and less tangible data (e.g. symbolic associations, user perceptions) to enhance effective product development. Designers can find themselves designing products for use outside their own experience, understanding and expertise. To a certain extent the designer can rely upon personal knowledge and experience, but this is often limited. Direct contact with users, to gain information about the potential users and the product in use, can provide a rich resource for the designing process (3). Hence the authors recommend that focus group methodologies may offer an effective form of data collection during the informative stages of the designing process (e.g. concept generation). Focus groups enable direct communication between designers and users. Employment of this method enables general and specific exploration of issues, as focus groups combine both context and depth. They provide backgrounds, reasons for individual opinions and experiences (2). Focus groups are suitable to retrieve data that is not readily available or experience not previously expressed. They provide depth for habit-driven topics (4).

It has been recognised that designers would benefit from the inclusion of design-research methods at undergraduate level. Recently graduated designers have been found to consider themselves as "poor researchers and now view researching as an important aspect of design education" (5). Designers need to immerse themselves into the design task specified by the design brief by collating a range of different types of information. We suggest that designers would benefit from the ability to carry out focus group research themselves. Designers often do not have access to the original material collected by market researchers. Through this division of labour, vital details may become lost through summarising the results. Moreover, designers may be particularly suitable to decide which are the most relevant questions to users regarding the design of a product. In focus group discussions, users provide feedback in their own words, thus helping the designer to understand user needs, including their aspirations and emotional bonds to products, as well as their cultural background.

The project currently undertaken at Loughborough University concentrates on the development of methods to support user-centred design. It is a one-year project funded by the EPSRC (Engineering and Physical Sciences Research Council) to develop focus group methods for designers. The method needs to be adapted to the requirements of designers. It may be combined with a variety of other methods from design practice and ergonomics (e.g. observation of product handling, rating scales, task analysis, or mood boards) to ensure an optimum output. A set of suitable techniques will be developed through a practical application to the design of a small domestic appliance. The final outcome of the project will be a resource of focus group material, a training guide for designers, as well as a range of conceptual products (e.g. kettle, toaster and coffee maker).

2. THE CASE STUDY

2.1 The aims of this case study

At this early stage of the project, the needs of designers require clarification, as they will be the users of the focus group guide and methodology. Hence, the working practice of designers needs to be understood – particularly to what extent design practice differs from the resources described in the literature. We also need to know whether, and under what conditions, designers perceive focus group techniques as appropriate to their work, and how the technique would fit into the methods commonly used. More information is needed about the formats of data that designers prefer, and how focus group data is currently being communicated to designers from external sources.

2.2 The interviews

A total of five designers were interviewed during this study, lasting between one and two-and-a-half hours. They included two freelance designers, one working for a design consultancy, and two working for a major manufacturer. Their ages ranged from 22 to 52 years. They were chosen to gain an overview over a broad cross section of design practitioners. The designers were interviewed using both open questions and a questionnaire. The questionnaire (refer to Figure 1) listed a range of different methods derived from the literature. It functioned as the basis for discussion, enabling coverage of many different design areas, as well as the different activities that designers engage in. It was also intended to identify which methods described formally in the literature (refer to Table 1 for a complete list), and taught at University level, differ from design practice. The list also included tools (e.g. CAD), and information resources (e.g. databases). All methods were briefly explained to provide clarification. Due to the extensive range of methods covered, a grid was utilised to indicate the different modes of method usage. Designers were invited to provide initial comments as appropriate and fill in the questionnaire.

	Feel free to add methods below	was it part of undergraduate training	frequency of use: never	rarely	often	always	type of project: minor improvements	modification of important details	major change, novel ideas	time line - please indicate when you use the methods: design planning	examine problem	generate ideas	evaluate and select	design of form/ details
	EXAMPLE	√			√			√	√					
26	performance specification method													
27	quality function deployment method (QFD)													
	idea generation													
28	constraint propagation													
29	product feature permutation													
30	orthographic analysis													
31	SCAMPER/ checklists													

Fig. 1 Extract of the questionnaire

The questions asked retrieved information regarding the product area that the designers work in, the types of tasks they are given, and the scope and circumstances of methods varying with different tasks. Questions regarding focus groups included the following:

- the degree of familiarity with the methods;
- details about how focus group data are communicated to designers;
- when focus group data would be useful;
- whether designers would find it beneficial to carry out focus group research themselves;
- views on suitable formats for presenting a training guide for designers.

2.3 Findings

2.3.1 The resources used by designers during the concept generation stage

The data were evaluated qualitatively. The questionnaire provoked a valuable discussion about the design process and the methods used. At times, designers preferred to offer comments, rather than filling the questionnaire in. Designers reported they tend not to use the formal methods described in the literature. This is mainly due to time constraints. Instead, an intuitive approach, using elements of most of the methods listed, is used – without applying any formal analyses. Freelance designers rarely carry out design planning activities that involve marketing activities. The more systematic the methods were, the less likely they were to be used (e.g. objectives tree method, evaluation matrices). There were considerable differences between the individual designers' preferences for methods. This is partly because design projects vary in content and context. Also, each designer seemed to have a personal set of adapted methods, adjusted to particular circumstances and requirements. All designers found it difficult to describe the creative design process and emphasised the iterative nature of designing – based on the repeated generation and evaluation of concepts in co-operation with the multi-disciplinary product development team.

Table 1 List of methods collected from the literature (including references)

No.	Method	Ref.
	corporate planning - design planning	
1	SWOT analysis	(6)
2	PEST analysis	(6)
3	tracking study	(6)
4	product maturity analysis	(6)
5	product development risk audit	(6)
6	systematic opportunity selection	(6)
7	competing product analysis	(6)
8	consider technological opportunities	(6)
9	style planning	(6)
	preparation/ understanding the problem	
10	Parametric analysis	(6)
11	Problem abstraction	(6, 7)
12	Product function analysis	(6, 7)
13	Transform to other area	(7)
14	Random input to stimulate thought	(7)
15	Counter planning	(7)
16	Objectives tree method	(7, 8)
17	Functions/ means tree	(7)
18	Alternatives tree	(7)
19	Task analysis	(6)
20	Life cycle analysis	(6)
21	Interview customers	(9)
22	Focus groups	(2)
23	Observational analysis	(10)
24	Benchmarking	(9)
25	Visiting	(9)
26	Performance specification method	(7, 8)
27	Quality function deployment method	(6, 7, 9)
	idea generation	
28	Constraint propagation	(9)
29	Product feature permutation	(6)
30	Orthographic analysis	(6)
31	SCAMPER/ checklists	(6, 8)
32	Analogies and metaphors	(6, 8)
33	Synectics	(6, 7, 9)

No.	Method	Ref.
34	Clichés and proverbs	(6)
35	Brainstorming	(6-9)
36	Brainwriting	(6)
37	collective notebook	(6)
38	boundary shifting	(8)
39	morphological chart	(7-9)
40	new combinations	(8)
41	lifestyle boards	(6)
42	mood boards	(6)
43	theme boards	(6)
	concept selection	
	evaluation matrix to rank concepts	(6)
44	utility function method	(7, 9)
45	Pugh's method	(9)
46	dot sticking	(6, 9)
	embodiment and detail design	
47	physical models	(6, 9)
48	computational models	(9)
49	"just build it"	(9)
50	rapid prototyping	(9)
51	process-driven design	(9)
52	part elimination strategies	(9)
53	assembly design	(9)
54	tolerance design	(9)
55	component design	(9)
56	manufacturability improvement method	(9)
57	elimination and simplification strategies	(9)
58	standardisation and rationalisation	(9)
59	standardising internal components	(9)
60	guided iteration	(9)
61	analytical optimisation	(9)
62	taguchi method	(9)
63	probabilistic design	(9)
64	failure modes and effects analysis	(6, 9)
65	design review checklist	(8, 9)
66	value engineering method	(6, 7, 9)

2.3.2 Perception on the use of focus groups

It was clear that the designers had a generalised understanding of focus group activity. None of the designers had ever carried out focus group research. For most designers, focus group data was a rare source of information. Two designers from the manufacturing company had once observed a focus group session.

Several designers were concerned about the significance of information that focus groups provide. They emphasised that focus groups would have to be well planned with clear objectives to avoid unwanted data being collected. There was concern that participants in focus groups may not be intuitive enough to contribute to new concept generation. Some designers expressed concern about the skills required in conducting focus group research – regarding issues such as the quality of questions, moderator skills, and efficient data analysis. The designers raised issues including: the individual has to be teased to convey honest information; designers are good with *things* – not necessarily with people; the method needs to be less time-consuming. Only one designer was confident that asking users about their perceptions of new designs, when using prototypes is easily done.

Designers, who are dependent on work given to them by a client, viewed the use of focus groups as almost impossible in their situation – as clients would not be prepared to pay for additional user research. This is because the client would have obtained the information already, and because clients do not view research activity as one that designers should carry out. Designers are anxious that researching user needs would be regarded not just as an additional cost factor, but also as a weakness, because the designer "should know". Designers working directly for a manufacturer emphasised the professional status of designers – "if you go and ask the public what to design you are not *flowing* – you are following culture, you are not *shifting*". The use of the method for generating new design was seen as constraining – "you end up doing what other people have in their imaginations...it's up to the designer to push a bit further". It was pointed out that users couldn't be expected to look into the future, as customers will rarely be prepared to reflect beyond the artefacts that they have interacted with, seen or heard of. One of the questions raised was "...what would then be the role of the designer?" It was suggested that attention has to be paid to the problem that focus group participants might have difficulties in focusing on the topic, as they need to get out of their usual mode determined by their daily concerns, into a more visionary one. One designer preferred a discussion with people of "similar beliefs", so that there is some "common ground" based on trust – since it was seen as important to make a focus group a "shared experience".

Designers had different views regarding the stage in the design process where the methods would be suitable. Some designers saw the importance of involving users early in the process, to assist in understanding cultural issues such as lifestyles and wider issues beyond functional details. Others preferred the use of focus groups for the evaluation of designs only, particularly as it would be most useful to be able to observe the product in use. One designer highlighted the difficulty of a group evaluation for a product designed by the person who moderates, as the personal aspects involved might cause an uncomfortable situation for both the participants and the designer. Interesting insights could be gained about the difficulties in integrating the information input into the creative process. It was described as an "intuitive learning process", during which ideas 'ripen'. The question was raised to what extent focus groups can be applied during this flexible process. There was resistance to data collected by other departments (e.g. marketing), as it was difficult to interpret and can prove restrictive.

Despite this negative reaction, designers conveyed their awareness about the need for a close consideration of user needs. Most designers regretted that they lacked detailed information, as it is often not provided by the client or market research departments. The need of immersion into the culture and aspirations of users and specific user groups was seen as particularly vital to design successful products. One designer emphasised the importance of knowledge about the product in use. Observation of product use was recognised as important, particularly where expert users have difficulties in reflecting about their routines.

2.3.3 Suggestions for the format of the training guide

The strongest message was that the guide should clearly show the benefits of the method to designers, inviting them to use it. Comments made included: present it as "I found it, it found me"; "send it as a birthday present"; "I would prefer somebody to talk me through it". Designers pointed out that the method would be difficult to communicate. The application of the method requires a paradigm change, possibly making it necessary to "link it more into education". Interactive presentations were recommended. A booklet outlining the benefits was suggested, in combination with a CD to explore the material according to what is most relevant. One designer preferred a book version because of the number of hours spent using a computer. Different types of media (e.g. book, website) with a good structure and a range of visual material was suggested. It was recommended to combine text and video showing many examples of use, to keep the designer interested.

3. CONCLUSIONS

This study has highlighted the reservations designers have to toward focus group activities. This may be due to pre-conceptions, but also due to practical limitations. The study revealed a series of insights into the requirements of designers regarding the adaptation of focus group methods for application by designers. The examination of design methods revealed the diversity of design practice, and the need for flexible design methods, demonstrating that methods are rarely applied in the formal ways suggested by textbooks. Designers appreciate the value of user involvement in order to gain increased empathy to enhance the designing process. However, the way in which the method is presented is of paramount importance. The next stage of our project will be concentrating on a designer-friendly guide that may incorporate visual, audio and two-dimensional material. Product success relies upon the needs of the user being satisfied on a number of levels. Focus groups offer the designer, as a new product developer, a user-centred design approach that can enhance the final design output.

REFERENCES

(1) J. Kitzinger, "The methodology of focus groups: The importance of interaction between research participants," Sociology of Health and Illness, vol. 16, pp. 103-21, 1994.

(2) D. L. Morgan, The Focus Group Guidebook, vol. 1. London: Sage Publications, 1998.

(3) D. McDonagh-Philp, "Gender and Design: Towards an appropriate research methodology," Proceedings of the 5th Product Design Education Conference, 1998.

(4) D. L. Morgan, Focus groups as qualitative research, vol. 16, 2nd ed. London: Sage Publications, 1997.

(5) S. Garner and A. Duckworth, "Identifying key competences of industrial design and technology graduates in small and medium sized enterprises," Proceedings of the International Conference of Design and Technology Educational Curriculum Development, Loughborough, P. H. Roberts and E. W. L. Norman, Eds., pp. 88-96, 1999.

(6) M. Baxter, Product Design: A practical guide to systematic methods of new product development. London: Chapman & Hall, 1995.

(7) N. Cross, Engineering design methods: Strategies for product design, 2nd ed. Chichester: Wiley, 1994.

(8) D. J. Walker, B. K. J. Dagger, and R. Roy, Creative techniques in product and engineering design: A practical workbook. Cambridge: Woodhead Publishing, 1991.

(9) H. W. Stoll, Product Design: Methods and Practices. New York, Basel: Marcel Dekker, 1999.

(10) N. Stanton and M. Young, "Ergonomics methods on consumer product design and evaluation," in Human Factors in Consumer Products. London: Taylor and Francis, 1998, pp. 21-54.

Design realization – a vital skill for product design students

C DOWLEN and **D COOK**
School of Engineering Systems and Design, South Bank University, London, UK

SYNOPSIS

Making things produces a vital and significant change in students' design performance and ability. This is to be expected by economic and societal contexts, by investigating students' goals and by looking at the feedback that can be obtained to the design process by the manufacturing phase and to the designer by manufactured objects. This theory is borne out in practice and by student feedback as courses seek to continue to provide design and make projects.

1. INTRODUCTION

There are a growing number of undergraduate courses aimed at producing product designers. In some we see trends towards the development of virtual products rather than real ones, and towards the development of non-functional models rather than functional products.

Professor Ken Baynes commented: "It is a common misconception that making things by hand is a thing of the past ... making things is going to be highly significant in the 21st century. ..."

This misconception is not useful, because it leaves students unprepared for the future as designers. Making things produces a vital and significant change in students' design performance and ability. This paper looks at some theoretical background and demonstrates results of getting students to make their designs.

2. WHY MAKE THINGS?

2.1 Making things is good economically.

Making things is about changing raw materials into products. This produces added value, with the worth more than the sum of its components. So the process of making things produces income and, ultimately, prosperity. Sir Terence Conran has said, "Put simply, without the skills of making there would not have been an industrial revolution. Britain's wealth today stems from the curiosity and inventiveness which applies hand and brain in tandem to understand the world and make a difference." Making things is good economically.

2.2 Making things is good for society.
Not only is making things good economically, but it is also good for society. The industrial revolution made a tremendous difference to our quality of life. People buy products to improve their quality of life, however quality is defined. It may be to overcome a problem they have, to fulfil a dream, or to make life more beautiful or comfortable.

2.3 Making things tells students about the real world.
Design is about changing the world. Not just in any way, but to fulfil the wishes of the designer in the designed products. So design is the link between the natural world of raw materials, the thought world of the designer, and the manufactured, human world.

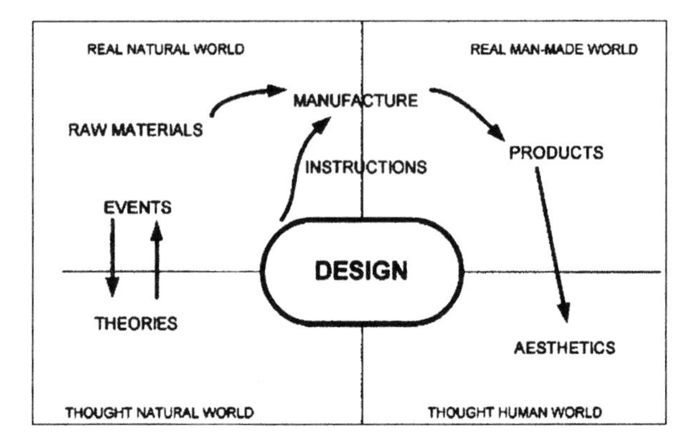

Figure 1: Overall Design Map

Figure 1 can be used to describe this. If we take the natural world on the left, the manufacturing process changes raw materials into the human-made world on the right of the diagram. The instructions for manufacturing are derived from design, which links the real world in the upper half of the diagram with the thought world below. See Hall (1).

As students' designs are realised, they develop awareness and knowledge about the way the real world's laws work. They could, perhaps, realise this through the medium of understanding of scientific theory, but until this theory becomes embodied within one of their own creations it will be hearsay, not experience. We understand the real world through doing. The interaction provides us with feedback about the way the world behaves, and this feedback informs our design processes, in the functional domain on the left of the figure and in the human emotional and aesthetic domain on the right. These feedback links have been added to develop the original diagram into figure 2.

A useful comment came in a British Television programme on taking a replica Viking boat across a boat-portage in the Shetland Islands. Robin Knox-Johnson was involved and his comment afterwards was "If we hadn't done it practically, everyone would have known the theory, but we started to learn the wrinkles: what you've got to remember, what you look for, what's necessary... All these little lessons, we could never have learnt unless we'd tried it." (2)

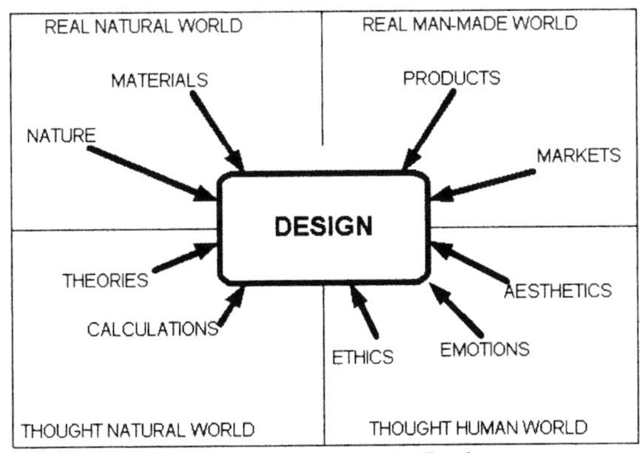

Figure 2: Feedback to the Designer

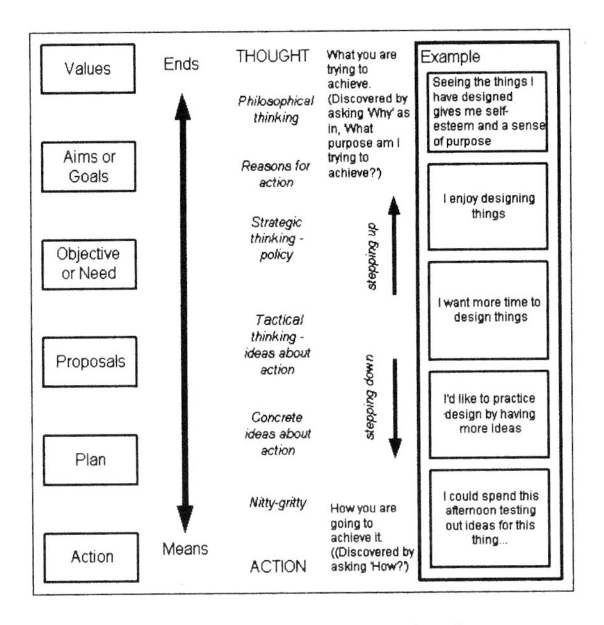

Figure 3: The Thought – Action Continuum

2.4 Making things is what students want to do.

In the end, students' ambitions are not content with conceptual thinking and finding good ideas, or even to turn these creative ideas into drawn or virtual objects. This is unsatisfactory. What designers want to do is to see their ideas becoming products in the market place. Seeing their designs becoming part of a successful change in society or meeting the need in the real world, or bringing them income is what satisfies. Getting products to market is critical to a designer's well being. This means that we, the educators, should get as close to helping students achieve this as we can. Students want to prove out their ideas in the real world and one of the best ways to do this is for them to be involved in manufacturing their designs.

Figure 3 shows the thought-action continuum as portrayed by Petty (3) and modified for the designer. We can use this to establish that the exercises and projects that we give students fit into the mould of thinking relevant to their intended profession – that of designer.

2.5 Employers want students to have ability in making things.

Potential employers of our students will be either manufacturing organisations or organisations supporting that process. The business and economy of the former relies on the standard of their manufactured products. If their products are better, they gain. This means that the products have been designed better; the student that can design knowing manufacturing requirements will be one who is more likely to impress the manufacturer. Organisations supporting manufacture such as design consultancies rely on the manufacturers for their income. A similar position exists. Students with manufacturing ability and knowledge make more useful employees. In fact, many companies who would previously have described themselves as design consultancies are now describing themselves as Product Development consultants, and this indicates that they are wiling and able to work with a manufacturer in a partnership rather than to be considered as a bought-in service, and this adds considerable value to their potential contribution.

2.6 Making things is part of the Total Design model.

This is a reverse reasoning process, more of an enabling one. Using the right model we will be able to understand better why students should be making things.

The Total Design model, developed in the 1980s by SEED (4) and others, started off from this manufacturing premise. Other design models, such as those from the German tradition, label the design process complete with the specification for manufacture.

One of the key things with all design models is that design is an iterative process. Feedback occurs at all phases and affects the way that a particular phase is carried out. The feedback loops from the manufacturing stage back into previous stages are significant. The incorporation of the manufacturing phase into the Total Design model means that the feedback from this stage becomes part of the design process, rather than being partially outside of it. This is an extremely powerful concept and one that allows the feedback to have a significant effect on design quality.

An additional comment about feedback is that it doesn't happen all at the same time, and neither is it related to the particular project. Feedback gained from the manufacturing phase of one project may be used for a subsequent project. Students who have developed manufacturing ability will be using this manufacturing ability unconsciously in the early stages of the design process. This means we should give students several design projects; their ability is seldom developed by one large one.

3. INCORPORATION WITHIN A COURSE

3.1 Design Evaluation

Our experience suggest that if students are not involved with the manufacture of prototypes then it is very difficult in most cases for them to carry out appropriate testing and evaluation of the products they design. There is therefore an important role for the prototype to perform at undergraduate level, i.e. does the design solve the problem set? This aspect of design development can be speculated upon but without a prototype it is not always possible to assess the success of the design. The prototype could be manufactured for the student but this is not usually possible and this would not be desirable as the manufacturing experience has so much to offer the undergraduate. It is recognised that to require students to undertake the manufacture of working prototypes is very time consuming for them and also that the manufacturing methods used for the prototypes may not exactly mirror those that would be used in production but the advantages far outweigh the disadvantages.

From experience, there is little doubt that a student or graduate with design and manufacturing experience is likely to have a good insight into what is needed to produce successful designs. It would of course be appropriate to think that designing is finished once the manufacturing drawings are signed-off but realistically this is not always the case. A Right First Time approach is to be encouraged and promoted and this is something that is constantly reinforced in the workplace and to undergraduates but the successful adoption of Right First Time can only be achieved after a great deal of experience has been built up. Experience of what works and what doesn't is gained by students actually manufacturing what they have designed and this can, and usually does, lead to modifications being carried out as the prototype is being manufactured. This 'designing on the job' may be relatively minor consisting, perhaps, of changes to make assembly more straightforward. Alternatively it may be major when the student finds that product cannot be manufactured in the way that was predicted or finds that the product is not going to function as required. This is not to suggest that the student should be pushed into manufacturing before complete design development has taken place but rather recognises that the experience gained in manufacturing related to design development is very valuable as part of the development of professional designers. It enhances their design competence and provides them with confidence in their design ability, as they are able to carry out a real evaluation of the products that they design.

3.2 Developing Employability and Professionalism

Potential professional designers need to be creative, innovative, and logical in their approach to designing if the products they produce are to be both desirable and functional. The development of creativity and innovation can be promoted in the production of 'flat work' and perhaps models but to mate them with the logical approach required for 'design for manufacture' the student needs to have the experience and understanding gained by being involved in manufacturing processes. Putting the student in this position also means that the creativity and innovation can be extended into the manufacturing requirements of detailed design, e.g. materials selection, manufacturing process choice, and assembly procedures.

To think about a design problem and to suggest solutions to it is part of the thought-action continuum of any designer but the requirement of the designer does not stop at this stage. The proposed solution(s) usually needs to be taken forward into a design that will fulfil its purpose from a functional stand point, is desirable, must be 'manufacturable', and be produced economically. For this to happen a designer must have an insight into and experience of manufacturing with the experience starting early in their career.

Prospective employers appear to recognise the usefulness of design graduates with 'making experience' and feel that it enhances what the graduate has to offer. It gives the employer the confidence to allow the graduate to work independently, at an early stage in their employment, on design developments from brief through to the production of manufacturing drawings and in many cases to become involved with production planning and supervision.

3.3 Aims and goals

Product design courses need to include both the achievement of a general level of competence in manufacturing and sufficient design and make design project work for students to experience the level of feedback that manufacturing their own ideas provides.

At South Bank University we see the overall aim of incorporating manufacturing within an undergraduate product design degree as being to enable students to become more creative and knowledgeable professional designers. On our BSc Course in Engineering Product Design the course focus of designing and making is on the final year project where the student is responsible for identifying a design problem and then taking it from initial brief through to the testing and evaluation of a working prototype. Shorter-term goals relate to the activities they will undertake in order to complete this successfully.

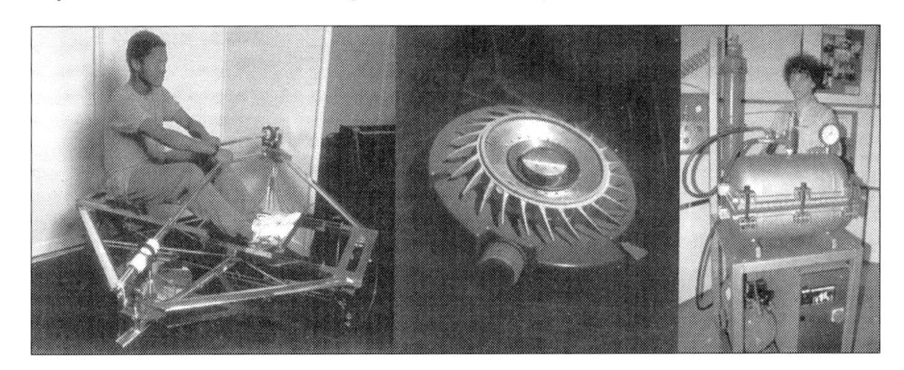

Fig. 4 Final Year Major Design Projects

3.4 Exercises

Thus in the first year, practical skills are developed in a traditional way and the students carry out a series of exercises to achieve competency. This introduction to initial skills provides a foundation for the further development of skills in the manufacturing of their designs as they progress through the design course. As they begin to manipulate materials successfully their manual dexterity starts to match their aspirations and their confidence begins to build. They start to gain a greater understanding of the needs of 'good design' from a manufacturing point of view and their knowledge of materials selection matched to both the design requirements and those of manufacturing begin to develop. This building of confidence is an important aspect of any student's academic career.

It should though be emphasised that it is not intended to produce students with highly developed manual skills in the field of manufacturing but rather to develop enough competence to produce successful prototypes that can be evaluated meaningfully. Students quickly discover that the manufacture of prototypes is a time consuming process and things can go wrong. This negative aspect of the exposure to manufacturing at an early stage of their undergraduate careers reinforces the importance of planning and managing the design process and the relevance of time management in design development.

3.5 Design Project Work

At the same time, we recognise that students must always have a design project on the go in order to focus their goals. This design project work is seen an integrating aspect of many design courses and it develops, so that from small projects in year one they develop complete functioning prototypes in year two of the course. As well as allowing students to bring design solutions to realisation they also allow them to apply much of the knowledge and information covered in supporting studies. The development of the students' communications ability is brought into play and engineering drawings are not only produced but also used to guide the manufacturing process. The theoretical work done on manufacturing processes and linked to aspects such as design for economic manufacture are applied in both the designing and making stages of prototype development. Work associated with product function such as applied engineering and materials technology is also used by the student in the development of prototypes. Such application helps to reinforce the relevance of the supporting studies and enables the student to use the knowledge and information provided in them to good effect.

Students are encouraged to develop designs as much as possible before starting to manufacture but it would be unrealistic to expect the design to remain unchanged as the manufacture proceeds. In many cases the design goes through developments as it is produced and in others major alterations occur. These changes and refinements are regarded as part of the student learning process helping to emphasis how design and manufacture need to be seen as very much interrelated and dependent when it comes to the final realised design.

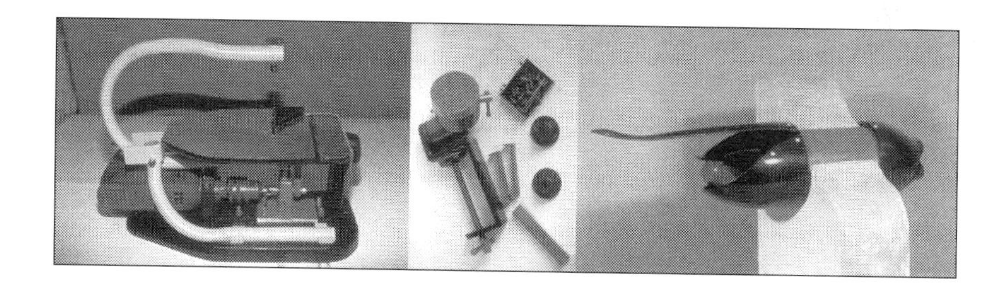

Fig. 5 Second Year Design and Make Projects

4. STUDENT FEEDBACK

How does this work out from the point of view of the student and their perception of the process? Following a first semester second year design and make project a questionnaire was handed to the students in order to see what their perceptions had been. This was arranged to be of the open format, so a statistical analysis could not be carried out on the results; its purpose was merely to obtain student feedback in terms of useful comments and perceptions about how the project had worked for them.

It was significant that few of the comments were irrelevant. Whilst many referred to understanding of specific manufacturing and design processes or to the development of time management skills, there were comments relating an increase in design ability and awareness through the process of manufacture. For instance "Modifying designs that I produced so that even the slightest change results in easier manufacture" illustrates that the feedback loops in the Total Design model are working, as does, "Designing with the manufacture process in mind for every single part of the project" and, "Modification is essential. Design on paper will change inevitably in the final prototype. Mistakes often lead to better solutions." "The importance of the relation between design, manufacture and economics" is a significant lesson to have been learnt and when asked about increasing confidence. Replies such as, "Seeing the design turned into a working prototype", "Gave me confidence as a designer by showing me that I can design something technical that works" and, "Having produced a working prototype increases confidence" indicate the positive nature of students' desires to be involved with making aspects.

5. CONCLUSIONS

The theory is one thing. The practicality is another. We think that both the theory and the experience come together to support the notion that making things produces a vital and significant change in students' design performance and ability, not just in their ability to manufacture. It is this interplay between the manufacturing and normally earlier phases of the design process that gives design confidence and competence. We shall continue to need products to improve our quality of living, even in the virtual age, and these products will continue to provide economic livelihood for their manufacturers, creators and designers.

REFERENCES

(1) Hall, A. G. Models & Mathematics: The Effect of Design on Theory in Engineering: Engineering Education Conference Increasing Student Participation, PAVIC Publications, 1994

(2) From: Secrets of the Ancients, BBC TV, UK broadcast on 2 Nov 1999

(3) Petty, G. How to be better at…. Creativity, Kogan Page, 1997

(4) SEED: Curriculum for Design: Engineering Undergraduate Courses, SEED, 1985

Reducing the risk – a partnership between academia and a small to medium enterprise

G TORRENS, A BRUSEBERG, and **D McDONAGH-PHILP**
Department of Design and Technology, Loughborough University, UK

ABSTRACT

This paper describes a User-Centred Design approach to new product development applied through a partnership between the Department of Design and Technology and a Small to Medium Enterprise (SME), Eze Drive Limited. The partnership enabled the SME to gain access to advice and technology. This provided an opportunity for academic staff and students to apply theoretical models and practices within a *live* project. This case study highlights the need for such collaboration and demonstrates the potential benefits that can be gained by both parties. The case study illustrates the use of value analysis and quality functional matrix. It shows the way in which the product was reviewed and redesigned ready for pre-production prototyping within one week.

1. INTRODUCTION

It is commonly acknowledged that a good working relationship between Academia and Industry is highly desirable – both to ensure that academic research and training maintains a close relationship with industrial practice, as well as to enable companies to have access to new methods, ideas, and insights. The Department of Design and Technology, Loughborough University, regularly undertakes such mutually beneficial collaboration.

The employment of a professional design consultancy often presents a major cost factor to SMEs. In this project, the company required a re-design of an existing automotive foot pedal extension to be fitted into a particular mainstream vehicle (Vauxhall Astra). Whilst previous products had been produced through small batch production (<2000 units), the new product needed to fulfil the requirements of medium to large batch production (>5000 units). The company therefore needed advice on material and process selection. Moreover, it was intended that the new product would facilitate convenient fitting and removal of the extension. It had to conform to the styling of the car manufacturer, needed to be visually unobtrusive, and appeal to the target market. The target users were people of small stature who required a safe distance between themselves and an air bag system.

There is a gap in support given to Industry by Government and Academia. When start-up companies have come up with concept proposals, they frequently lack resources to improve and develop the product ready for manufacture. The design concept is often underdeveloped for investors to risk taking it through to the next phase of development. Moreover, companies face major financial risks in carrying out research and development from a concept and idea stage through to a commercial realisation. Universities can *kick-start* the early stages of concept development – by helping the company through advice. The aim is to facilitate the development of a product that attracts capital investment that will take the product to the market. Through undergraduate and postgraduate student work, Universities can provide a resource that is unavailable within an SME's new product development budget. For undergraduate industrial design students it is vital to have access to design problems, through which commercially based experience can be gained. Research and lecturing staff can provide support to students.

The Department of Design and Technology is involved in supporting new business through shared risk. SMEs gain a new product within a limited Research and Development (R&D) budget. The company will provide only basic cost (e.g. materials, temporary employment of a student) which is a small percentage of the commercial rates for a consultant design and development team. In return, the company must agree to academic publication of the results within a six-month period. The Department gains current industrial information for research and training purposes whilst maintaining a presence within the field. This mechanism for collaboration enables the Department to be involved in developing efficient and effective methods of new product development from concept to market. It should be noted that this form of collaboration is reserved for small businesses and start-up companies. The philosophy behind this initiative is to introduce new companies or individuals with an idea or invention to appropriate methods of new product development, in order to ensure sound economic decision making when reviewing the commercial viability of the product. This activity does not fit into commercial R&D rates, and so does not compete directly with a design consultancy, and complements business advice from business link and innovation centres. SMEs have to fit within academic time schedules, which can be very restricting for a larger company but is less problematic for a start-up company.

2. THE AIMS OF THE PROJECT

Eze Drive Limited is an SME with a very limited budget for R&D. A product opportunity was identified by the company, who was not in a position to take it further. Loughborough University was approached with the task to develop and modify an existing product – an automotive control pedal extension set that can be fitted into cars, designed for smaller members of the community who commonly experience difficulties in reaching the pedal controls. It provides additional distance between the driver and the steering wheel mounted air bag. The need for a safe distance between driver and the airbag has been highlighted through the published research and popular media, where drivers have been injured due to the rapid inflation of the airbag. The product is specifically aimed at people below 5th %tile of stature of a documented UK population, and a certain disabled groups of people.

The company Eze Drive Limited was familiar with the market and could pass on information about users to the Department, which formed the basis for a product design specification (PDS). The market information was based on the main author and the company's experience in the area. The company had already gained feedback from users on a prototype, but required the re-design of the existing configuration. The Department was able to assist the company with the production detailing within the context of a user-centred approach, with its knowledge about all the different stages of the process to get from concept to commercial conclusion, including design ergonomics and production engineering.

The product was being manufactured using small batch production methods. Sub-contractors produced some of the parts. The combination of fabricated steel, fasteners and turned steel components made it heavy and expensive to produce, when compared with injection moulding or die-casting. Because the pedal extension is part of the control interface of a potentially hazardous product, it should not slip off the pedal, bend or break. It therefore had to be robust due to of the forces placed onto it, and remain firmly in a plane using a semi-permanent connection. To fit the extension to a range of car models, the original design required multiple parts due to foot pedal shapes being variable – making the design more complex, more expensive, and adding weight. The combined weight of the original pedal extension and the standard pedal exceeded the strain put onto the pedal spring, thus keeping the pedal depressed without further force from the driver's foot. The company currently fits their own design with a modified spring on the end of the foot pedal leaver to compensate. In order to avoid the need to replace pedal return springs, and to provide an off-the-shelf fit, a reduction in weight was required.

To decrease the weight of the extension, it needed to be specifically designed for a certain car model. If designed for a particular foot pedal shape, the number of parts for the extension kit could be substantially reduced. Consequently, the extension set needed to be modified to fit the foot pedals of a particular main stream manufacturer, chosen for its popularity. The aim of the original brief was to further develop the existing control pedal extension set, ready for large batch production, and a target manufacturing cost of £30. The brief included the following objectives:

- benchmark existing products; apply value analysis methodology to reduce the number of components;
- identify a number of manufacturing options for the production of the solution, and associated materials; identify original equipment manufacturers or sub-contractors who can supply or manufacture required components;
- provide a design solution to assist in the quick release of the extension, which has visual empathy with the styling of the mainstream car manufacturer for which it was intended, and that is visually unobtrusive.

All dimensions, location over the existing foot pedal and the configuration of the extension foot pedal for the new design, were to be based upon the given existing product. A set of foot pedal controls of the mainstream car manufacturer, provided by the company, were used to detail the exact dimensions of the new extension.

3. PEDAL CONTROL CONCEPT DEVELOPMENT

3.1 Benchmarking

A literature search and review was conducted, primarily using the Internet. An initial review of the world market did not uncover many similar products. There were two companies found to be producing comparable products:

- Gary E. Colle Inc., Poway, California, USA (1);
- Easy Rider Pedal Extensions, Glendale, California, USA (2).

Both products used a metal fabricated assembly, off the shelf fasteners and basic adjustment within the designs. Both manufacturers claim that their product will fit most cars and vans, with extra adjustment in the case of the Easy Rider extensions. The American market primarily caters for automatic-drive vehicles with a large brake pedal (approximately three times the size of a manual brake pedal). Both US products rely upon fitting over the existing foot pedal for location. One of the manufacturers was marketing within the UK. From available literature, it seemed that the US products were to be fitted by the user. The US system of location assembly and methods of production were very similar to the current UK product, being universal but relying on a skilled fitting. The scope for a market extension into the US with a similar philosophy needed to be considered. This was based on the idea to specialise on a particular model, to produce a lightweight extension, and to make it easy to fit for home use.

3.2 Value analysis of the existing design

Value analysis method is used to analyse the components of an existing design regarding both the cost and the value of each element regarding its functionality. By rating cost in comparison to value, the method aims to systematically eliminate parts that are not needed to simplify the construction, minimise production cost, and improve the product use (3), (4). The method was chosen because one of the main concerns of the project was the reduction of parts and the use of alternate materials. Hence, the method assisted a user-centred approach to the product evaluation, providing information to be used within a quality-function matrix.

Fig. 1 Existing foot pedal extensions

The existing design (see Figure 1) is fabricated from sheet steel, using off the shelf fasteners and replacement rubber control pedal pads. The total cost, including powder finishing, assembly and packaging is £40.00 per set. The set retailed at £89.99 through Eze Drive Limited, with around 2,000 units sold each year. There are seven main components within the existing product, not including fasteners. The brake and clutch pedal extensions weighed 680g and the accelerator extension weighed 637g.

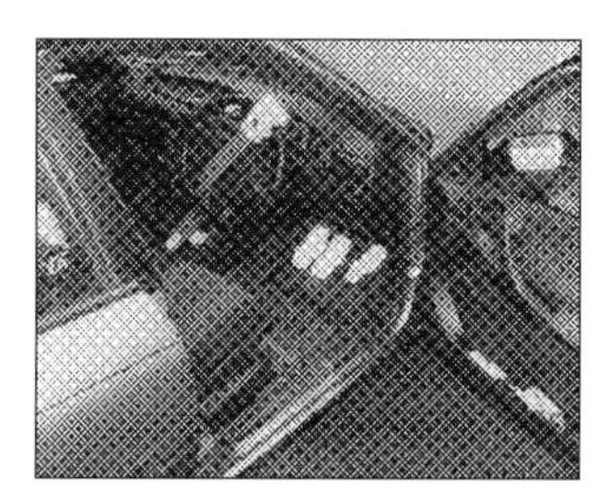

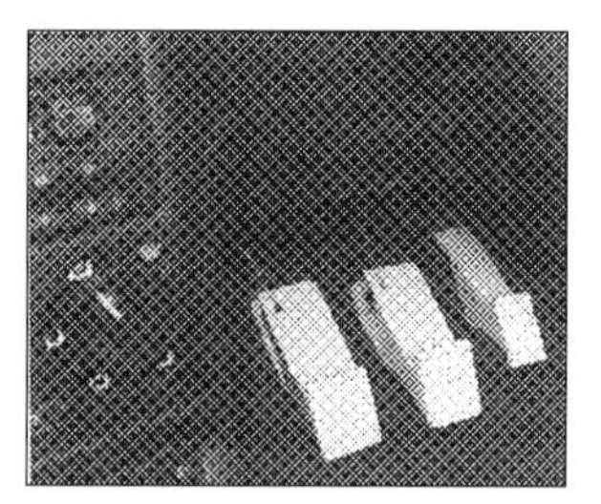

Fig. 2 and Fig.3 The concept pedal extensions in position within the specified car model

To reduce weight and number of components, a design solution was developed that incorporated most of the fabricated bracket parts into one component (see Figures 2 and 3).

A single location and clamping pin was used to lock the extension over the existing pedal unit. The proposed concept solution has three components per pedal extension. There are two configurations of extension units, one for the brake and clutch, and one for the accelerator. The hardened steel clamping pin and lock nut could be used on all three designs. The weight of the new extension unit, without clamping pin and lock nut, is calculated to be 330g. The accelerator pedal extension unit is calculated to be 261g. Both are around half the weight of the existing design. The use of a single clamping pin enables easier location and fitting in comparison to the construction of the existing design. The location and clamping function of the steel fastener and lock nut provide a secure fastening of the extension unit onto the pedal. Over-tightening was avoided be specifying a fastener that provides visual (pressed down on to a washer) or audible (click) feedback to the user when the optimum pressure has been applied.

To test the new extensions, a sketch model of the proposed design solution was fitted to the specified car model. Fitting was easier than using the existing units. It was recommended to the company that a detailed survey of the control pedals of the model range of the car manufacturer should be conducted. This should be completed before committing to the next phase of product development – to ensure the dimensions used in the prototype would fit across the manufacturing tolerance of the Vauxhall Astra pedals.

The market potential was reviewed in terms of safety, and regarding the ease of fitting. The time to fit the device was drastically reduced, as the construction was now held together by only one element. The extension could now be fitted in approximately 2-5 minutes, where the currently available designs take about an hour. As it was vital that the device cannot slip off or break, the forces on the pedal were reduced – in particular taking the load at the top and on the back section of the pedal into consideration. The extension was designed so that the mechanical loads are taken in both directions. The pin merely assists in securing, without taking any load. The configuration of the new design also allows for wear of the rubber pedals, by the wedge shape of the fixing bolt pushing the pedal support bar into the extension housing. The extension enables an easy release that takes account of safety issues. A quick release system was not used to ensure there could be no accidental release. Hence, a semi-permanent fixing was specified, although an increase in fitting time was incurred.

4. MANUFACTURING OPTIONS AND ASSOCIATED MATERIALS

In order to produce a relatively complex bracket arrangement, four manufacturing options were investigated: (1) aluminium sand casting; (2) aluminium gravity die-casting; (3) structural foam injection moulding; (4) Computer Numerically Controlled (CNC) machining. The suggested aluminium grade to be used was LM25TF. This grade has sound mechanical properties, is lightweight and strong.

The concept design was produced to enable the use of a two-piece mould. If structural foam injection moulding was utilised, a side core needed to be used on one side of the moulding – adding to costs. The hole needed for the clamping pin could be drilled into the sand cast option. A side core would be required for the gravity die cast option, adding cost to the mould tool. The side core may be removed manually, rather than being set for removal via the mould machine. This would reduce set up costs, but would increase handling costs per unit. Costs have been obtained from three local companies, for comparison.

A number of issues remain for further investigation by Eze Drive Limited, that depended upon information from suppliers and specific tests – including the cost and final specification of the clamp pin and lock nut, and the identification of the specific grade of aluminium to be used. Further testing would be required regarding the mechanical and structural evaluation, to ensure the design would meet and surpass relevant safety standards and industry guidelines.

5. THE OUTCOMES

The outcomes of this work were documented in a report and presented to the company. They included:

- one set of 'sketch' models, made of card and foam, that were used to demonstrate the ease of attachment to each control pedal;
- CAD/Engineering drawings of the design solution;
- an exploded view of the design solution;
- a component list;
- a list of appropriate suppliers and subcontractors to supply components.

6. SUMMARY AND CONCLUSION

Universities can assist start-up companies in reducing the risks taken when investing into a new concept. The expertise available through student work and guidance from University staff can contribute substantially to the development of new products where companies lack the resources for professional consultancy in the earliest phases of development. Assisting small companies with further development of ideas to gain evidence about the success of their new product in the market encourages investment from other sources. The rapid turn around of this design project was only possible through close collaboration between stakeholders (i.e. manufacturer, user and designer). The outcome of this project has benefited both the company and a student industrial designer, which has provided a valuable case study teaching material. This paper, and other academic publications derived from this project, have provided a 'technical story' for the company. This can be used as a marketing tool to be shown to potential customers by the company, as the research and publications behind the project increase the credibility of the product. The success of the product developed to date ensured further co-operation between Eze Drive Limited and the Department of Design and Technology in the area of new product development.

REFERENCES

(1) G. E. Colle, "Gas and Brake pedal extensions, Gary E. Colle Inc.", http://www.garyecolleinc.com/extensions.html, (February 2000).

(2) C. M. Secor, "Easy Rider Pedal Extensions", http://www.pedalextenders.com/index.html, (February 2000).

(3) M. Baxter, *Product Design: A practical guide to systematic methods of new product development*. London: Chapman & Hall, 1995.

(4) H. W. Stoll, *Product Design: Methods and Practices*. New York, Basel: Marcel Dekker, 1999.

Sustainable Development

Suppliers to multinationals – opportunities for increasing manufacturing capabilities in less-developed countries

E E CAHUA, S P MAGLEBY, C D SORENSEN, and **R H TODD**
Department of Mechanical Engineering, Brighma Young University, Utah, USA
W C GIAUQUE
Marriot School of Management, Brigham Young University, Utah, USA

ABSTRACT

So far, efforts to reduce poverty in less developed countries (LDCs) have generally focused on development of infrastructure. They pay little attention to the methods by which most wealth in the world is created, namely participating in manufacturing value chains and capturing part of the value to be retained locally. This paper proposes the development of local manufacturing by establishing additional local suppliers for transnational corporations (TNCs). However, some questions arise. What specific items should local suppliers venture to manufacture? What are the management and engineering challenges? Which approaches and methods might best be implemented to help improve design and manufacturing capabilities in LDC's? This paper proposes some ideas that might help answer these questions.

1. INTRODUCTION

Poverty throughout the world is of obvious concern to every thoughtful person. Humanitarian considerations alone dictate enormous concern, but added to that are considerations of wasted human potential, wasted economic potential, and the prevalence of hunger, sickness, inadequate housing, illiteracy, poor sanitation, unemployment and, increasingly, environmental degradation.

Most research and resulting recommendations for alleviating poverty focuses on one of two areas – governmental policy making or micro enterprise, the latter dealing largely with home production, small scale infrastructure (water supplies, schools, etc.), and agriculture. As helpful and laudable as these efforts are, they pay little or no attention to the methods by which most wealth in the world is created, namely participating in manufacturing value chains and capturing part of the value to be retained locally.

For many products, particularly those produced by transnational corporations (TNCs) (corporations with operations in more than one country) manufacturing is a worldwide activity. Products are manufactured in many locations, including in LDCs, for marketing, political and economic reasons. However, we have observed many plants located in LDCs which utilize only local labor, importing the design and the vast majority of raw materials and supplies used by the enterprise. We hypothesize that the percentage of needed materials sourced from the local economy could be increased substantially. If this is true, a significantly larger portion of the value chain could be captured by the local economy, which would exhibit a multiplier effect far greater than the direct monetary amount.

This paper proposes that additional local suppliers for TNCs could be established by working with universities, multinational corporations and smaller local manufacturers in LDCs to develop methods, approaches and models that will enable local manufacturers to supply support goods needed by larger manufacturers. Such items could include, for example, spare parts, product add-ons, packaging, manufacturing jigs and tooling, and factory consumable products. We believe that once a relationship has been established, the manufacturing ability of the local manufacturer will continually improve, thus allowing an ever-larger range of products to be designed and manufactured locally.

2. LITERATURE REVIEW AND BACKGROUND

We believe that the best way to create wealth and improve the standard of living in LDCs is through the development of manufacturing, and in particular, to begin with certain items manufactured as supplies to TNCs. Brandt (1999) declared for Industry Week, "The fact is that manufacturing has become more important to the global economy than ever." He pointed out that manufacturing conducts 90% of all private research and development, invests 74% of all information technology dollars, pays salaries 89% higher than those in retailing, generates 4.5 times more jobs than retailing, represents 80% of all world trade, and has achieved annual productivity gains that are three times higher than those in the service sector over the last decade. Brandt added, "No other sector has done as much to create innovation, general prosperity and increased productivity than manufacturing."

However, most work on poverty deals either with broad strategic and policy issues at the governmental ministry level, or with local self-help initiatives, such as micro credit, enhanced agricultural and home production methods, and education.

Battat et al (1996) have conducted relevant research in this area. They discussed the relationship between local suppliers and foreign companies operating in developing countries, or what they call "backward linkage." In addition, they discussed the importance of institutional support, both public and private, for upgrading local supply industries in developing countries. Three areas of support were identified: (1) technological improvement, (2) managerial training, and (3) financing. They also described some programs that have been successfully implemented in countries such as Singapore, Korea, Ireland and Taiwan, and the impact on actual backward linkage formation.

Other researchers have focused on the area of international entrepreneurship. Lado and Vozikis (1997), for example, proposed a conceptual model that provided an enhanced understanding of the efficacy of technology transfer to promote entrepreneurship and stimulate economic development in LDCs. They argued that the efficacy of the technology-transfer mode will depend on: (1) identifying the level of development of the country to which the technology is transferred, (2) the technological elements being transferred, and (3) the recipient firms' absorptive capacity. An important conclusion was that TNCs may not be effective in engendering economic development in the LDCs through technology transfer unless the transfer of such technology also has the potential to stimulate entrepreneurship in the form of new business ventures, improved product and service quality, and greater organizational learning by local business firms in the recipient country.

Some researchers argue that the supplier base in LDCs is both a serious problem for manufacturing and a key determinant for foreign investment. Austin (1990) discussed that LDC suppliers often are not equipped technologically or managerially to meet the demands of manufacturers and processors, which causes a multitude of quality and cost problems. Lawrence and Lewis (1993) argued that weak suppliers hinder the application of the JIT philosophy, which requires frequent and reliable deliveries of small lots of quality parts, ideally from local, single-sourced suppliers. Wilson et al (1995) argued that cheap labor is not a key determinant for foreign investment anymore, mainly due to the growing sophistication of global organizations and modern methods of production. They argued that multinationals are increasingly more interested in procuring local inputs.

Although the United Nations have also carried out research in the area of TNCs, practically no studies were done about the relationship between local suppliers and TNCs. According to Peter Hansen, former Executive Director of the former United Nations Centre on Transnational Corporations (UNCTC), the literature about TNCs was limited to (1) examine the factors and motivations that prompt transnational corporations to invest and produce abroad rather than to supply foreign markets through exports, and (2) investigate host country characteristics, including elements of public policies that attract foreign direct investment (Pearce et al, 1992).

In summary, very little attention has been paid to such issues as determining what percentage of design materials and supplies used by TNCs in LDCs is sourced locally, the nature and magnitude of obstacles to greater local sourcing, or whether efforts to identify and overcome those obstacles is feasible. Existing efforts are generally short-term and opportunistic, whereas this proposed model is designed to be long-term and developmental.

The potential to develop additional local suppliers to TNCs and thus improve manufacturing capabilities in LDCs is high, as recent studies show that there is an increasing trend toward transnationalization. According to the *World Investment Report 1998: Trends and Determinants*, a survey carried out between July and November 1997 among 300 managers of TNCs and international experts around the world, supplemented by about 100 direct interviews, suggest that the trend towards a further transnationalization of firms will continue in the medium term, independently of the size, sector and location of the TNCs. More specifically, between 1996 and 2002, the contribution of foreign activities to respondents' business is expected to rise from an average of 47 per cent to 56 per cent in sales, from 35 per cent to 45 per cent in production, from 34 percent to 42 per cent in employment, from 36 per cent to 42 per cent in gross investment, and from 32 per cent to 41 per cent in assets (see Figure 1).

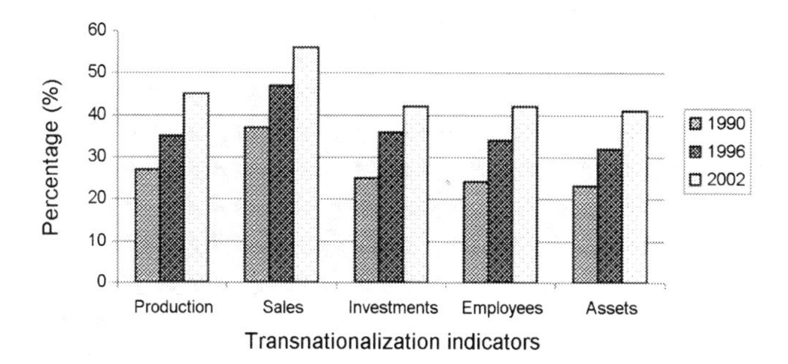

Fig. 1 Respondents agree that the contribution of foreign activities to TNCs' business will continue to increase *(Source:* Invest in France Mission, Arthur Andersen and United Nations, 1998).

As a result, a rising number of companies will establish genuine transnational production and sales networks, compared to a minority in the early 1990s. Only 33 per cent of the respondents to the survey considered their companies "completely global" or "highly coordinated internationally" in 1990; this proportion had risen to 56 per cent in 1996 and could reach 78 per cent in 2002, while the proportion of those considered "little coordinated internationally" had fallen from 67 to 45 per cent between 1990 and 1996 (see Figure 2).

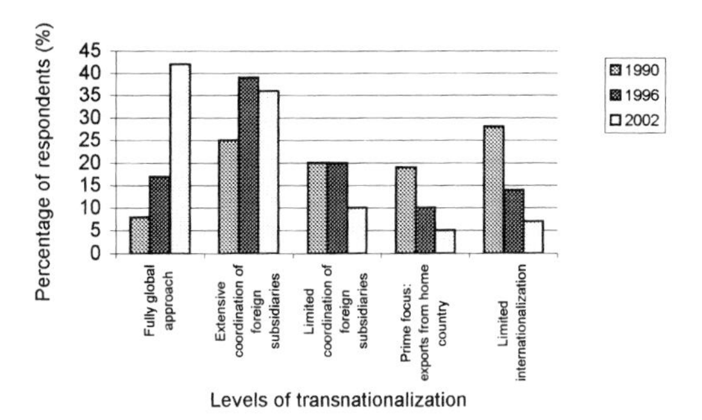

Fig. 2 Respondents also agree that there is an increasing trend towards transnationalization

(Source: Invest in France Mission, Arthur Andersen and United Nations, 1998).

From a recent visit to Peru, we could feel a strong need to develop local suppliers of materials and components. For example, Goodyear del Peru must import 98% of the rubber they need for their local operations. They have conducted a preliminary study for a natural rubber processing plant, and are willing to sign a 10-year contract with anyone interested in investing in and implementing such a project. In Indonesia, foreign-owned producers of consumer electronics buy only about 25 percent of their intermediate inputs from local suppliers. In the Philippines, foreign car assemblers depend on local suppliers for less than 15 percent of their intermediate parts needs. In Mexico, the maquiladoras, companies producing mainly for export markets, buy less than 2 percent of their inputs from Mexican companies (Battat et al, 1996).

In summary, there is a great potential for the development of manufacturing in LDCs given the fact that TNCs are increasing their operations in those countries. But how can local companies improve their design and manufacturing capabilities so that they can offer competitive products in terms of quality and price? Which products should they focus their efforts on? The next section proposes a model that might help answer these questions.

3. PROPOSED MODEL

Numerous TNCs now operate manufacturing facilities in third world countries, often to take advantage of favorable material and labor rates, and sometimes to satisfy governmental local content requirements or other political imperatives. We propose working with such firms to develop additional local suppliers for specific items that are currently imported. This would involve:

- reviewing purchasing records for likely items (ideal items would be heavy, bulky, or damage-prone items expensive to ship in from the outside, items consisting of locally available raw materials, and items not requiring major technological investments);
- determining existing and potential capabilities of the local economy and manufacturing environment, given the constraints on infrastructure existing locally; and
- brokering purchase agreements between the two parties, consulting with the local firms as necessary to help them design the products appropriately, and develop and implement quality and manufacturing procedures needed to compete effectively.

Preliminary discussions indicate that efforts should be initially focused on such categories as spare parts, product add-ons, manufacturing jigs and tooling, and factory consumable products. Components going directly into finished products may be subject to restrictive specifications, which would make it difficult for even superior products to gain immediate acceptance.

Early study results have shown that for a developed economy, there are four key ingredients for success in manufacturing:

- a knowledgeable workforce;
- a market for the manufactured goods;
- sufficient capital, and
- an infrastructure to support the manufacturing operation and enterprise.

In LDCs, these four ingredients are also critical and yet may not exist in conventional forms available in developed countries. However, they may be created in non-conventional forms by the indigenous population if useful methods, approaches and models are available to the local population which are appropriate to their local circumstances. The objective is to develop those methods, approaches and models, which when used by the local population with appropriate external support will enable them to develop manufacturing enterprises which will successfully produce goods needed by TNCs or other distributors located in developing countries.

We envision a four-way partnership among universities from developed nations, universities in LDCs, transnational corporations and local suppliers. Each partner's role and benefits, as we see them, are outlined in Table 1.

Table 1 Partner's Role & Benefits

Role	Benefits
Universities from developed nations	
• Provide Leadership o Identify promising client companies o Work with clients - identify promising products to source locally o Build upon relevant literature to develop new thinking o Identify research outlets & formulate publication strategy o Take the lead in raising funds • Create appropriate knowledge o Product development & engineering o Process development & engineering o Management approaches & systems	• Fulfilling university missions o Educating students in areas contributing to useful, fulfilling careers o Advancing knowledge in poverty alleviation o Being of service to the world o Enhancing international skills & presence • Funding fort faculty & student research/publication • Fulfill faculty & university academic & research goals o Academic publications o Classroom projects o Cases & source material • Funding for student research, &internships • Learn through partnerships from other universities & companies
Universities in LDCs	

• Provide local expertise & contacts o Governmental agencies o Locating potential suppliers o Share in finding potential client companies • Partner with foreign university o Writing & publishing in academic outlets o Developing & writing course materials o Curriculum development o Encouraging faculty & student research o Finding & developing funding sources • Provide follow-up on ongoing projects • Provide education & training in areas of expertise o Industry-specific knowledge (if appropriate) o Sources of vocational training (if appropriate)	• Source of funding for faculty & student research • Internship & career opportunities for students • Contacts for research & consulting • Publications & research • Chance to learn from partners to update curriculum, course materials
Transnational Corporations	
• Work with researchers to develop information o Materials & supplies purchased o Logistics of current supply chain o Economics of current supply chain o Local supplier needs to be competitive • Funding & support o Project support o Internship opportunities o Co-researchers as interest & opportunity allow	• Develop capable local suppliers • Faster, leaner supply chain • Political, humanitarian benefits of developing the host country
Local suppliers	
• Fulfill contract with buyer o Work with researchers to develop management & technical capabilities o Work with researchers on product & process Design o Implement manufacturing plan • Educate partners on processes, local conditions • Provide internship opportunities	• Additional business, probably substantial, with world-class customers • Enhanced manufacturing capacity & capability • Training & enhanced professionalism for management, personnel • No cost consulting

4. IMPLICATIONS FOR RESEARCH AND CONCLUSION

The proposed model would constitute a valuable addition to the literatures of development and international entrepreneurship. In addition, new research opportunities will be developed in the areas of design and manufacturing in LDCs. However, prior research needs to be done to determine what percentage of materials is currently imported by TNCs, what specific items should be produced locally, etc.

The potential to create additional local suppliers to TNCs and thus improve manufacturing capabilities in LDCs is high, as there is an increasing trend towards transnationalization. Initially, supplies should include items such as spare parts, product add-ons, manufacturing jigs and tooling, and factory consumable products, as components going directly into finished products may be subject to restrictive specifications. To successfully produce goods needed by TNCs or other distributors located in LDCs, new methods and approaches need to be developed or enhanced. We propose a partnership between universities from developed countries, universities in LDCs, TNCs, and local suppliers to develop such methods.

We would like to conclude this paper by inviting researchers and practitioners to share their ideas and suggestions regarding our proposed model, as we believe their feedback will be a valuable contribution to improve our research approach, and faster indigenous design and manufacturing in LDCs.

REFERENCES

(1) Austin, J.E., 1990. Managing in Developing Countries, Free Press, New York.

(2) Brandt, J., 1999. "Leadership for the New Manufacturing," Industry Week, vol. 248, no. 9, pp. 47-50.

(3) Battat, Joseph Y., I. Frank and X. Shen, 1996. "Suppliers to Multinationals: Linkage Programs to Strengthen Local Companies in Developing Countries," The World Bank, Washington, D.C.

(4) Lado, A.A. and G.S. Vozikis, 1996. "Transfer of Technology to Promote Entrepreneurship in Developing Countries: An Integration and Proposed Framework," Entrepreneurship Theory and Practice, vol. 21, no. 2, pp. 55-73.

(5) Lawrence, J.J. and H.S. Lewis, 1993. "JIT Manufacturing in Mexico: Obstacles to Implementation," Production and Inventory Management Journal, vol. 34, no. 3, pp. 31-35.

(6) Pearce, R.D., A. Islam and K.P. Sauvant, 1992. "The Determinants of Foreign Direct Investment: A Survey of the Evidence," United Nations, New York.

(7) United Nations Conference on Trade and Development (UNCTAD), "World Investment Report 1998: Trends and Determinants" (WIR98).

(8) Wilson, S.R.,R. Balance and J. Pogány, 1995. "Beyond Quality: An Agenda for Improving Manufacturing Capabilities in Developing Countries," E. Elgar, Aldershot, Hants, England; Brookfield, Vt., US.

Ecodesign

J LLOVERAS
Department of Engineering Design, Technical University of Catalonia, Spain

SYNOPSIS

The ecodesign as a join of technological design and the ecological preservation, is an important question of our time. It is an environmental need for humanity to resolve to adopt a respectful technology.

The basic concepts of ecological preservation are: environmental impact, life cycle, ecobalance, sustainability, reducing material, recycling, reuse, etc. These should be in the mind of the designer or the designing team, when they develop new products or reengineer old ones.

Technological innovation design has actually sense in an ecological frame.

"Ecodesign" is a module of "Technological Innovation Projects", a PhD program of the UPC.

1. INTRODUCTION

The initial aim of this doctoral module, "Ecodesign", was to introduce ecological considerations in the conceptual design phase, and its name reflects this in general.

1.1 UPC academic references

"Ecodesign" is a module of PhD program named "Technical Innovation Projects" which initially was part of "Engineering of Technical Systems" PhD program with the name "Technological Innovation". This was presented for first time in May 1995. In December of same year, started the teaching of the "Ecodesign" (ref.2Q44E), once every two years.

When in 1997-98 "Technological Innovation" became a separate PhD program with the name "Technical Innovation Projects" ("Projectes d'Innovació Tecnològica" ref. 53), Ecodesign continued to be a module of it with the 53013 reference.

According to the new Spanish doctoral act (Real Decreto 778/1998), the structure of the teaching period has changed, and the modules of "Technical Innovation Projects" PhD program are offered since 1999/00 in an annual base. Now "Ecodesign" module has 3 credits, the equivalent to 30 hours of taught time. This PhD program (1), (2), is offered by the Department of Engineering Design ("Projectes d'Enginyeria") of UPC ("Universitat Politècnica de Catalunya").

1.2 Ecological design as a real need

Few years ago, the engineering design books were saying nothing for the product life after being used by the consumer. Normally the used products were being thrown to the landfill where they were disappeared as useless material. This reflected the social impression of product life.

Nowadays in the new design books appears the whole cycle of the products life, from prime material to waste residues management. Scientific studies demonstrate that some severe environmental impacts of human economic activity and as a consequence the generated problems over some important physical, chemical and life parameters of our planet.

In fact the wild life is in equilibrium on the Earth for millions of years and generation after generation leaves clean the planet because all biological material is recycled -except some fossil residue-. Life is responsible for example: to create and maintain the atmospheric composition of gases and to preserve the salination degree of seas. However, there are certain limits imposed by the planet concerning the life capacity, which should not be exceeded. Life and its sustainability action is a wise teacher for our conduct regarding the use of unpolluting and recycling materials. These are key words for the general human activity.

In recent years the quick evolution and acceptation of the environmentally sound concepts by the society are due to different sources of information concerning environmental problems like scientific studies, United Nations programs, some Non Governmental Organisations (NGO), and political green parties. These with their actions sensitize and press ahead for changes in the attitudes around environment problems, so that the ecological conscience in world culture increased in the last decades.

In general people are receptive and accept the diverse voices, which speak with unanimity about preventing environmental impacts. On the other hand that means new eco-laws, eco-labels and discussions about eco-taxes.

One of the answers to these problems is that human produced goods should have a minimal environmental impact, which could be achieved with adequate materials and energy management. These are key issues for successful ecodesign.

These concepts of sustainability, recycling, etc. are an important global need and at the same time a good opportunity for technological innovations, since many products should be redesigned taking them in account. Furthermore, the products which daily are used, packaged, transported, etc. and afterwards should be recuperated and recycled, reveal a new area for product reengineering and create new facilities.

The designer or the designing team, as the technical responsible to create new products should keep in their mind these general principles, because they are them who can do these technological changes.

The ecodesign objectives to be reached are distributed in a scale of different levels. In the recent years, new ecoproducts have had an unprecedented evolution.

2. THE ECODESIGN MODULE

2.1 Module objectives
Ecodesign module initially had had an objective to wake up the conscience and also to discuss with the students the general aims about ecological problems and their design related aspects.

A positive feedback occurs when teacher transmits ideas and thoughts to the students aiding him to deepen in the existent knowledge and to detect new one.

The general level of eco-consciences raised gradually since 1995. Now the teaching is emphasising upon a more practical and applied studies.

The ideal objective would be that when some students could carry out a real exercise or dissertation of technological innovation by redesigning or by creating, an ecoproduct or an ecoprocess of high ecodesign class.

2.2 Ecodesign module contents
Initially the sources to implement this subject was of a diverse background about these topics and specially a new library section about Life Cycle Assessment (LCA) founded by a commission of Industrial Engineering Association of Catalonia. Next it was the very interesting International conference, "Technology, Sustainability and Imbalance", (3), (4) in Terrassa near Barcelona, during 14th - 16th December of 1995, with important speakers from all over the world, which structured and reinforced the sustainability concept background. Finally it was the International Standards Organisation (ISO) publication of the 14.000 series norm about LCA.

The main topics of this module are sustainability concepts, design with minimal environmental impact, design for easy dismantling and recycling and Life Cycle Assessment (LCA) (5), (6).

The first part of module generalise, aware and discuss with the students, the real state of the world and its limitations, the environmental impact of human activity, the application principle of material reduction and recycling and finally the energy problem and possible solutions in the future.

The second part of it consists of the study of LCA, that includes different definitions of concrete concepts like ecobalance, product ambient strategy evaluation, design change evaluation, etc. and ISO 14.000 (7), (8), (9), series, finally all the above are applied by students using taught estimation techniques and they hand an assignment. Moreover demonstration of the most updated available software for LCA estimations helps students to understand the use of computer facilities in this area of research. The difficulties of doing complex LCA are mentioned.

2.3 Some Ecodesign examples
At the final year of industrial engineering studies there is a dissertation to be done. The students must complete this Final Thesis ("Projecte Final de Carrera") to obtain the industrial engineering degree. Normally the topic is a product design or an industrial installation that it can be proposed by the student, by the teacher, or by a company (if the student has a contract between industry and university to make a specific work). In all the cases the work plan and

course are supervised by an UPC teacher. Last few years UPC strongly recommends that each final Thesis work would have one part of it dedicated to an environmental impact study.

The author leads some industrial engineering final dissertation and some doctoral thesis upon topics related with ecodesign applications in diverse levels of engagement. Same examples of studies or projects which have been developed by students of our department, are the following: A device to compact and store materials in rubbish collecting points in the streets, Dustbin with rubbish separator, Composting plant for a organic residues in "Vallès Oriental", Solar water heating and photovoltaic electricity production in detached houses, An offshore wind power park in the Mediterranean sea near "Delta de l'Ebre" (Catalonia). The last two ones are related with the renewable energy production. Other studies as: Sea desalination with minimum energy cost, Electronic faucet for water saving, Used water purification and reutilization in a house, Electric vehicle, Electromagnetic radiation avoid, etc.

2.4 The future enlargement of Ecodesign module

The Ecodesign module prepared for the next 2000-01 doctoral course is remodelled to include the Waste Engineering module and has increased its credits extension to 4.

Waste engineering means separation, recycling, sub-product reuse and their implementation problems. It is an intensification of the last part of product life and the material regeneration. This addition completes the Ecodesign module with the sustainability concepts, minimal environmental impact, LCA, easy dismantling and recycling and material reuse.

The teachers of the new Ecodesign module are the author (Dr. J. Lloveras) for the general ecodesign vision, Dr. M. Gonzalez (since 97-98) for detailed LCA and Dr. D. Curcurull (since 98-99) as specialist of waste engineering management.

3. ECODESIGN FUTURE

3.1 Design Opportunities

Gradually the products and processes will be designed or redesigned with increasing ecological considerations.

It is an environmental necessity and at the same time technological innovation opportunity, for example, the treatment of ordinary domestic rubbish is one main area for innovative solutions.

3.2 Universal products

The eco actions to come to ecoproducts have several degrees of performance. In the conceptual design phase there are several possible levels of ecodesign. The lowest level is occupied by products that have been undergone some little superficial changes. In turn Ecodesign attempts to prevent and minimise environmental impact; in a subsequent higher level Ecoeficient Ecodesign attaches also an economical analysis; and finally Sustainable Ecodesign adds the promise of an appropriate technology for the future generations.

In my point of view, if we could classify design products, the higher category would be occupied by these ones which offer an universal usefulness meeting the claims of sustainability (sustainable ecodesign) while at the same time the great majority of earth's population could be considered as potential buyers.

Certainly there are only few "universal products" or "universal ecoproducts" with the previous features. However, I hope their number will increase with the time becoming available for more and more people in all over the world.

In summary, I expect the future "universal products" would implement at least the following characteristics: easy fabrication, low cost, subsistent, usefulness, respectfulness, not being attached to fashion, easy maintainable, easy reparable, without discriminating nationality, religion, culture, sex, age, etc. "Universal products" is a new research area.

4. CONCLUSION

Ecodesign is a module of doctoral program "Technological Innovations Projects" of UPC and attempts to rise the environmental conscience teach and investigate about eco technical design.

These general concepts consist a very important background for the future designers in order to form an ecological sense, since possible environmental impacts, sustainability and decommissioning facilities for recycling issues must be in their mind during the conceptual design phase.

If the old technology have a part of the blame for the environmental impacts, now the technical designer can bring about a new respectful technology. This change which affect all technology gives also a great opportunity for economic activity. It is a long way.

The new technology has acquired an ecological aspect and the designer have to be conscious of the ecological problems and skilful enough to confront them. “Green technology” has already been aligned between the Hi Tech.

"Universal products" as a definition for ecological sustainable and for all people available products, is a research area.

Generally the student has not a very good background in these issues, so that in the UPC recently there is an institutional effort to raise the environmental conscience.

REFERENCES

(1) http://www.upc.es/3er-cicle/Cast/Doctorat/Programes/dreta.htm

(2) Lloveras, J. “Technological Innovation Projects. Structure of a PhD Program”. Proceedings of the 12th International Conference on Engineering Design (ICED 99). Munich. WDK 26. Vol. 2, p. 863-868, Ed. Lindemann, Birhofer, Meerkamm, Vajna. Pub. Technishe Universität München. München (Germany), 1999.
ISBN 3-922979-53-X

(3) Sostenible? Tecnologia, Desenvolupament Sostenible i Desequilibris.
UPC, Dep. Medi Ambient Generalitat de Catalunya. Ed. Icaria. Barcelona 1997. ISBN: 84-393-4255-1, 84-7426-318-2.

(4) Lloveras, J. "Tecnología avanzada respetuosa" (Respectful advanced technology). Comunication (abstract) to "Congreso Internacional de Tecnología, Desarrollo Sostenible y Desequilibrios". Universitat Politècnica de Catalunya. 14 - 16 Diciembre de 1995. Terrassa (Barcelona). 1995.

(5) SETAC, Guidelines for Life-Cycle Assessment: A "Code of Practice". Report from the SETAC Workshop held at Sesimbra, Portugal, 31 March- 3April 1993.

(6) Rieradevall, J. & Vinyets, J. "Ecodisseny i ecoproductes". Generalitat de Catalunya. Departament de Medi Ambient. Ed. Rubes. Barcelona 1999. ISBN: 84-393-4992-0.

(7) Environmental management systems. Specification with guidance for use (ISO 14.001: 1996).

(8) Clements, Richard B., Complete Guide to ISO 14.000. Prentice Hall, Inc., Englewood Cliffs, N.I. 1996.

(9) Robinson, G. & Roberts, H., ISO 14001 Implementation Handbook.Butteworth - Heinemann a division of Reed Educational & Profesional Publishing Ltd. UK. 1998.

Design Education
Philosophy and Practice

The spectrum of design culture

C LEDSOME
Imperial College of Science Technology and Medicine, London, UK

ABSTRACT

This paper examines the tensions between different aspects of design, by looking at its history, the current situation and the way forward. Design, as a multi-faceted process, did not exist in classical times. Only with the great cathedrals of the Middle Ages did tensions between architects and master masons bring a foretaste of problems to come. Various efforts to bring some unity of appreciation and co-operation have been made. Design, by its very nature, grows and changes. If we are to achieve a more co-operative approach, the author argues for a broader understanding of the unity of the field.

"Design" encompasses those activities which lead to the complete description of a product, project, process, or system to satisfy a market need (including the design of a system which may itself manufacture another product). It includes the management of those activities, and the necessary instructions for realisation, maintenance and use. .... Engineering includes those design activities where functional safety, reliability, quality, efficiency, and economy must be assured, no matter how they are realised.

"Attaining Competence in Engineering Design", Design Council, 1991

1. INTRODUCTION

Let me tell you a story.

Back in the 1850s, a young man by the name of Sam Clements left his native Missouri to seek his fortune. He travelled on a paddle steamer down the Mississippi and was amazed by what he saw. Tremendous vistas of grasslands and forest appeared round every bend in the river and the wide rushing water seemed to have a life of its own. He had a number of jobs and later came back to the Mississippi to train as a river pilot on those same steam boats. More years passed and he became a writer adopting as a pen name the call of the leadsman sounding a two fathom depth of water, Mark Twain. He later described how he realised one day that as a pilot he could no longer see the magnificence of the river. His eyes sought the ripples denoting a sandbank or a submerged log. He watched the motion of the trees and the grass for warnings of a sudden stray wind. He looked for new channels opening up as the river bed shifted. He saw far more and now knew the river intimately, but he had lost that magnificent vision of the river as he had first seen it.

A similar thing happens when someone takes on the role of an engineer. They are attracted to it by its most visible products, such as cars, aircraft, bridges, or spacecraft. However, when they have learned what they need to know about material properties, control systems, energy transfer and structural requirements, not to mention the processes of design and manufacture and their business and management, they have lost that awe of the mysterious power of the machine. Those designers, who are not privy to the arcane crafts of engineering, strive to maintain and enhance those very aspects of the human appreciation of design elegance, which current engineering formation inevitably tends to suppress. Conversely, a lack of appreciation of the engineering aspects of design can lead other designers to produce products with less than satisfactory performance.

2. ORIGINS

I design, therefore I am human[1].

"The capacity to design, that is the intellectual ability to explore and compare alternative potential courses of action in order to select the one which is the most desirable, within practical limitations, is a fundamental human characteristic."[2] This is the factor which has distinguished humanity from the rest of the animal world and enabled us to evolve the civilisation we have today. Design is such a fundamental part of our mental abilities, that it did not begin to be recognised as a single coherent field until the Victorian era, but we can recognise different aspects of it from our earliest activities. We begin to see direct evidence of specific people with more than usual design expertise in the Babylonian and Egyptian civilisations. Arches and vaults met the aesthetic pressure for wider wall openings and high roof spaces. Trial and error allowed the Pharaohs to lie beneath ever larger pyramids. The names of some of the architects and masons from those times are recorded in wall inscriptions.

Then came the Greeks, who first took a broader view of our intellectual activities and separated philosophically "pure" science and art from more pragmatic activities. Greek science was carried forward by logical argument rather than experiment and often bore only a superficial relationship to reality. Mathematics was held up as the ultimate of logical thinking. Greek art was concerned with perfection with the Golden Ratio governing proportions and shapes. With a scale related to the lengths of strings and pipes, music was a branch of mathematics. The real world was regarded with some disdain as being less than the perfection illustrated in poetry, painting, sculpture and the ideal world of the scientific hypothesis.

The Romans were much more pragmatic. For example, π was set at 3, or rather III, which is as good as it gets in Roman numerals. Their buildings were, and even in ruin remain impressive with central heating and baths, and their extensive road systems and aqueducts show very large scale design and systems thinking. The equipment and weapons of the Roman Armies were well designed and produced in large quantities. Right through the Middle Ages it was the legacy of the Romans, rather than the Greeks, which eventually produced the gradual resurgence of technology, architectural design, and improvement in the general quality of life early in the second millennium.

The rise of the Craft Guilds set standards of workmanship, professionalism, and training, which still find echoes today. These operated alongside and interacted with the artists, such as painters and sculptors, and the natural philosophers, astronomers and other forerunners of today's scientists. Indeed they were both dependent on the Guilds for many of their materials and had their own craft secrets such as mixing paints to their own recipes or the methods of grinding lenses. There was no separation between the perception of these different interests, indeed many were proficient in several fields. Leonardo da Vinci is perhaps the supreme example, a consummate artist (both painter and sculptor), an imaginative engineer (if not always pragmatic in his details), and an inquisitive scientist (particularly in biology). This was not seen as a contradiction, each was simply a different aspect of the world we live in.

The early forerunners of today's industrial design began to be more apparent. Furniture became less heavy and utilitarian and began to take on an elegance which developed over the centuries. Tapestries gave way to wallpaper; china replaced earthenware; architecture took on the great houses adding fountains and gardens. The printing press produced magazines and newspapers spreading the news of new styles and fashions along with advertising with ways to spend the growing wealth of at least part of the population. One minor area of tension gave a foretaste rifts to come. The only really large projects in the middle ages were palaces, fortifications and churches. Here there were differences between what the architects wanted and what the master masons could build. The heights and spans of cathedral roofs, as an obvious example, became a matter of pride with vaults and flying buttresses multiplying in all directions as the masons tried to keep up with architectural and ecclesiastical aspirations.

Generally though, the broad interactive approach worked right up to the early years of the industrial revolution. Then, as weaving began to be organised into factories, mines got deeper and first the canals and then the railways began to spread, industry was seen to be two separate activities, design, linked to the intellect, and the "mechanical" work of production and construction, linked to manual skills. Such skills were seen as separate from crafts, which began to be seen as more akin to the arts. "Mechanical" in this sense disparagingly implied working to instruction with no creative input. (This was an accusation later made by a civil, i.e. canal, engineer when referring to railway engineers, which contributed to the split which set up the Institution of Mechanical Engineers and perhaps helped to give them their title.)

It has been said that design was a way for gentlemen to be involved in manufacture without getting their hands dirty[3]. For a while the basic simplicity of the broader aspects of canals and railways made it possible for almost anyone to contribute to their planning without being too concerned about the technical intricacies. Then the need for bigger pumps, more complex locks, more powerful locomotives and longer bridges to meet the demand began to leave the "gifted amateur" behind. A few boiler explosions and collapsed bridges soon revealed the science of the day as grossly inadequate to analyse or explain the workings of the new technologies and engineers began to carry out their own research and later to set up their own degree courses, often with little content outside engineering specifics.

Greek philosophy became influential and emphasised the growing separation of "pure" science and the arts from the applied sciences and engineering. The separation of conception from creation became more marked. The Great Exhibition of 1851 was the last time in this country that all creative activity was portrayed together, although the seeds of fragmentation had begun many years before. Prince Albert's vision was breathtaking in its scope and years ahead of its time. If he had not died soon after, the exhibition might have heralded a new beginning, but instead it became a swan-song, with all of the more exuberant design of the early Victorian period suppressed by the long 50 years of mourning. Even today, here in the UK, any lavish decoration or overtly unconventional design is regarded with some suspicion. Thus, although engineering matured in the late 19th century, with the railways and the resultant ease of travel and transport becoming an established part of life, industrial design was stifled awaiting the stimulation which came from abroad.

It took the first World War to shake the UK establishment out of a complacent assumption of superiority dating back to Napoleon with the realisation that other countries were catching and even overtaking us. Then came the political shock of the Russian revolution on top of the devastation of much of Europe and the appalling loss of a high proportion of a generation of men. This left Britain and the rest of Europe looking for a new direction. It came from the USA. Relatively unscathed by only being involved in the last part of the war and with the profits from arms sales, America breezed in with bright new music, fashions and pressure for change. Fashion took off in completely new directions. Art Deco was the style for architecture and interiors, and a resurgent Germany produced Bauhaus. With industrial design now exploring its new potential, engineering hit the doldrums. The debts from the war (the UK was the only country to pay all of its arms bills to the USA, who had to build Fort Knox to hold all of the gold) followed by the stock exchange crash of the twenties and then the depression left the UK little to invest in new engineering infrastructure or advances in technology. Then came World War 2, destroying much of the old infrastructure.

It is to Britain's credit that after all that, it emerged from the war optimistically and began a sustained period of growth and economic stability, which with a few hiccups, still continues. A battered London staged the Olympic games in 1946 saving the games from oblivion. The "Britain Can Make It" exhibition of 1946 and the 1951 "Festival of Britain" gave people a view of a new future, which sustained a new demand for both industrial and engineering design. C. P. (Lord) Snow's Rede Lecture in the late 50s proposed the existence of two cultures with little interaction, one based in the humanities and the other in the sciences. This emphasised the differences between the human and technical aspects of design and held back moves towards a broader view of design for twenty years. The Council of Industrial Design had been set up in 1944 to help turn the engineering advances of the war into desirable peacetime products. In 1971 its remit was extended to include engineering and it became The Design Council. It expanded its educational role and gave advice to industry. Even so, successive governments have failed to appreciate the need for well designed products from a broad industrial base. One sector after another has dwindled to near extinction or become dominated by foreign owners. Much touted support for design has proved to be a chimera. In 1994 the Design Council itself was reduced from a peak of 350 staff spread round the country only five years earlier to a rump of just 30.

Increasing professional responsibility in engineering led to degree accreditation by the engineering institutions and initially split the education design provision in two, those deemed worthy of accreditation and those who were not, opening up a divide. Then that gap began to fill. The Royal College of Art - Imperial College Industrial Design Engineering joint course for post-graduate engineers had a high profile start in the early eighties. Undergraduate courses at South Bank, Bournemouth, De Montfort and now many other places soon became popular, with some opposition, in must be said, from reactionaries at either end of the design spectrum and in both industry and academia.

SEED, Sharing Experience in Engineering Design, was set up following a conference in 1979. Although based in engineering and primarily mechanical for much of its existence, it has always been receptive to input from other areas of design. The Institution of Engineering Designers has evolved over the same period and now provides recognition of engineering, product, and software design expertise. The Royal Society of Arts, the Royal Academy of Engineers and many other bodies are now also taking a more broad based approach to design.

3. CURRENTLY

Where are we now?

In 1977, the Carter Report[4] illustrated the relationship between industrial and engineering design with the following diagram:

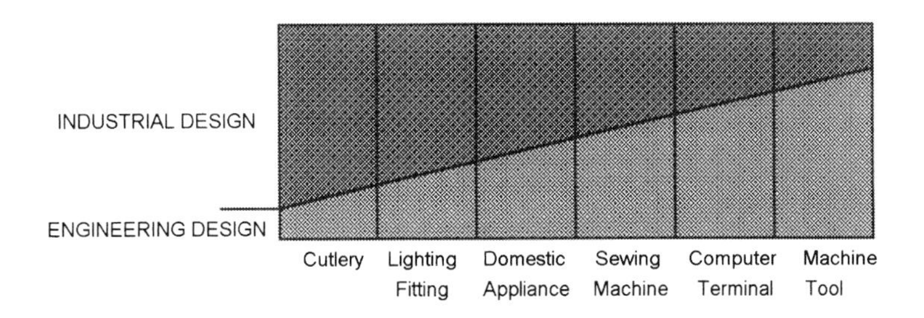

Fig. 1 Relationships between industrial and engineering design

This attempted to show how a typical spectrum of products depended to different degrees on the design skills of both engineering and industrial designers. Thus cutlery designers had a wide range of design options provided they stayed within the capacity of the existing manufacturing technologies. Machine tools, on the other hand, showed some concern for ease of operation, but were primarily inspired by the engineering requirements of their operation. Domestic appliances and sewing machines fell somewhere between. The interesting point is that computer terminals (this was before the PC) are shown near the engineering end of the spectrum. Today we would probably take the performance and reliability of the computer more for granted compared to its ease of use, matching to existing equipment and even its appearance in an office. (Recent advertisements for Macintosh machines have shown them in different colours and asked what flavour you want.) Clearly this product has moved along the spectrum towards a higher industrial design content since 1977. I believe all product types tend to move in that direction but at very different rates. If we try to illustrate that in a new diagram we get something like this:

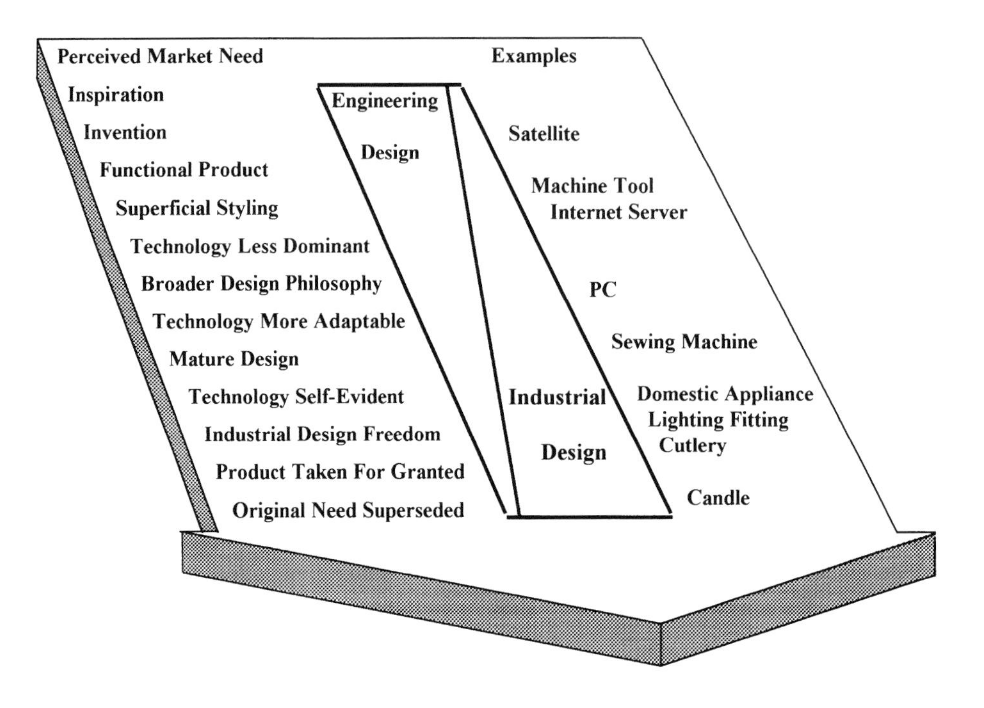

Fig. 2 Relationships between engineering and industrial design

Here we can see how a perceived functional need may initially be unfulfilled until technical inspiration leads to invention, then a viable, but perhaps rather ugly product, which sells because of what it can do. The first televisions in the 30s were the size of a small wardrobe, early lawn mowers looked like miniature combine harvesters, without the covers. New generations of the technology become smaller, lighter, cheaper, more compact and easier to operate. This gives industrial designers the freedom to make the product more acceptable to the user, and the market expands.

Eventually the functional components become so easily incorporated that the industrial designer can turn the product into such a familiar part of our lives that its existence is taken for granted. Even when their function has been superseded, familiar products may live on, for example: half-timbering that has no structural function, a moulded candle which will never be lit. This simple linear progress is probably rare. New perceptions of the product type, new technologies, splits in areas of application or combinations with other products can cause it to skip back in the process if the change is seen as a desirable improvement. Most probably progress will be in a series of iterative loops familiar from design models. Occasionally the combination of an efficient engineering solution and a sound industrial design concept coupled with convenient manufacturing methods will produce a classic design which satisfies the market. This can stop further product development for decades. An example is the diamond frame bicycle of the thirties, which stifled change for forty years.

If all products do tend to move this way, and my own limited knowledge of design history would seem to confirm it, we might be able to predict design changes over longer time scales.

This might enable companies to invest in appropriate material stocks or manufacturing facilities ahead of the market demand, provided the forecast was reasonably reliable. Of course technological development often moves in unexpected ways and rarely at the pace we expect. There is always the spectre of the classic design lurking in the wings.

We could re-draw the Carter diagram to illustrate a wider range of relationships:

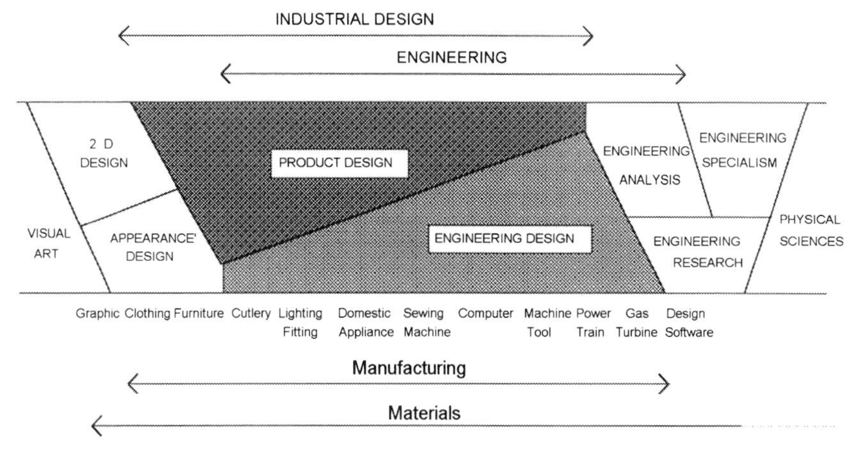

Fig. 3 Modified Carter diagram

I'm sure we could discuss at length the significance of how far the various arrows and areas should extend and overlap, but it is only intended as an illustration to provoke some thought. In drawing this diagram I am also aware that we all have a tendency to see our own field as being at the centre of everything so I am loath to draw any conclusions from it whatsoever.

So, at the start of a new millennium, (sorry, it just crept in) when so many other things are being reviewed, the whole concept of design seems, at last, to be moving towards a more unified pattern. A recent book by Bill and Gillian Hollins[5] includes services and longer term company strategies within the design ambit. Design processes are now being perceived as closely related to similar management procedures at all levels, and the scope for a new understanding in both fields is only just being explored.

4. FUTURE PROSPECTS

Lay in a course Mr Data!

The advent of massive computing power with internet levels of communication has clouded our view of the future. It has become fashionable to have a computer. Just as with the Rubic Cube of 20 years ago, everyone wants one but most don't know how to use it properly. Information technology is not an end in itself but a very powerful tool for doing things. As with all new technologies, we have first used it to do better what we did before. Indeed the initial spurs to developing computers stemmed from engineering analysis, particularly on the space programme. Then, as its capabilities became apparent and our expertise grew, we have explored its potential. We have developed computer graphics, then 3-D modelling, then direct links to machine tools and rapid prototyping. At the same time many control systems have now been taken over by the silicon chip, even popping up in the humble toaster.

Now we are beginning to use it to do things we only dreamt of doing before, things which were impossible because of the sheer size of computation. As it becomes more familiar as a tool, we are beginning to stretch our imaginations and do things we would not even have conceived of before. Virtual reality (just like real reality only you get to wear a hat), GPS, digital film stars, and internet banking are just a few examples. We are already beginning to take IT for granted, just as we have with so many other technologies. What we use computers for is much more important than what they are.

Computers allow us to "intellectually explore and compare alternative potential courses of action in order to select the one which is the most desirable, within practical limitations" much more easily and confidently. In other words, whatever else the computer can do, it is the best new design tool since the invention of the drawing board and slide rule. It will allow us to combine the best engineering and industrial design concepts into truly elegant designs for many more of our products than ever before. I feel optimistic that the efforts of many people and organisations over recent decades to produce a more unified approach to design are at last bearing fruit. Who knows, one of these days we may actually get the design message through to those who's policies seem to have been most influenced by Snow's divisive message, the politicians and civil servants.

REFERENCES

(1) Colin Ledsome, 'Engineering Design' is a Tautology, 'Engineering Science' is an Oxymoron, PDE '98.

(2) Colin Ledsome, Not Just a Pretty Face, Talk at an OU residential school, 1992.

(3) Quoted by Professor John Wood, Royal Academy of Engineering Workshop, 1997.

(4) 'Industrial Design in the United Kingdom' (Carter Report), Design Council, 1977.

(5) 'Over the Horizon', Bill Hollins and Gillian Hollins, Wiley, 1999.

Creativity in engineering education

P JARVIS and **K HAUSER**
School of Manufacturing and Mechanical Engineering, University of Birmingham, UK

SYNOPSIS

This paper offers a brief review of the main articulations and viewpoints of creativity, as presented in the psychology-based research and a framework relating together all the key factors. The authors have attempted to clarify some of the basic concepts through scientific analogies, in order to render them more accessible to engineering educators. This is done in the hope that this valuable material might be more easily conveyed to engineering students.

1. THE PLACE OF CREATIVITY IN ENGINEERING EDUCATION

There is a great deal of psychology-based research on the subject of creativity, which is not fully exploited by those in engineering education. There seem to be a number of reasons for this. Firstly, the application of creativity is predominantly associated with the arts rather than with engineering. Furthermore, its investigation is associated with psychologists, and therefore not something to which engineers can contribute. It is often understood as an innate talent, rather than a skill that can be improved. Perhaps most significantly, creativity is difficult to define and to quantify. In summary, the generally inadequate understanding of creativity by engineers leads to its neglect in their education. The authors believe, however, that the education of engineering students would be greatly enhanced if greater attention were paid to the understanding and teaching of creativity.

2. CREATIVITY AND ENGINEERS

While the term "creativity" is popularly associated with fine arts, the trend in business is towards an association with the concept of competitiveness, where creativity is required at all levels of any enterprise in order to produce unprecedented thoughts, efforts or performance. An example of this view comes from Cropley: (1) "Modern life is marked by rapid technological advancement as well as rapid change in many other areas (e.g., social, political and economic) and one important aspect of creativity is that it offers prospects not only of promoting change, but also of turning it into progress.". Creativity is therefore not just a valuable asset for engineers, it is a necessity, as engineers are essentially concerned with progress.

Interestingly, from linguistics, one can find similarities between the concepts of engineering, ingeniousness and creativity. On the one hand, the words "engineering" and "ingeniousness" have a common Latin root: "gegnere", to produce or generate, while this same Latin verb often strongly emphasizes the idea of birth, just as is found in some translation of "Creare", the Latin root for our English word "creativity". The concept of birth is therefore a central theme to both engineering and creativity, expressing the ultimate creative act.

3. THE MULTIDIMENSIONAL NATURE OF CREATIVITY

3.1 The "four P's of creativity"

A fundamental characteristic of creativity is that it is multidimensional. Rhodes (2) reviewed over 60 definitions of creativity and identified four significant factors. These are sometimes referred to as "the four P's of creativity": the *Person*, the *Process*, the *Product*, and the socio-cultural environment otherwise known as the *Press*. In figure 1, the authors have attempted to address all the dimensions involved in the creative act, by combining the 4 P's of creativity and Cropley's model of psychological constellation (3)

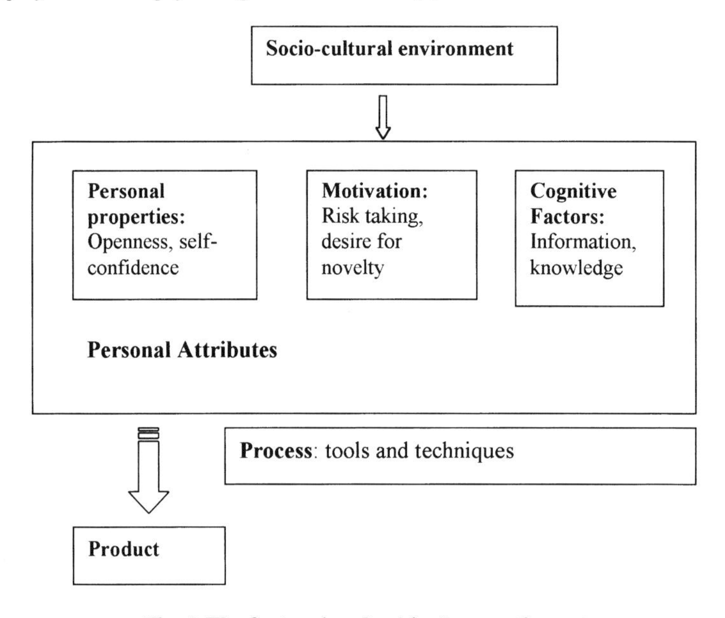

Fig. 1 The factors involved in the creative act.

The socio-cultural environment, or *press*, which the authors call the *place*, refers to the combination of factors external to the person, such as technology, social trends, fashion etc.

The *person*, or personal attributes, are sub-categorized in three groups. The *cognitive factors* are a very important prerequisite to any creative act, and encompass all the knowledge and experience of the particular problem or research field. Without such expertise, progress and novelty is difficult to achieve, since the recombination of knowledge to produce novelty is precluded. Motivation and drive are partly due to psychological dispositions, and partly due to external influence. *Personal properties* are the more intimate characteristics of a person, such as ones openness, flexibility, and curiosity.

The creative *process* describes the way that the creative act takes place. The researches may be oriented towards the cognitive process or towards more practical applications such as tools and techniques.

Creativity as pertaining to the *product* is mainly concerned with the signature of creativity in the product. It is universally accepted that the creative process results in novelty and value. There is admittedly some disagreement regarding the meaning of value and novelty, but without going into philosophical considerations, the product issued of a creative exercise "must be relevant to the issue at stake and offer some kind of genuine solution" (4). Some definitions include further criteria such as ethics (5) or the actual real production of the idea (6), but all suggest the combination of value and novelty.

Although there are four main dimensions to the problem, this paper will concentrate on the person and the process. These dimensions are reflected in the variety of viewpoints to be found in the relevant literature.

3.2 A variety of viewpoints

Creativity is addressed by authors from diverse disciplines and viewpoints, which is reflected in their writing. Elsewhere (7) four viewpoints are identified.

"The first such [viewpoint] is that of the academic psychologists. For example Bolton (1976), and Finke *et al.* (1996). They tend to concentrate on the attributes of the person, as they pertain to creativity, and the process of being creative at the cognitive level.

The second viewpoint is that of what might loosely be labeled the pragmatic practitioners, for example Von Oech (1983), Basadur (1995) and de Bono (1969, 1971, 1992, 1994). They tend to approach the subject from a pseudo-psychological viewpoint, but the emphasis is on improving performance in creativity, rather than understanding the phenomenon as the academic psychologists seek to do. The emphasis tends to be upon the process of creativity and particularly on techniques to enhance this process.

The third [viewpoint] is that of architects and industrial designers, such as Lawson (1990), Lawrence (1986) and Baxter (1995). These may be referred to collectively as the art based designers, to distinguish them from the engineering designers. This group tends to emphasise the process that gives rise to the product, and to a degree the creativity of the person.

The last [viewpoint] is that of engineering designers. The emphasis of this group is mostly on the process of creativity but also on the creative person, though the psychologist's distinction between personal qualities and abilities is rarely made."

3.3 Different psychological approaches

The approaches to creativity, previously adopted by the psychologists, have been put into 4 categories.

1) The case study approach, based on introspective reports by eminent creative personalities.
2) The psychoanalytic approach, which explains "creative expression in terms of the sublimation of unconscious conflict." (8)

3) The psychometric approach, which concentrates on measuring abilities through various tests.
4) The pragmatic approach, in which practical approaches are adopted. This is the group most commonly referred to by engineers. Examples of techniques based on this apprach are the attribute listing technique, by Crawford (9), morphological analysis by Allen (10) and Zwicky (11), the famous brainstorming technique by Osborn (12) and lateral thinking by deBono (13-16).

The scope of the current paper is to try to explain the process element of creativity by drawing on what academic psychologists have said about the person, rather than by offering new tools or techniques. i.e. prescriptions for the process. Indeed, the authors would contend that a better understanding of this process would help students to better use the creativity techniques which already exist.

4. UNDERSTANDING CREATIVITY

4.1 The struggle of defining creativity

Creativity is a puzzling concept mainly because it is so difficult to agree on a definition, both at the public and expert level. Everyone thinks they understand what "creativity" encompasses, but definitions range from "doing something artistic" such as painting or writing, to general problem solving; whether dealing with mundane or highly specialised problems. It is essential to mention at this stage that the multidimensional nature of the subject (the four P's of creativity) is partly responsible for the confusion.

Also, the understanding of creativity has been very dynamic through the ages. "From earliest times until the Renaissance, it was widely believed that all desirable innovations were inspired by the gods or by God. During the Renaissance, this view began to give way to the idea that creativity is a matter of genetic inheritance. In the beginning of this century, the debate turned to an argument over the relative contributions of nature versus nurture. In recent decades, there has been growing acceptance of biospychosocial theory, that is the belief that all creative acts are born of a complex interaction of biological, psychological and social forces."(17)

4.2 Novelty and value

Novelty and value are the common denominator of all definitions of creativity. Regardless of the context, a creative act involves these two elements. The difference between the creativity involved in a work of art and in an engineering product lies solely, in the authors opinion, in the issue at stake. In the first case, emphasis is put on the aesthetic quality of the product. The issue at stake here would be to please the eye of the observer. In the second case, commercial and technical aspects would probably dominate. In other words the essence of creativity can be uniquely defined, regardless of where it is applied. This leads to a first analogy to represent the process.

5. CREATIVITY AND THE ELECTRICAL CIRCUIT ANALOGY

Reflecting on the problem solving aspect of creativity, it was proposed that what bound a problem to the person trying to solve it was the *resistance* that the problem posed for the individual. This resistance is precisely what was earlier referred to as "the issue at stake". The nature of the resistance could be categorised in the following manner: technical, commercial, human, environmental and aesthetic.

This resistance to be overcome is at the core of the relationship between the person and the task. The resistance of the problem varies according to the personal attributes of the would-be creative person. For example, Jean-Paul Gautier would probably be more at ease with a problem of which the resistance is mainly aesthetic than would a NASA engineer, and conversely, the resistance of a technical engineering problem would probably be lower for this same engineer than for M. Gautier.

To explore the concept of a creative task having resistance, an analogy was drawn with the overcoming of resistance in a simple electrical circuit.

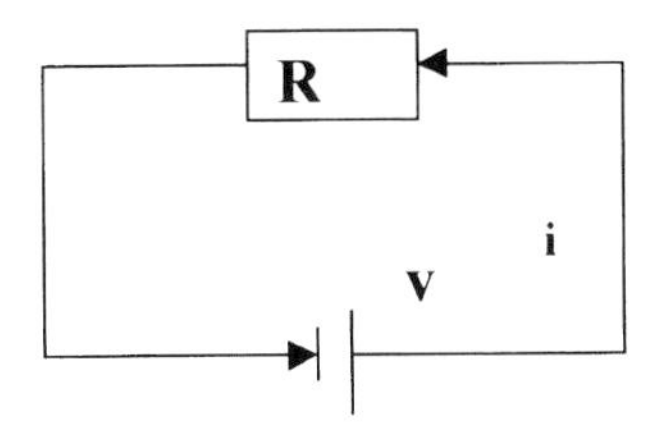

Fig. 2 Diagram of an electric circuit.

In an electrical circuit (figure 2), a battery with a certain electrical potential (V) overcomes a resistance R to produce a flow of current (quantity of charge per unit of time). Similarly, a person with a certain creative potential overcomes a problem with a certain resistance to produce a flow of ideas (number of ideas per unit of time).

6. THE COGNITIVE PROCESS

6.1 The Trinity model

The next step in the analysis of the creative act is to look at the cognitive level or, by analogy, the internal working of the battery of an electrical circuit. Finke *et al.*, academic psychologists suggest a model called the Geneplore model (figure 3), which explains the process in terms of 3 elements which are the generation of pre-inventive structures, the pre-inventive exploration and the application of product constraints.

The Finke *et al.* model is described in the following manner: "In the initial generative phase, one constructs mental representations called pre-inventive structures, having various properties that promote creative discovery. These properties are then exploited during an exploratory phase in which one seeks to interpret the pre-inventive structures in meaningful ways. These pre-inventive structures can be thought of as internal precursors to the final externalised creative products and would be generated, regenerated and modified throughout the course of creative exploration." (18)

The Geneplore model offers a rare explanation of the cognitive process of creativity, based as it is on experimental evidence. However its terminology is not readily accessible to engineers.

Elsewhere (19), the Geneplore model has been adapted, and in doing so, this problem has been addressed. The resulting model has been called the Trinity model (figure 4).

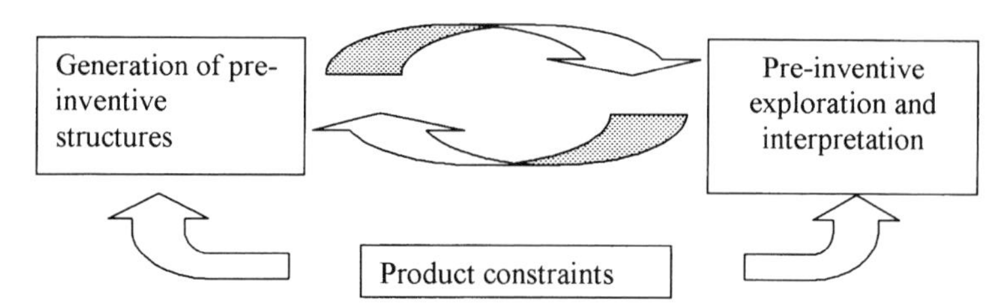

Fig. 3 The Geneplore model after Finke *et al.*

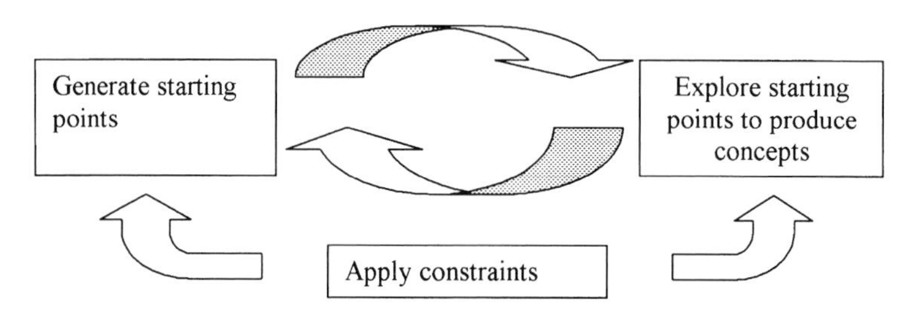

Fig. 4 The Trinity model of creative cognition

6.2 The Geneplore model and the surface assessment analogy

6.2.1 The surface assessment analogy

One of the authors has suggested an analogy in order to clarify the use of the Trinity model.
"It is proposed to visualise design as being akin to searching across an undulating surface. The X-Y position of a point on the surface represents a two dimensional abstraction of the multidimensional attributes of a concept. The height of a point on the surface represents the fitness or utility of the concept corresponding to that point." (20)

This visual analogy suggests that the creativity process is analogous to searching for low points on an undulating surface. The value of a solution to a problem is represented by its depth on the surface.

Thus, the three phases of the Trinity model are about generating new areas on the surface to look at; exploring these thoroughly, and accurately determining the value or "depth" of the solutions under evaluation; neither undervaluing nor overvaluing them.

6.2.2 Surface generation

One of the phases of the engineering design process where creativity plays a key role is the actual formulation of the problem, or the understanding of the surface one is going to explore. The mental formulation of a problem's structure has tremendous implications in terms of the potential to generate novel and valuable solutions. The criteria that are chosen as an organising principle of the surface can be the difference between a revolutionary break through and a dead end. For example, if James Dyson, when exploring the surface of vacuum cleaners, had generated an initial surface organised according to the shape of the vacuum cleaners rather than their suction systems, dual cyclones would probably not have entered his mind. The logical structure we choose to make sense of the problem we are tackling is arbitrary, but some ways have much greater potentials than others. A key part of the generation phase is to make that decision.

6.2.3 Levels of magnification

Any real-life undulating surface has some level of roughness or surface texture. Roughness on the design surface corresponds to significant differences in the value of designs, which only differ very slightly from one another. That is points which are very close together (in X and Y) can be at significantly different depths. This reinforces the importance of detail design which corresponds to homing in on the microscopic irregularities of the design surface.This process of employing increasing magnification in order to understand a surface is employed in surface metrology. It is represented in the Trinity model by its iterative nature that can be understood as increasing the level of magnification in order to perfect the details.

For the exploration to be efficient, it is important to target the level of magnification correctly for every phase of the design. Failing to do so leads to one of two problems. If the search takes place at too low a level of magnification, there is a lack of focus on the real task. Conversely, if the search takes place at too high a level of magnification, the work focuses unnecessarily on details, which are not yet important and so is inefficient.

6.3 Surface dynamics: visions and discoveries

A further characteristic of the surface is believed to be its dynamism. The value of a solution to a particular problem varies with changes in the level of technology, social needs, trends, fashion etc. Consider, for example, the surface representing solutions to the problem of stirring hot drinks in the most economical way in cafes etc. In 1950, plastic spoons were probably the best solution. In the 1990's, sugar is provided in individual sachets, rather than in bulk, and so the scooping attributes of the spoon are no longer needed. The plastic stirrer therefore has effectively replaced the spoon in such circumstances.

There are a lot of factors affecting any design surface, making it dynamic rather than static. The whole of human knowledge is interconnected like a spider web. When something happens in one field, i.e. a specific point on the web, it potentially affects a lot of other fields, both those closely and distantly related. Conversely, if an invention which might have occurred, does not actually occur, it will prevent the development of a whole series of related discoveries. For example, if plastic had not been invented, the world would be a very different place today. There is consequently not one surface, which represent the solutions to a problem, but two:

1) The *reality surface* representing the actual state of knowledge.
2) The *potential surface* representing solutions which could be achieved if the knowledge and technology to date was fully exploited.

Using these concepts, one can identify two sorts of creative act. The first and easier one is to find discrepancies between the two surfaces. This would mean, for example, to apply an existing technology in a new way, such as using dual cyclone technology for vacuum cleaners, or to spot sociological trends, which is what came into the equation with the introduction of the walkman. This process will be called *discovery*. The second one, which is more difficult, is to generate a novel and valuable solution, which lies below the potential surface as of today. This means that the design is invented but the technology to concretise the idea is not yet there. Leonardo di Vinci's helicopter design is a good example. This process will be called *visioning*.

7. CONCLUSION

There are many prescriptions available to help in the creative process, and no attempt has been made to add to this list here. Instead, a framework has been presented encompassing all the aspects of creativity (figure 1). This, and the supporting analogies, would suggest certain general approaches that one can usefully adopt.

Although engineers tend to prefer the views of pragmatic practitioners because of their practical approach, the rest of the material on creativity, and more specifically the psychology based research, can help shed some light on the bigger picture, drawing some logic out of such a widely misunderstood subject.

REFERENCES

(1) Cropley, A.J. (1999). Definition of Creativity. *Encyclopedia of Creativity.* Vol I, Academic Press, USA. p. 513
(2) Rhodes, M. An Analysis of Creativity. *Phi Delta Kappan,* April 1961, 42
(3) Cropley, A.J. (1999). Education. *Encyclopedia of Creativity.* Vol I, Academic Press, USA. p. 633
(4) Cropley, A.J. (1999). Definition of Creativity. *Encyclopedia of Creativity.* Vol I, Academic Press, USA. p512
(5) Cropley, A.J. (1999). Definition of Creativity. *Encyclopedia of Creativity.* Vol I, Academic Press, USA p.511
(6) McKinnon, D.W. (1975) *An Overview of Assessment Centers.* Center for Creative Leadership
(7) Jarvis, A.P. (1997). *Exploring Design.* Phd Thesis.
(8) Finke, Ronald A. Ward, Thomas B. and Smith, Steven M. (1996). *Creative Cognition: Theory, Research and Applications.* Cambridge Mass.; London, MIT Press. p.9
(9) Crawford, R.P. (1954). *The Techniques of Creative Thinking.* New York, Hawthorn.
(10) Allen, M.S. (1962). *Morphological Synthesis.* Englewood Cliffs NJ, Prentice Hall.
(11) Zwicky, F. (1948). The Morphological Method of Analysis and Construction. *Courant.* Anniversary Volume. New York, Wiley Interscience.
(12) Osborn, Alex F. (c1963). *Applied Imagination.: Principles and Procedures of Creative Problem Solving.* 3rd revised edn. New York, Charles Scribners.
(13) de Bono, Edward. (1969). *The Mechanism of Mind.* London, Jonathan Cape.
(14) de Bono, Edward. (1971). *The Use of Lateral Thinking.* London, Penguin.
(15) de Bono, Edward. (1992). *Serious Creativity.* London, Harper Collins.
(16) de Bono, Edward. (1969). Water Logic. London, Penguin.
(17) Dacey, John. (1999). Concepts of Creativity: a History. *Encyclopedia of Creativity.* Vol I, Academic Press, USA. p. 310
(18) Finke, Ronald A. Ward, Thomas B. and Smith, Steven M. (1996). *Creative Cognition: Theory, Research and Applications.* Cambridge Mass.; London, MIT Press. p.18
(19) Jarvis,A.P. (1997). *Exploring Design.* Phd Thesis. p. 159.
(20) Jarvis,A.P. (1997). *Exploring Design.* Phd Thesis. p. 164

A problem-based learning approach to the formation of electronic design engineers

D KING and **R SHUTTLEWORTH**
Manchester School of Engineering, University of Manchester, UK

ABSTRACT

Introducing circuit design into the early stages of an electronic engineering degree is a highly desirable activity but it poses problems because many students lack the abstract concepts and background theory to understand the circuits they are dealing with. Lectures and examples classes are a good way of teaching the analysis of circuits, but the gulf between analysis and the synthesis skills required by the design process is considerable. For a number of years a project-based approach has been used as a means of improving students design ability in the Manchester School of Engineering, but experience has shown that this has to be done in such a way that it builds confidence and a willingness to explore design ideas.

1. INTRODUCTION

The electronic design module introduced into the second year of the degree course at Manchester has been developed to satisfy a number of educational requirements. A primary function of the module is the development of electronic design skills and the harnessing of well-developed analytical abilities to facilitate the circuit design process. Combined with learning about electronic design, students are exposed to working in groups having 4 or 5 members so that the problems of group organisation must be addressed if the project is to be brought to a successful conclusion. Only a part of the background theory required during the design work has been covered in any detail during lecture modules at the time when the design module is undertaken. As a result there is need to either learn the necessary theory or adopt a heuristic approach to some of the factors required to finish the work successfully. A result of this type of approach is that all students gain a practical understanding of electronic design, which can not be developed through lectures alone. Also, they are exposed to problems and given time to reflect on what they have learnt. It is found that this process helps considerably with the understanding of related topics that are encountered in subsequent lecture modules.

Throughout the module emphasis has been placed on developing confidence in electronic design ability that comes from making systems which work. Design and development projects are very rewarding when the system being produced works and satisfies an identified need. A successful outcome to a demanding project increases the motivation to succeed in subsequent projects. Without a successful outcome design projects are highly de-motivating and for this reason it is important that all the students taking part in design experience some success in their work. As a result steps have been taken to try and ensure that the design process is demanding, but provides an acceptable level of success for all the students involved.

2. CHOICE OF PROJECT FOR THE DESIGN MODULE

The system chosen as the subject for the design work determines what the students will learn through undertaking the activity. It will also have an influence on whether the project can be brought to a successful conclusion. Several projects have been tried and the one that has met with the greatest level of success is the design of an audio link using an infra-red transmitter and receiver. The system was identified as being useful as a means of communication between a lecturer and deaf students in a lecture theatre as well as in other similar situations. As far as the students are concerned the project has obvious applications and the requirements are easily understood. In terms of learning about electronics a number of different types of circuit are required to build the infra-red voice link and this broadens the experience of the students.

A major problem for designers when approaching a project of this type is fracturing the design into a set of sub-systems that can initially be designed and developed independently. Fracturing a design requires more experience than most students possess so the basic building blocks for the infra-red link are defined in the laboratory manual for the module. Figure 1 shows the sub-systems required in the audio link. It is not the intention to enforce a particular solution for the design problem and any student group wishing to go its own way is permitted to do so. There is a small number of groups that do go their own way, but they are usually populated by students who have had previous extensive experience of electronic design.

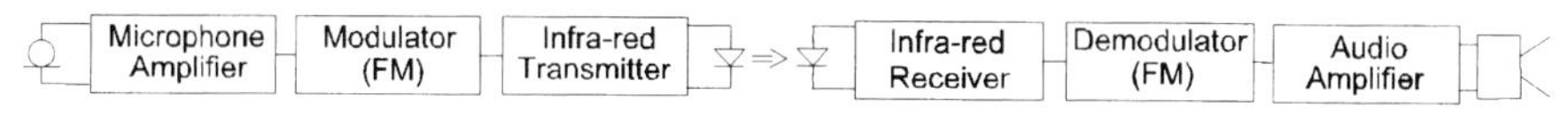

Figure 1
The infra-red audio link

The variety of different types of circuit required to meet requirements of the sub-systems is extensive. Audio circuits are needed which are capable of working with small signals from the microphone as well as supplying relatively large currents to the loudspeaker. Most of the knowledge required for the audio circuits has been covered in lectures prior to commencement of the design. The frequency modulation (FM) circuit uses a voltage-controlled oscillator and the demodulator suggested for the system is a phase-locked loop (PLL) circuit. There are numerous data sheets for PLLs and the students are expected to work from these to develop their designs since this subject matter is not covered in lectures until the third year of the course. Pulse shaping and amplifying circuits are needed for the infra-red transmitter and receiver and these partially build on experience of designing a video amplifier covered earlier in the second year of the undergraduate programme.

3. THE PROCESS OF CIRCUIT DEVELOPMENT

It is good design practice, and inherent in the embodiment process, to build and test sub-systems independently before they are finally assembled into the whole system. The students are encouraged to build the audio amplifiers first, and this they often do by dividing the group into pairs, one pair designing the microphone amplifier and the other pair the audio output stage. At this point the obvious way of testing the amplifiers is to connect the two in series, as shown in figure 2a, and observe the result. Adopting this test strategy the group soon realises the need for co-operation in defining the interface between the amplifiers, that is the signal level required and the importance of input and output impedances.

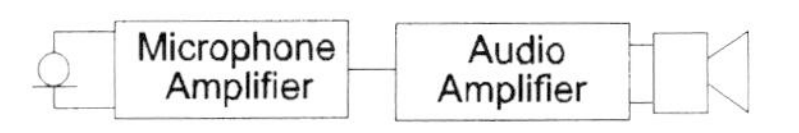

Figure 2a
The first stage in the development process

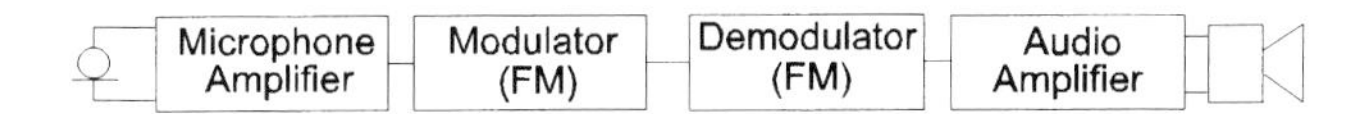

Figure 2b
Testing the FM modulator and demodulator circuits

Following the successful completion of the audio stages the FM modulator and demodulator can be designed and tested using the system configuration shown in figure 2b. Finally the infra-red transmitter and receiver stages are designed and the full system tested using the arrangement shown in figure 1. Progressively adding functions to the system, which can be tested with blocks already developed, gives confidence that the design is progressing satisfactorily and that the final goal can be achieved. Providing a path that shows the way to progressive development raises motivation levels in the students and greatly assists in promoting good organisation in the student groups.

4. THE SUB-SYSTEM CIRCUITS

For some of the student cohort the problem of selecting circuits to satisfy the requirements of the subsystems proves quite difficult. To overcome the problem there are suggestions made about the types of circuit that might be tried. Figures 3a and 3b give examples of suggested circuits for the audio stages of the design. The circuits are not complete as they stand and their performance is only just adequate to fulfil the minimum requirements of the system. It is expected that most project groups will produce infra-red links that perform better than the barely acceptable performance attained if the suggested circuits are used. This process of suggesting less than ideal circuits proves useful because the student groups build the circuits and then go through a process of improving them to the point where the circuits work well. The work involved in this process of improvement builds a practical knowledge and understanding of how the circuits work which is not obtained from lectures. It is also a useful lesson because the same path is followed in many embodiment design activities where improvements on an initial concept are needed to meet the requirements of a specification.

A range of components is available in the laboratory for students to use on the project. The range is restricted to devices that are relatively inexpensive so that destroying components during development work is not an issue. At first sight it might be thought that this would lead to the students being careless with components but in reality it is soon realised that a broken component costs time and is to be avoided in an activity which runs to a relatively tight time schedule

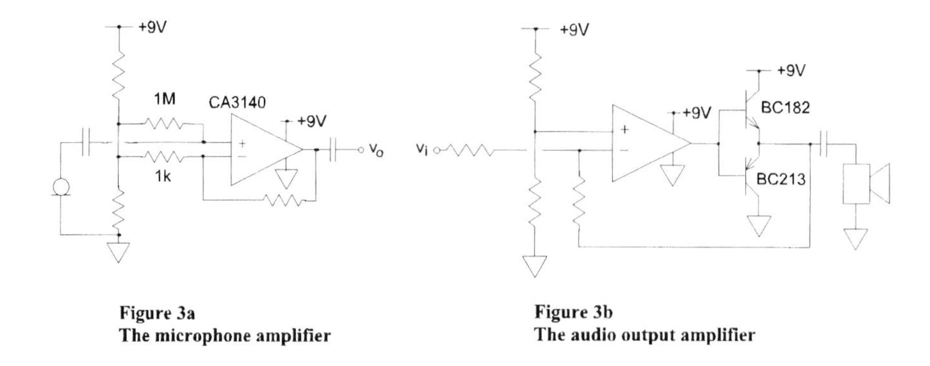

Figure 3a
The microphone amplifier

Figure 3b
The audio output amplifier

5. ASSESSMENT STRATEGY

A frequently used approach to assessing design projects is the submission of an extensive final report that describes the work and achievements of the project. This approach is often counter productive in long term student projects because the design work tends to get left to a late stage and then everything is done in too short a time scale. Insufficient work is put into the design, so usually the results are poor, and the lesson taught is that report writing should be used to disguise the fact that less than enough work has been put into the project.

A different approach has been taken to the infrared audio link design in that pressure is put on the student groups to work throughout the project period. A set of short progress reports is required from the groups, each report representing approximately 100 man-hours work for the whole group. It is expected that every student will put 100 hours into the module so the progress reports are required at the end of every quarter of the project. A strict limit is placed on the length of the report so the work involved is kept to a minimum for both students and staff. Experience has shown that allowing the submission of longer reports turns the design process into a report writing activity and this is not the objective of the project work. Only a single side of A4 paper is accepted as the written report, covering the achievements of the previous quarter and the projected work for the following quarter. Any number of circuit diagrams and results graphs can accompany the report. A Gannt chart showing the timetable for the next quarter is also a deliverable item as part of the report. The inclusion of circuits and graphs in the report is useful because they should have been produced by the students for their own notes, and they are items that contain a large amount of information that can be easily assessed by academic staff.

When the quarterly reports are due for handing in there is a review of the progress being made with the practical work in the laboratory. This takes the form of an interview where demonstrations of working circuits are required and the notes being kept in the design log book are inspected. The aim in these quarterly reviews is to judge the progress being made as well as providing help with solving circuit problems identified during the interview. Also, group organisation problems are identified and suggestions made to overcome difficulties that might be developing. Every attempt is made during the review to be constructive and help with the design process rather than being critical of the work done.

The overall assessment scheme and the marks allocated for various elements of the module are shown in the table below.

Table 1 Assessment scheme

First Quarter Project Review	5%
Second Quarter Project Review	10%
Third Quarter Project Review	10%
Peer Group Assessment	15%
Final Group Presentation and Report	25%
Individual Report and Practical Work	30%
Catalogue Advert	5%

The three quarterly reviews account for 25% of the total mark and, combined with the final group presentation and catalogue advert, accounts for 55% of the total. Individual assessment is based on the part of the final report that the individual submits, as well as a peer group assessment making up a total of 45%. An important feature of the project is that the group of people have to work together to make the design successful and the proportion of marks awarded reflect this objective. A peer assessment is included to provide a sanction that can be applied by the group to members who fail to deliver work they have been assigned, and it also supplies information about the work individuals contribute which is hard to assess by other means. The catalogue advert is meant to advertise the product that has been designed and sell the product in the market place. This activity is treated as a competition by the students and results in surprisingly attractive adverts from some groups.

6. STAFF INVOLVEMENT WITH THE DESIGN WORK

Contact between staff and students is an important part of undergraduate electronic design projects because seeing an experienced person solving a design or development problem conveys a lot of information which is important, and cannot be taught successfully in other ways. The assessment process is geared to promoting student-staff contact through the design reviews undertaken at regular intervals. In the periods between assessments further assistance is provided by staff running design surgeries where students can come along with problems and a solution, or a means for finding a solution, is suggested. Usually one hour a week is spent by each member of staff in the laboratories running the surgeries.

The work load for both academic staff and students with design projects is high. Student cohorts of up to 80 students (20 groups) have been taken through the design process of the type outlined, and the activity has been run by two members of staff. The staff loading is equated to one lecture module each by the School of Engineering but this is an under-estimate of the work involved in both running and organising the project work.

7. THE NEED FOR INITIAL CONFIDENCE BUILDING EXERCISE

Several years ago the electronic design project was organised and run as outlined in the previous sections. The outcome for many of the students and the staff was less than hoped for, given the work that had to be put in by all concerned. Observing what was happening with the students it became apparent that they were not confident enough in their own abilities to build working electronic circuits using manufacturers' data sheets for design information. All the students had been taught to solder during the Engineering Applications module in their first year and had some experience of building circuits.

To improve confidence levels in circuit construction and development the design project work starts off with a short individual design and build exercise. Each student is given the specification for an astable oscillator and asked to use a 555 timer circuit to build the oscillator. The specification for the frequency of oscillation and the mark-to-space ratio for the output voltage waveform is made different for each student to prevent plagiarism. Using the data sheet for the 555 timer circuit the exercise takes one laboratory session (3 hours). Students are not allowed to progress to the infrared link design work until they demonstrate a circuit working to the specification given to them. As a reward for successful completion a mark of 5% is added to their individual mark for the module. The improvement in confidence with building circuits following the initial exercise on the 555 timer is considerable and repays the time and effort spent on the activity.

8. CONCLUSION

The electronic circuit design module using the infrared audio link has been running in the form described for two academic years and the second cohort of students has now finished the project. Judging the success of project work in teaching the required engineering skills is difficult. Perhaps the only measure of success is whether a higher level of ability is demonstrated in subsequent projects. In the case of most degree students there is the final year project which is undertaken, and indications are that improvements in practical ability have occurred with the present final year students. Furthermore, the students' knowledge of electronics and the ability to accept some of the more difficult concepts of phase-locked loops and high speed switching circuits seems to have been enhanced.

When changes are made in the method of teaching students the outcome is not always apparent over short time scales. The final success of the electronic design module implemented in recent years will only be determined from experience gained over a number of years with graduates which become practising design engineers.

9. ACKNOWLEDGEMENTS

Many of our colleagues in Electrical Engineering at Manchester have contributed ideas through discussions about project work, and the authors are grateful for their help. The visiting Design Professor, Glenn Birchby, has made very helpful and constructive suggestions which have been built into the project. Finally our thanks go to Andrew Morris, the technician in the projects laboratory, whose enthusiastic help and work has made the project run smoothly.

Introducing interplays between product, process, and supply chains to engineering undergraduates

A MCKAY, J E BAXTER, and **A DE PENNINGTON**
School of Mechanical Engineering, The University of Leeds, UK

ABSTRACT

A key challenge in the education of undergraduate engineers is to provide them with an appreciation of three important concepts that influence much of what a practising engineer experiences during a product realisation process: namely, product, process and supply chain. This is especially important if the functionality of Computer Aided Engineering (CAE) systems is to be introduced in a realistic context. This paper introduces an assignment that is part of a module that introduces CAE systems for design and manufacture. The assignment is intended to allow students to gain first hand experience, albeit in an artificial situation, of the interplays between product, process and supply chain.

1. INTRODUCTION

The success of the majority of products is governed not by the quality of the product itself but by the time that it takes, and the costs incurred, to deliver a concept to the marketplace. Further, the long term prospects of manufacturing organisations are frequently governed more by their ability to deliver a fit for purpose product to time and cost rather than their ability to deliver a supposedly ideal product late and outside its budget. Hence it is often process and supply chain issues that lead to failure rather than weaknesses in the product itself. Graduate engineers must appreciate these issues if they are to be effective in business environments.

This paper reports on a case study-based assignment that is intended to allow students to experience for themselves these dependencies. It is part of a module that introduces the functionality of Computer Aided Engineering systems to level 3 and 4 mechanical engineering undergraduates. It is included in this module to give students a better appreciation of what CAE systems can and cannot do to support practising engineers in their work. The students participate in a design process, for a relatively simple product, that is conducted across a supply chain. Each student works in a small team and each team participates in a chain that starts with an initial high-level requirement and ends with an electronic product definition of the product that they designed. The module is delivered to approximately 60 students per year and the assignment has been run in its current form for the past two years.

2. RATIONALE

It is generally accepted that the majority of engineering products are complex. Analysis of such products, and consideration of complex systems thinking, indicates that it is the number of different kinds of relationship that create this complexity (1). Similar conclusions can be drawn about the processes that engineers use -- there are many people in different roles using different tools in different ways. Extended enterprise thinking is emerging as a way of considering not only individual organisations but also the extended enterprise within which it participates (2). An extended enterprise, for the purposes of this paper, can be defined as the collection/network of organisations that together design and deliver a product. Process chains can be seen as paths through the network along which things flow. For example, physical artifacts flow along a supply chain. A general issue addressed by this paper is related to how complexity might be introduced to Level 3 and 4 undergraduate Mechanical Engineering students in a way which will benefit them in their future careers. Whilst both engineering products and engineering processes can be described as being complex, it can be argued that it is only the extended enterprise that is a complex system because it is the extended enterprise that is composed of a number of self-determining individuals which interact with each other in such a way that the extended enterprise evolves.

The assignment described in this paper was designed to provide students with a concrete example and experience through a real process and supply chain, and a realistic product. In this setting, students have an opportunity to use at least one CAE system, experiment with the functionality that is available and so experience some of the limitations.

3. STRUCTURE OF THIS PAPER

The intent of the assignment is to allow students to learn through experiencing a situation and then reflecting upon it. In this way students gain an opportunity to make at least one full circuit of Kolb's learning cycle (3). Subsequent lectures draw upon their reported reflections of their experiences and relate it to theory and current best practice. In Section 4 of this paper the details of the assignment that is given to students are outlined and in Section 5 the requirements of the students and their teams are summarised. The use of a web-based bulletin board for electronic communication within and between teams is discussed in Section 6. Finally, in Section 7, key learning points for students, based upon our experiences running the assignment are highlighted.

4. ASSIGNMENT REQUIREMENTS

The starting point for the students is a predefined product specification, a predefined collection of teams and guidance on requirements for the process that they are to use. Students are told what is required of them but they are not forbidden to do anything else.

4.1 Product requirements: a fishing game

The product that is used is a children's magnetic fishing game. The fishing game has a number of fish that are located in a rotating disc and which move up and down on a cam surface. As they move up their mouths fall open. The object of the game is to catch the fish using a fishing rod with a magnetic end that attaches to a metal insert in the open fish mouth. A sample is provided for the students and the design brief is as follows:

> *The customer requires a fishing game that is approximately twice the size of the sample that has been provided. In all other respects the game should be the same as the sample. As such, the required product structure is as given. However, the game is difficult to store when it is not in use. For this reason, the customer also requires some form of package that can be used to store the game when it is not in use.*

4.2 Process requirements

The first task for each team is to select and/or define its own process. As a minimum, the process must take (as input) some product requirements and produce (as output) an electronic product definition of a product that meets these requirements and information that a design team would pass to a manufacturing planning department. The process must create the outputs in a timescale that fits in with the required timescales of the entire supply chain process. A number of industrial processes are provided for students to use either directly or as a basis for their own process definition; these include the Rolls-Royce Integrated Product Development process (4), the Boeing Corporation's core business processes of Acquire, Define, Produce, Support, the Pahl & Beitz design process (5) and the core processes from the Supply Chain Council's supply chain operation reference model (6).

4.3 Supply chain requirements

The supply chain is really a design chain since no physical artifacts flow. Figure 1 shows the

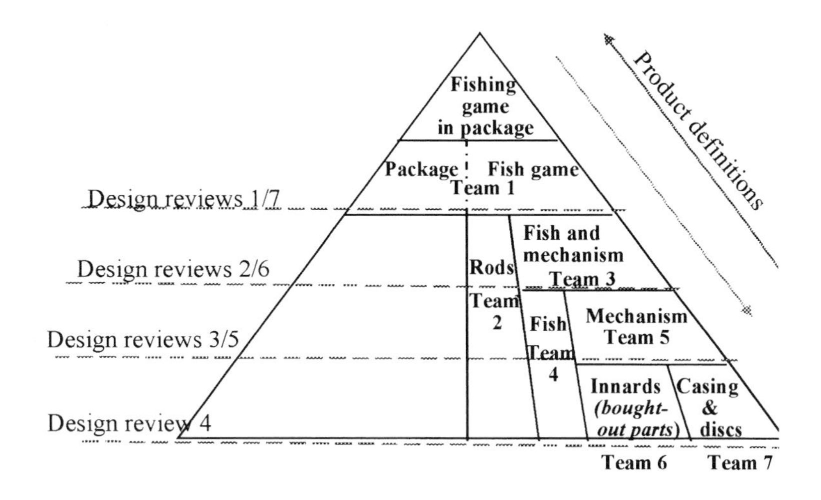

Fig. 1 The "Fishing Game" Supply Chain

supply chain that is set up to supply a fishing game and package design. The module leader acts as the customer for the design. Team 1 is responsible for supplying the whole design, Team 2 is responsible for supplying the *rod* design to Team 1, Team 3 is responsible for supplying the *fish and mechanism* design to Team 1, and so on. Each team is responsible for presenting their requirements and completed designs at one or more design reviews. For example, in week 3 of the assignment Team 5 present their requirements for the *innards* and *casing & discs*. In week 5 of the assignment they present their design for the *mechanism*. Each of the presentations at the design review is assessed. Because of the number of students taking the module two independent supply chains are set up, pyramid A and pyramid B. In this way students have opportunities to compare the operation of their own pyramid with another.

All teams have a number of tasks to complete within two days of starting. These are as follows:

1. to name team members who will carry out the following roles: initial contact point for the team's customers, initial contact point for the team's suppliers, product owner, and other roles/responsibilities that the team wishes to identify;
2. to outline a high level process that the team agrees to use (at least initially);
3. to post the agreed process, people and roles in the virtual assignment room.

These tasks are essential for the successful operation of the supply chain and the making of these decisions helps the teams to form.

5. OUTPUTS/DELIVERABLES

Through the first half of the assignment each team is required to post requirements for their suppliers on the module bulletin board which is then used to support a short presentation at a design review. The design reviews provide opportunities for suppliers to question their customers and for all students to observe what is happening in both pyramids. In the second half of the assignment electronic product definitions are posted on the module bulletin board for use by each team's customers.

Soon after the last design review each team is required to submit a final report that includes the following information:

- details of the product that was worked on and the process that was used;
- details of the teams with which the team interacted during the assignment and the nature of those interactions;
- examples/print-outs of the inputs and outputs to/from the process and/or team;
- the way in which computer-based tools were used (highlighting the way in which they helped the process; discussing any limitations that were found (how they constrained the process and why the limitations exist); suggesting improvements/additions to the computer-based tools that the team might have found useful);
- anything else that the team thinks is interesting.

Marks are allocated as follows: 25% for contributions to the bulletin board, 25% for presentations at scheduled design reviews and 50% for the final report.

6. NATHAN BODINGTON BUILDING

The Nathan Bodington Building (NBB) (7) is a web-based service provided by the Flexible Learning Development Unit for the University of Leeds. It is a software system created at the University of Leeds that can be used to create a highly interactive web site for teachers and learners using the metaphor of a building for easy navigation.

6.1 Discussion room design

The objective for the use of the NBB discussion rooms was to act as a record of the fishing game assignment interactions, both customer-supplier and other, and to provide an open place where students could communicate without needing to meet in person. One NBB room is used for each supply chain pyramid. Each discussion room is monitored by a postgraduate demonstrator who is also responsible for posting messages to give guidance as necessary to the students. No structure is given to the NBB discussion rooms, it was decided that the students needed to experience the problems caused by the lack of a structure for their data and interactions. Some teams then imposed a structure. To ensure that students make use of the discussion rooms marks are allocated for its use (25%). The students are also required to make their design review presentations from the discussion room; these, too, are assessed (25%). This ensured that the details of their product requirements and their electronic product definitions, CAD models, of their designs are recorded in the NBB.

The NBB software does not allow the deletion of submissions to discussion rooms or their editing after a message has been seen by someone other than its author. Consequently submissions cannot be altered by students to improve their marks after the deadline for submission. The security access to each discussion room is set up so that only those students in teams belonging to a given supply chain can see that supply chain discussion room. All staff and postgraduate demonstrators involved in the assignment are given access to all rooms. The NBB software also keeps a record of who has read any given message.

6.2 Experiences of discussion room usage

Students posted a range of messages from just text to pictures, animated pictures and HTML controlled information. Some posted images on their own WWW pages and made reference to them in their NBB postings to enable them to include more complicated material in their design reviews. At any one time the major activity in the discussion rooms tends to revolve around the teams approaching a design review. During the first half of the assignment the teams post their requirements and ask questions about them. In the second half the teams post their designs. In the supply chains that have been experienced by the authors, the team number 1 have interpreted the doubling of the size of the game differently. For example, in one year one team changed the physical size of the game by a factor of two and the other doubled the number of fish in the game. So from the outset evidence of the need to agree product requirements with the customer was available.

In one pyramid, thirty-three message threads were started in one discussion room. After a while the students started to complain that there was no structure to the threads that were started. One person in that supply chain took it upon himself to start a number of threads to structure the discussion. Threads were started for each of the team interactions up and down the supply chain (vertical interactions) and each team's design reviews. However this did not prove sufficient and as a result of debate during a design review more threads were added to deal with interactions between teams at the same level in the supply chain (horizontal interactions). This again provided examples of what happens in reality: namely, the benefits of structuring interactions and associated information in a supply chain so that the appropriate information can be readily found, and the drawbacks encountered in not structuring (or inappropriately structuring) the information. Also, even when a structure is developed, it may need to be upgraded to deal with unforeseen situations. The major problems that the teams encountered were associated with the over or under specification of requirements and the late delivery of designs. As a result there were delays in communicating requirements down the supply chain and digital designs up it. This is typical of what happens in reality.

6.3 Discussion on the use of NBB Discussion Rooms

In general the use of NBB discussion rooms has been successful. The exercise has allowed students to experience the problems and tensions that happen during the interactions between suppliers and customers. The students have not reported any major problems in using the NBB discussion rooms. This indicates that the NBB technology did not prevent them from meeting the requirements of the assignment. The one issue that did arise were those students who wished post images and did not know how to do it or did not have a WWW area to put the image files in. The lack of knowledge on how to post images was filled by the support of the postgraduate demonstrators and student interactions. The staff and postgraduate demonstrators involved with the module did have WWW areas that they owned. The number of image files that the students needed to use was small. Thus students without WWW space were encouraged to send the images to staff and postgraduate demonstrators to be made available on the WWW. This was done on an ad hoc basis but did not take up a lot of time. However, it is intended to provide the students with a more formal mechanism and WWW space for the posting of images in the future.

More students read messages in the MECH3330 NBB discussion rooms than posted them. It is suggested that this might result for the following reasons: one person in a team was assigned the role of posting the team's input to the discussion rooms; all students had the knowledge and skills to read messages in the NBB discussion rooms but fewer people had the skills to post more than just text messages. It should also be noted that few teams requested help and guidance in posting HTML submissions to the NBB. The authors have also had the opportunity to interview a number of students after the module has finished. One of the observations made was that teams often preferred to have informal meetings than conduct the entire debate through the NBB. This is what happens in industry; there are often informal meetings and telephone conversations between customers and suppliers to iron out details. However, it is important to document the final outcomes of such debates to ensure that everyone is clear about what has been agreed.

In general the use of the NBB discussion rooms was successful. They act as a permanent record of the supply chain interactions associated with the "Fishing Game Assignment". Examples from the NBB discussion rooms can be used to support follow up lectures. The success in terms of take up and use of the discussion rooms was in part due to the association of module marks with its use. The students applied the technology successfully although more attention could have been usefully spent on the content and structure of the design review presentations.

7. LEARNING POINTS

In this section students' learning and experiences are outlined using the ideas of product, process and supply chain as a framework.

7.1 Product-related learning points

The focus of the module is the role of shape-based CAE systems in design and manufacturing. A key and early learning point related to the lack of geometric information and allowed students to form an opinion on the suitability of contemporary shape-based CAE systems for the support of the early stages of product definition processes. In terms of CAE system functionality some of the underlying assumptions and limitations soon become apparent; for example, how do you model the string in the rod, knots in the string or the magnets?

7.2 Process-related learning points

On more than one occasion teams have asserted that their process was being affected by the teams with which they interacted. Complaints relating to late delivery of requirements or electronic product definitions were relatively easy to deal with. More challenging was teams' abilities to deal with poor quality. In defining their processes very few, if any, teams have yet considered [from the outset] including steps to check either the quality of the material that they pass to others or the quality of the material that is delivered to them. This problem is especially highlighted when poor quality CAD files are exchanged because, as the poor quality moves up a pyramid, the associated difficulties increase dramatically. Discussions relating to the quality of the requirements and CAD files that flow prompts the following kinds of questions.

1. What, if any, is the team's release procedure for Electronic Product Definitions?
2. Who ensures the quality of what flows?
3. How does a team determine whether or not the quality was acceptable?
4. What are the characteristics of a good CAD file? [1]
5. What happens if the quality is not acceptable?
6. What happens when flows change -- are teams sure that they were always using the correct data?
7. Would there have been an advantage in adopting quality control procedures and who or what would have benefited?

7.3 Supply chain-related learning points

Two key points have been highlighted in the context of supply chains. One is that there are two kinds of process, those that are internal to a team and those that are external across a chain. The other is the importance of different forms of communication. The value of face to face contact, the need for common data formats, file sizes and the quality of information are all identified as factors that influence successful communication. Some teams also note that there is benefit in communicating across a level of a pyramid as well as along customer-supplier relationships. In this way students are beginning to see and experience the benefits of different kinds of extended enterprise relationship such as partnering.

8. CONCLUSIONS

Before the introduction of this assignment the interplays between product, process and supply chain were covered through lectured material. The assignment was introduced because a significant proportion of students did not see the relevance of the material and did not perform as well as one expect in examination questions related to it. A detailed analysis of student feedback showed a correlation between students' industrial experience and their performance in this aspect of the module. That is, the students with little or no industrial

---

[1] Meaningful [to the recipient] names, clear product structure and no spurious elements are three characteristics of a good CAD file that students are encouraged to consider.

experience did not appreciate the problems and issues associated with working with suppliers and customers. It was therefore decided to introduce an assignment to the module to give the students experience of working in a supply chain. As part of the assignment students were required to attend design reviews and present their designs. Overall the assignment has been well-received and student performance in examination questions related to this subject has improved. Student feedback on the fishing game assignment indicates that they consider it to have provided them with experiences that will be of value in the future. Krause, Rampach, Kimura, and Suzuki identify time, cost and quality as key factors that influence the performance of product development processes (8). A fourth factor that is commonly considered is risk. Through this assignment students gain invaluable insights into each and experience of their consequences.

ACKNOWLEDGEMENTS

Thanks are due to the staff and students from the School of Mechanical Engineering at the University of Leeds who have participated in the fishing game assignment.

REFERENCES

(1) Peter Conveney, Roger Highfield, *Frontiers of Complexity: The Search for Order in a Chaotic World,* Faber and Faber, 1995

(2) C. Fine, *Winning Industry Control in the Age of Temporary Advantage*, Perseus Books, 1998.

(3) D. Kolb, *Experimental Learning: experience as the source of learning and development,* Prentice Hall, 1983.

(4) P.C. Ruffles, *Project Derwent process - A new approach to product development and manufacture*, published by the American Institute of Aeronautics and Astronautics, 1995 .

(5) G. Pahl and W. Beitz, *Engineering Design: A Systematic Approach*: Springer-Verlag, 1988.

(6) Supply Chain Council, URL: *www.supply-chain.org*

(7) Nathan Bodington Building, URL: *www.tlsu.leeds.ac.uk/nathanbodington.html*, University of Leeds,

(8) F.L. Krause, C. Raupach, F. Kimura, H. Suzuki, *Development of Strategies for Improving Product Development Performance*, Annals of the CIRP, Vol.46/2/1997.

Innovation, design, and CAE in new product development

H GILL and **E UNVER**
School of Design Technology, The University of Huddersfield, UK

ABSTRACT

There seems to be a need for clarification on a number of issues that are seminal for the design community: for example, the constructs of invention, of creativity and of innovation (or innovative) appear to be used interchangeably. They are all a vital interest for design but they are different things. Furthermore there is a failure to distinguish between what design has a ***legitimate interest*** in or should ***contribute to*** and that which is its ***direct responsibility***.

The authors of this paper will seek to clarify these and other issues. One of these interests is the increasing demand for sustainable practice: design has a crucial role here and this paper will explore how computer aids can be a powerful tool in this area and also more generally in both design and manufacture.

1. INTRODUCTION

At its simplest innovation can be defined as:

> ***'the initiation and implementation of change'.***

Other definitions such as:

> ***'Innovation – the successful exploitation of new ideas – is essential for sustained competitiveness and wealth creation. A country aiming to keep ahead of its competitors needs companies which innovate'. (HMSO 1996)***

stress the business, economic, political and technological dimensions of a concern for innovation that can obscure the fact that -**'innovation is for everyone'.**

All spheres of human endeavour benefit from a steady flow of new ideas and their subsequent implementation. So innovation can of course be technological and much of the literature centres on technological innovation but it also has applications in social, political, educational and other programmes. Organisations have to develop a creative dynamic that permeates all levels of the enterprise rather than it be the responsibility of one or two 'gifted' individuals. To establish such an ambience within an organisation can be difficult but some very large companies, 3M Corporation, Sony and Unilever have formal structures for *'unlocking the potential of its people'. (DTI Winning Report 1996)*

The word ***innovation*** is often confused with ***invention*** and ***creativity*** : ideas are described as being ***Innovative or inventive or creative***

It is now generally accepted – and certainly for the purposes of this work – that the term innovation is used to describe the process that

> ***Begins with an idea (preferably inventive or creative) and takes it through to implementation or widespread diffusion.***

Space does not allow for a fuller treatment of the innovation process other than what follows since the focus is to be in the role of design within the process.

2. THE INNOVATION PROCESS

Interest in innovation has grown steadily but especially over the past ten to fifteen years. More recently still the associated issue of entrepreneurship has been the focus of attention. Rothwell (1992) has described a growth in understanding of innovation from earlier notions of it being driven by 'Technology Push' – the market being a passive recipient of the products of the technological infrastructure through the 'Market-Pull' and 'coupling' model to what he calls the 'fifth generation' model. In this the process is pursued by teams who are networked electronically rather than physically and international (global) boundaries apply.

Other authors have prepared models as explanations of the process some of which can be used as operational guides. A comprehensive model is proposed by Gill (1999) and this gives some idea of the complexity of the process, the major components, the principal players, the way the direct focus moves and to provide a structure for identifying where IT input can be beneficial (see section 5).

3. DESIGN AND NEW PRODUCT DEVELOPMENT

Many authors both in books and in learned papers have confused the design process with the innovation process: the example Wright (1998) is typical. In his book 'Design Methods in Engineering and Product Design' he illustrates ***'The Design Process'*** as including – 'The determination of customer requirements'. He also includes the model from Pugh (1990) 'Integrated Methods for Successful Product Engineering' in which the ***Design Process*** is seen to include ***Market Analysis.***

> ***Market analysis is a part of the innovation process but if designers are analysing the market what are the marketers doing?***

The confusion stems from failing to distinguish between the:

> ***Legitimate concern for, contribution to, interest in the whole process including market analysis and indeed manufacture and the direct responsibility for.***

Various stages in the process of innovation require specialist input and although decision-making at various strategic points is collective, there comes a point – and design is one of these points – when the focus is on a particular activity: only the designers do the actual designing although the specification which guides the design input can be (and usually is) contributed to by other players. The same is true of manufacture and of marketing. It is a fact – not sufficiently emphasised in design education – that a lot of work has been done and very important decisions have usually been made before design input is engaged.

WHAT will be designed is often decided without recourse to design: Design is asked to concentrate on the ***HOW*** it will be designed.

Many in the UK see the flaws in this and seek to persuade companies to access design skills much earlier in the process and to use design much more as a strategic tool as opposed to a purely functional one. There is plenty of evidence e.g. Peter Drucker in a NEDO report (1978) writes of companies

> ***'seeking to do better that which they shouldn't be doing at all'***

i.e. striving to improve manufacturability and other features of products nobody wants.

Hence the design community have to work at convincing business of the crucial importance of design but, to return to the confusion between design and innovation process, we will not do this successfully if we don't understand the nature and position of our input in the overall process of innovation. Design has an interest in – a concern for the whole process including user response etc., but its 'direct responsibility' is more circumscribed. Market analysis may precede and/or follow a strategic decision on ***what*** to design. There are some exceptions and the Sony Walkman is a case in point – 'gut-feel' (sensitivity to market possibilities) substitutes for market research – but in the vast majority of cases a rigorous study is needed to ***inform*** design. Design responds (hopefully creatively) to this and in turn ***informs*** manufacture. Project management structures determine the precise nature of the interaction between the players.

3.1 Conception to Consumption

3.1.1 Before Conception

There is an understanding proposed by some commentators that conception and consumption are the beginning and the end – the alpha and the omega! – of design. Consumption suggests a product (or service) in use and hence --depending on the product – can be a long way from the ***end*** of the design: manufacture, sale, etc., have been completed. But is not the ***end*** of the innovation process and this point is addressed in section (3.1.2)

The concept of the ***'concept'*** is clearly a well understood part of design: there is the 'concept design' stage; conceptual thinking is a recognised skill exhibited by the designer and concept designs are evaluated (preferably systematically) before one or more concepts are taken forward for development; French (1971) has written a book '.../ Design The Conceptual Stage'. But non of this is at the ***beginning*** even of the design process. Indeed the concept which follows an informed and rigorous challenge of the brief is the response to the ***concept*** – the original idea which is the ***basic prescription given*** to the designer – it is the raw material the designer has to work with. This challenge might question the whole notion of the product or maybe some aspect of it as ***prescribed*** by the brief. This is especially true if the brief is brought to the designer from a client who has not conducted any meaningful research.

The point is that very often (more often than not) a substantial effort has preceded the ***input*** to design. The conception – the idea – of ***what*** to design predates and is often decided remote from the design activity; it may even be past its sell-by date by the time it reaches design. It is crucial to a creative response from design to understand this reality

3.1.2 Beyond Consumption

Some of the issues that arise after the promotion and sale of a product: customer response; competitor activity; service reports and so on are part of the life-cycle of a product that can be extended by successive 'incremental improvements'. Indeed the vast bulk of innovative activity in the consumer products market is incremental and radical change is relatively rare. So consumption of the product can be sustained over a long period; choosing when to make a radical rather than an incremental change is a very difficult decision for businesses to make since the difference in investment between those choices can be enormous. Companies have folded because they have got this wrong. But there is a growing challenge for design both in the life-time of the product in use and after this useful life is ended – in its retirement. These requirements have, of course, always been present but environmental pressures –***the need for designs to be greener*** - is a growing pressure and it will have an influence on designs now and in the future much more so than in the past.

Life-cycle costing (LCC) has shown that it is in their use that products have their most serious environmental impacts. Hence it is we – the users – not the manufacturers who are the major polluters. But although this is an enormous challenge to the creativity/ingenuity of designers it is also, perhaps, the way in which businesses can be persuaded of the power of design.

Economy in use, longevity[†], ease of manufacture and ease of disassembly for recycling-re-use (retirement) can only be achieved by design. It is true that companies still mostly see environmental sensitivity as cost-incurring rather than cost-saving but there is a growing body of case material to demonstrate effectively that:

Greener Products are Better Products

Hoover's 'New Wave' washing machine was conceived, designed, developed and manufactured using a team approach to project management and life-cycle costing (albeit to a limited extent). It was the first to receive an eco-label for environmental performance and hence enabled Hoover to break into the German market for the first time.

Swedish Railways (SJ) adopted LCC in the procurement from suppliers of the X2000 tilting high speed train. Unlike the UK's own APT (advanced passenger train) which was a failure the X2000 is being evaluated by railways throughout the world. SJ accepted delays in the overall project time so that the commitment to LCC could be sustained. The result is a train far more reliable and up to 30% cheaper to maintain and run than its predecessors.

[†] There is a debate surrounding this issue. New products deplete resources so the longer they last the less the depletion. However more efficient technologies will emerge which the 'old' product does not exploit. Change might be environmentally beneficial.

Increasingly there is pressure for companies to accept a buy-back approach and to be responsible for the products they sell from ***'cradle to grave'***. Design has a central role to play in this process so it must project its thinking ***beyond consumption.***

4. THE DESIGN ACTIVITY

Design is often described as a

Problem-Solving Process or as

A response to a recognised need.

The need recognition – the problem identification – is often taken for granted.

But to fully comprehend design we need to recognise it as a – ***problem-processing activity***

rather than problem solving because as the design converges from concept to detail many sub-problems (lower-level problems) need first to be identified before they can be solved.

Problems not identified at the design stage identify themselves during manufacture, or worse, in use, with the attendant loss of goodwill.

If design is a problem-processing activity what do we need to know about:

- ***That which is processed – namely problems and***
  The processor – the designer. He/she may be the single most important variable in the 'equation'. A designer's personal value system can materially affect the outcome.

Neither of these points is given due attention in the literature nor in design education.

Given due attention to the foregoing design can be seen as follows:

- There is an organising strategy that is inseparable from
- an accompanying mental process that assembles, organises and transforms the information required,
- there is a range of tools that can be deployed as appropriate and the designer has to be the master of these tools including an ever-growing array of IT tools
- there are a number of constraints that must be observed and which act to evaluate and filter decisions about solutions.

These four principal components of the activity interact in an indeterminable way to assemble, evaluate and convert information inputs to very specific outputs that are a prescription for

Manufacture, sales, operation, maintenance, and retirement of the product so described.

5. APPLICATION OF CAD IN THE DESIGN PROCESS

The University of Huddersfield has invested in CAD laboratories in which Alias WaveFront, Catia, Solidworks, CosmosWorks FEA software are all run on NT based Silicon Graphics machines. This development has brought us even closer to industry standards than before. CAD has empowered students with weak visualisation skills. The technology has improved these students understanding of their designs. Traditional drawing skills, which have been essential in the evolution of form through analysis and development of sketch visuals, have not yet been completely supplanted. However, the technological alternative has already removed the drawing board from the industrial setting and certainly will change the landscape in design education. Traditional studio teaching and assessment methods will also change if more CAD-based and the student is to be given equal access to staff advice during the development phase of their design projects. Jagger and Unver (2000)

Another significant development in CAD is the trend towards collaborative engineering. Industrial CAD users can produce innovative solutions to complex problems through the medium of the Internet. In the past, each user would require access to the same CAD system and connections to a local area network, to enable an effective collaboration.

5.1 CAD Modelling Techniques and Finite Element Analysis and Rapid Prototyping

Solidworks as our main 3D Design software can be categorised as Feature-based, Parametric Solid Modeling and CATIA as 3D Hybrid Solid Modelling and ALIAS as 3D Wireframe/Surface Modeling software.

Finite Element Analysis (FEA), as related to the mechanics of solids, is the solution of a finite set of algebraic matrix equations that approximate the relationships between load and deflection for static analysis as well as velocity, acceleration and time for dynamic analysis heat transfer, fluid flow, electrical and magnetic phenomena and acoustics. We use Solidworks to create the physical data necessary for analysis by creating a mesh of elements and CosmosWork FEA to analyse, FE Solvers and Post Processing.

Rapid prototyping is also a relatively new field that is gaining speed due to its accurate creation of three-dimensional models exact to the designers' CAD models. This has already started eroding the model-making profession, but it has given the designer more power and responsibility over the model making.

While CAD software is evolving, peripherals are changing too, with Virtual Reality (VR) and input devices. Some developments may not be as successful as CAD and this has been due to the constraints of the office environment, but this (the office) itself may also be challenged, changing the work environment for those in it. Product design has always needed a three dimensional form to give the manager, designer or client an idea of how the final production unit will look, feel or operate. Traditionally the task was carried out by a model-maker, who would receive initial plans and dimensions and make the prototype or visual model to the required finish. Gradually as time has progressed so too have the tools at the craftsman's disposal improving output and quality. This has developed from simple hand tools to more advanced items such as CNC lathes and milling machines and Stereolithography.

The Internet capabilities have already been added to most of High-End 3D Design software. In the future, collaboration over the Internet will be possible with manipulation and editing of live CAD models in peer-to-peer browser sessions. As the Internet and intranets improve in reliability, it may also be possible to execute more robust engineering design functions through a combination of browser and Java technologies. Enhancements in Java performance and security will also increase the client-based capabilities, as will continued extensions to Java, HTML, VRML, etc. As STEP matures, STEP-based browsers will emerge, breaking down the barriers between proprietary CAD systems while maintaining the intelligence within the models.

6. CONCLUSIONS

It is not mere pedantry to want to distinguish between Design and Innovation. Design is part (perhaps in many instances the most important part) of the innovation process but it is not the whole thing. ***Market intelligence is crucial input to inform design but it is not its direct responsibility; just as manufacture which is, in turn, informed by design is not its direct responsibility.*** In project management based on a matrix structure the boundaries are purposely made less distinct and concurrent decisions making pursues the team approach. This allows for more balanced decision making but designers are the only ones able to do the essential designing and the other players have their direct responsibilities in the same way.

If design is part of the process that precedes its direct input it will not have to spend time challenging the brief/specification. It can devote its creative energies to interpretation of the specification knowing its relevance to market needs.

This interpretation will increasingly take account of beyond consumption in pursuit of more environmentally responsive (and responsible) designs. A more precise understanding (for some, I think, a realisation) of its role in the innovation process and recognition of its ability to produce greener designs can be the way in which design finally convinces industry/business of its real value as a strategic tool. Design education has to make sure that these issues are part of their programmes. It is clear also and indeed already widely recognised that IT will play an increasingly significant part in both Design and the overall innovation process.

Product Design over the past 15 years has seen a tremendous change in the way in which the designer works, this is largely due to the increasing role that computers and software are playing. Not only have computers become cheaper, more reliable and faster, but CAD software has taken great leaps in development in this period making sophisticated 3D graphics available to most designers.

The future of product design is set to undergo an even more radical change than that of the CAD revolution that has affected the product design field in the last ten years. The software is set to become more accurate, faster, affordable and easier to use. The past implications of the advance in technology has been wide ranging, from the loss of draughtsmen and the arguable increase in freedom of the designer, through the decreased restraint of technical drawing. Manufacturing has also been evolving generally towards faster, more accurate machines (an example is injection moulding), whilst there has been a reduction of human reliant forming processes and an increase in automation and robotic assembly.

CAD, in principle, could be applied throughout the design process, but in practice its impact on the early stages, where very imprecise representations such as sketches are used extensively, has been limited. There are some new software programs currently available which are trying to fill this niche such as ALIAS Studio Paint. It remains to be seen how effective they will be and how widely they will be implemented. The advantages of CAD modelling as a result of its links with rapid prototyping technology will eventually have an effect on the model-making workshops of most Universities. At the present time, however, the cost of RP hardware is beyond the reach of most schools of design.

Finally computers can only enhance a good concept, and in a commercial environment, it can be expected that the CAD user has already developed a sound grasp of these basic conceptual skills. Using CAD will speed up the design process, help to visualise the product etc. but will not transform a bad designer to a good one.

REFERENCES

(1) Corfield KG (1978) Product Design: A Report for NEDO, London

(2) French M (1971) Engineering Design: The Conceptual Stage, London, Heinerman

(3) Gill H (1999) Public Address – unpublished

(4) Rothwell R (1992) Successful Industrial Innovation: Critical Factors for the 1990s. R+D Management Vol.22 No.3 pp.221-239

(5) Wright (1998) Design Methods in Engineering Design and Product Design, London, McGraw Hill

(6) Clarke C (1998) Finding Your Way in the Mid-Range Market, CAD CAM Nov 98 pp. 15-19

(7) Jagger B and Unver E (2000) A Case Study on the Effects of CAD on Design Education (to be published in Design 2000 Conference Coventry University)

(8) Matthews C (1998) Solid Works in a Solid Market, CAD / CAM, Aug. 98 pp. 13-16

(9) Moon A and Jamieson R (1999) Benefits of Digital Prototyping for Small Businesses.

(10) Time Compression Technologies, Dec 99 Vol. 7, Iss. 6, pp 28-36.

(11) CATIA V5 Release 3, CAD / CAM Magazine, Jan 2000, pp 16-18.

Information from selected websites, eg. www.catia.ibm.com, www.solidworks.com, www.aw.sqi.com

Benchmarking of MCAD systems in design

A P MONGEY and **W F GAUGHRAN**
Department of Manufacturing and Operations Engineering, University of Limerick, Ireland

ABSTRACT

Remaining competitive is the key to success in today's highly paced, ever changing marketplace both in industry and education. In attempting to evaluate the latest advances in *Mechanical Computer Aided Design (MCAD)* systems, the end user is often confused by advertising hype and industry jargon in choosing an appropriate MCAD system. Acknowledging the fact that MCAD systems are becoming the industrial and educational standard for design and manufacture the task is now set to decide which modelling system best suits the needs of the design engineer. The research shows that *benchmarking* is a valuable selection tool. Benchmarking should not be solely dependent upon fact-based predictions but also on which system is most likely to emerge as the 'standard' in the marketplace. For designers, time and development factors are inseparably linked. In the transition from two-dimensional to solid modelling or parametric systems as well as in the evaluation and selection of a new system, training is a key factor. *Learning Curves* for such systems are discussed, as are benchmarking standards.

Key Words: *Mechanical Computer Aided Design (MCAD), benchmarking, Learning Curve.*

1. WHICH MODELLING SYSTEM

The underpinning choices for selecting a CAD system for mechanical design and manufacturing fall into a technology hierarchy beginning with choosing 2-D only or 3-D. However the limitations of 2-D become obvious very quickly. Traditionally 2-D drawings have been prone to errors as illustrated in figure 1, "the impossible part", also the user needs to input 2-D drawings of the intended product model, needing more than one view to describe the intended 3-D product. This means when downstream changes occur the same data needs to be re-entered for each view. With 3-D, though users only input data once. The 3-D system assumes the burden of the user in converting the product profile for each drawing view. It understands the difference between model geometry and drawing geometry. However 2-D drafting still remains critical in design, this is ensured by its inclusion within most 3-D products, " *the simple line remains a mighty tool*". (1)

Selecting an MCAD system opens a wide choice of possibilities: surface modelling, solid modelling, explicit input, parametric modelling, variational, feature-based, or a mix thereof. Engineers adding computers and software to their design/manufacturing process have never had a greater opportunity to make great advances but also to make mistakes. The wide variety of MCAD systems available today can pose the design engineer with the problem of selecting that system which best suits his/her needs.

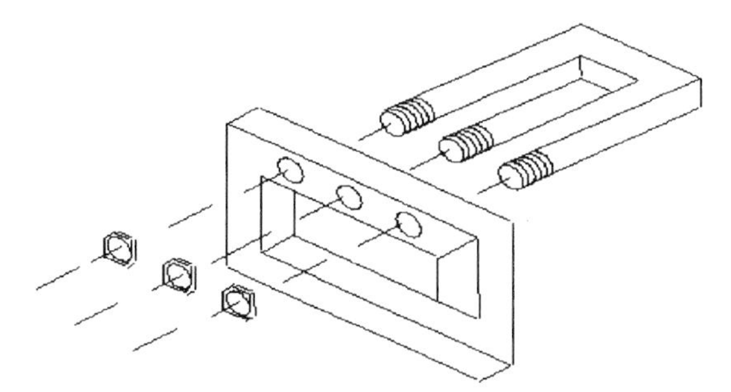

Fig. 1 '2-D' representation of impossible part

MCAD is one of the fastest growing computer aided design software market sections, hence new advances seem to occur daily with the introduction of new capabilities and improvements on existing systems, fuelled by dramatic advances in hardware technology. With these new advances come claims from vendors that their system is more advanced than others. So it is essential that one understands how the latest MCAD systems compare. This poses the question how does one compare such complex systems. Vendors often obfuscate the differences between systems; well-trained personnel that can quickly glaze over any shortcomings of their software usually carry out demonstrations.

Proper planning is critical in the selection of or transition from 2-D to 3-D modelling. In many ways planning is even more important than when design engineers gradually converted from manual to 2-D computing in the late 1960's early 70's. Many users may feel sophisticated in the use of the their current systems but with selection process for a replacement system being every 5-7 years or sometimes even longer but they are much less so. With the future of MCAD seeming to move towards data translation between separate systems, users need to select a primary system best suited to their needs.

2. STEPS IN SELECTING A MCAD SYSTEM

Before purchasing a MCAD system, a careful, well- thought-out plan for selecting a system should be developed and followed. This plan can be divided into five phases:

1. Establish the need for a MCAD system.
2. Survey and Select Objectives for the System.
3. Develop Benchmark Tests based upon Objectives
4. Benchmark selected Systems.
5. Evaluate, Select, Purchase and Install a MCAD system.

Let us now discuss each of these phases in greater detail.

2.1 Establish the need for a MCAD system

During the next few years competitive pressures will force many users to consider modern MCAD systems. The first consideration in the selection process is to determine whether a

MCAD system is needed. Evaluating what is needed to keep the edge is ongoing, or should be. A wise business continually examines its design-through-manufacturing (DTM) processes and associated tools/systems to insure optimum working to decrease costs while increasing profits.

Due to the increasing integrated nature of the manufacturing environment in which electronic information is created, the decision on which MCAD system to implement is a major strategic decision and should ultimately be made at managerial level. However all potential users of the MCAD system (Fig. 2) should be consulted regarding how, when, and where a system would be used and whether it would be cost effective in their particular operations.

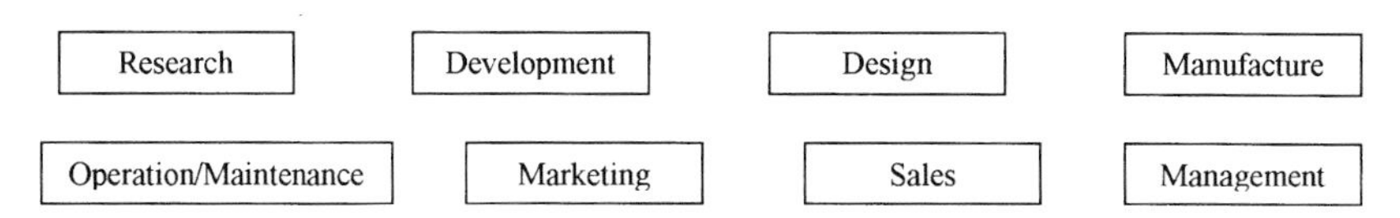

Fig. 2 Spectrum of Engineering Functions (2)

A well-chosen MCAD system facilitates successful product development by promoting a design process in which multiple departments can work concurrently to bring a product to market. Never purchase a system based on the consideration that this is the first step into the future and the company wants to project a progressive image. The process of determining the need for a MCAD system can be time consuming and frustrating. Nevertheless, speed should be sacrificed to careful deliberation in this phase.

2.2 Survey and Select Objectives for the System

All of those people outlined in figure 2 above need to evaluate the entire DTM process to determine whether or not it is concurrent with modern engineering technologies and whether processes need to be developed/improved. Once the entire process has been evaluated and a new model developed or the old model improved upon, objectives can now be set upon which the selection of a new system can be set. Much of the success in the following steps relies upon the foundation work being properly completed here; this is in fact the most important section. Users evaluating MCAD systems also need to assess the importance of objectives set within their DTM process. For instance assembly modeling capabilities will have a higher importance with companies whose products contain 500 or more components than those that contain very few.

2.2.1 Considerations to take into account during this step include

Investigate how well a system will exchange information or interface with other MCAD or CAM systems. Many systems do not have this capability. One important consideration is the CAD system's ability to exchange information with other MCAD systems and other engineering applications. One standard for such exchange is the initial graphics exchange specification (IGES). There are also a number of other common formats. Make sure that the system you select can export common file formats, particularly to other applications you are planning to use. Consider whether the system will be multipurpose. Will other office operations, such as word processing, database generation be used in conjunction with the system? Is OLE (Object Linking and Embedding) important in your DTM process.

It is generally agreed that software should be selected before hardware. However, some MCAD systems are *turnkey* system that is, a total system with software and hardware combined and inseparable. Therefore, you should examine all the features of any given system very carefully before being attracted by spectacular hardware. For example, some systems use dual monitors. The chance of being impressed by this feature may over- shadow the question of whether one really needs the two displays. So a checklist for hardware requirements must also be developed.

2.3 Develop Benchmark Tests based upon Objectives

MCAD systems demonstrate enormous complexity. Paper analysis are incomplete, too simplistic and are not useful for comparing systems in terms of design functionality, usability and performance. Such a determination can only be made by performing similar tests on different systems. These tests will simulate the needs of the user. The benchmark tests should be representative of how it is envisaged the designers/engineers will be using the system in the DTM process. The tests should be designed to represent the tasks that will be performed and the sequence in which they are to be performed.

In addition a comprehensive set of procedures must be put in place to evaluate the performance of the software, ideally these procedures must be formulated by the different personnel involved in the DTM process and also observed by these personnel. Tests will be of little use if observed by personnel who do not now what is required of the system. However although this is the ideal situation it is often impractical. Hence criteria must be set based upon the objectives. These criteria form the basis for the benchmark development and evaluation. This will give a true indication of the value of different systems. Benchmarks can be a very time consuming process and users will quickly discover benchmarks present their own set of problems. The benchmark's factor of difficulty should be such as to test for the desired results, while at the same time allowing the test to be completed in the given time.

In setting out procedures for evaluation one aspect that needs to be taken into account is the importance of different elements of the DTM process in the engineering environment. This can be overcome by applying different weightings by engineering discipline (e.g. Design, Drafting, Analysis, Manufacturing, etc.). This allows the different disciplines to assign more importance to functionality used frequently and less to functionality used less frequently. Other elements that must also be taken into account are staff training (discussed in section 3), cost and vendor support.

2.4 Benchmark selected Systems

Once objectives and criteria have been set and the benchmark designed, before carrying out the benchmark, demonstrations of various systems should be requested. It is necessary to request initial demonstrations as it is unrealistic to carry out the benchmark on all of the systems. Vendors should be told of your expectations beforehand, if the vendors are completely aware of the requirements, they will be able to give a more realistic presentation. During vendor demonstrations initial questions can be posed based upon the objectives set above. With each succeeding demonstration, the questions will be more effective. Moreover, the answers received will be more meaningful. It is a good idea to select your own project that you would like to have demonstrated. Having a project which is used consistently across the systems helps to evaluate each software package fairly

It is important at this stage that all of the collected information be organised and carefully reviewed. A list should be made of only those systems that merit further serious consideration. Once a list of MCAD systems has been determined for benchmarking a team of personnel must be selected to carry out the tests, this team should be comprised of at least one person from each section of the DTM process who will actively use the system, it is also necessary to have an experienced operator of the system carry out the actual benchmark i.e. a representative of the vendor.

2.5 Evaluate, Select, Purchase and Install a MCAD system.

For each system under consideration, evaluate the extent to which the systems complies with each objective and criteria as observed during the benchmark tests. For instance a system that cannot address any aspect of a particular requirement may get a rating of "0%" for that requirement, while a system that can address all specified aspects and even provide additional functionality relating to that requirement may get a rating of "100%". It is then possible to compare the total system scores to see which system best satisfies the requirements.

3. LEARNING CURVES

In today's market time is of the essence, development of any new product is inexorably linked to time. When changing from 2-D to Solid Modelling or simply selecting a new system, training is a key factor in the selection process. The shorter the learning cycle for a MCAD system the better. However it can often be difficult to quantify learning.

One method first developed for the Aerospace industry during World War II was the concept of Learning Curves (3), however these deal with repetitive tasks and rating operator performance.

Learning Curves can be applied two ways to computer applications:

1. The modelling of a specific component on different software packages and the time taken to model it.
2. Time taken to learn how to use a software package proficiently.

However it can be quite difficult to apply a specific curve to any particular system as the users aptitude, previous experience and ability affect the rate of learning. Therefore again similar tests with the same user on each system is one solution, however when testing a second system the user is more familiar with the benchmark hence it may be easier. Another solution would be to use two different users carry out the same tests on different systems, however this leads to the problem of previous experience on such systems.

Learning curves can be based on complex mathematical models, however the simpler the learning curve model the better (4). It is expected to show that the methods incorporated in general learning curve principles are not only limited to the rating of operator performance on specific tasks, but can also be utilised for analysis and prediction of the time spent learning particular computer applications.

4. CONCLUSIONS

By following the steps outlined above a prospective user of MCAD system software can have confidence in their evaluation process, they can be assured that the selection process is up-to-date, comprehensive and un-biased. Therefore producing valid results that best meet the companies needs and represent the ability of the target systems to perform in their environment. Purchasing system software without following the steps outlined above will probably result in the selection and implementation of a system that fails to give the maximum benefits in the DTM process.

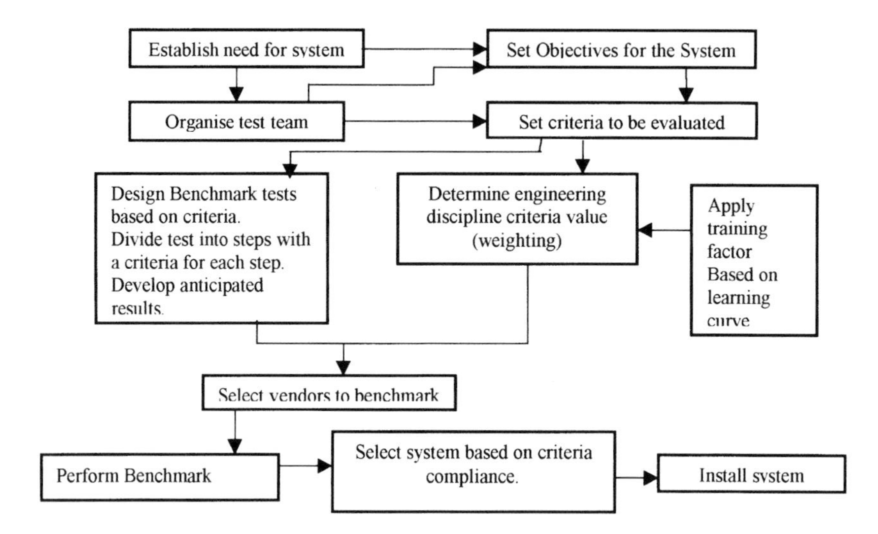

Fig. 3 Steps in the process of choosing a system

At the University of Limerick a research project is currently in progress which intends to evaluate the learning curves as well as benchmarking the functionality of several MCAD systems. It has been found that following the steps outlined in figure 3 will lead to the successful installation of MCAD system software in an engineering environment. The project is in the process of developing a sample benchmarking test suite that will be adaptable to most engineering environments, and to develop learning curve models that are applicable to MCAD system software.

The results should provide guidelines to design engineers which will be based on scientific survey and test methods, rather than anecdotal 'evidence' and vendors claims that their system is the best.

REFERENCES

(1) Versprille, K., 1999 Which CAD is right for you., Integrated Manufacturing Systems, Society of Manufacturing Engineers.

(2) Dieter, G., 2000 Engineering Design: A Materials and Processing Approach, McGraw-Hill, Singapore.

(3) Wright, T.P. "Factors Affecting the cost of Aeroplanes", Journal of Aeronautical Sciences 3, pp 122-127, 1936.

(4) Towill, D.R. "The use of Learning Curve Models for Prediction of Batch Production Performance." International Journal of Operations and Production Management. Vol 5. No.2. 1985

Estimating in engineering design

R W SIMONS and **P R N CHILDS**
School of Engineering and Information Technology, University of Sussex, UK

ABSTRACT

Estimating a task or project is second only in importance to the specification. Without a good estimate, there is no appreciation of the time that the task will take, the range and quantity of resources required, the costs involved, or a means of planning, monitoring and controlling the implementation. This paper discusses various estimating techniques that are used in a number of situations such as bidding for a contract and typical university projects, and describes a method used to impart estimating skills to inexperienced university students.

1. INTRODUCTION

Historically, estimating was associated with stop watches and time measurement and was the basis for rate fixing on the shop floor. It arose out of the desire for the workforce and employers to have a record of process times and other activities, against which wages could be set. An important by-product of the establishing of rate fixing, was the need to prepare manufacturing information in detail, allowing every piece part to be individually assessed. As a result, there was a significant reduction in the skills required in production units, as all the processes had to be clearly defined and nothing left to the discretion or expertise of the operator. One of the major professional engineering institutions in the UK (IEE) for many years regarded estimating as a 'technician' activity, not in any way qualifying for chartered status. With the increasing complexity of products and projects and the recognised need to be able to control every aspect of a task in order to be cost effective and efficient, organisations now place a very significant emphasis on the ability to estimate accurately.

Estimating is a difficult task and relies heavily on the experience of those given the responsibility. In all cases they will have insufficient information on the specification of the product or project, much of which will not be defined until the implementation is under way. Some of the information they have will be inaccurate and some will be changed as time passes. Notwithstanding all these hazards and many more, it will be from the initial estimates that a firm will bid for a contract and the price quoted will commit that organisation to the risk of profit or loss, depending significantly on the goodness of the estimate.

2. BIDDING FOR A CONTRACT

Every business organisation, irrespective of size, bids for contracts on a regular basis. Each bid made commits that company to risk which could involve severe financial penalty.

However in every case each organisation is hoping to make a useful profit and establish a reputation for the service or product being offered. An estimate is required of every feature that it is to be delivered. Firstly the specification of the delivered equipment or service has to be studied in great detail. The content of the specification has to be broken down into work packages, each of which has to be assessed for difficulty and content. The foundation for a planned structure is provided by the set of targets, associated tasks and estimates of time, manpower, materials and services required. Poor estimation of these tasks is a foundation for disaster. The greatest hazards are the complete overlooking of some things that have to be done and the underestimation of time and cost for all tasks.

A consideration in planning is to determine what would happen if something goes wrong. Some of the anticipated pitfalls and some of the unanticipated will in fact occur. Best and worst case timing and cost for each task should be estimated and from them best estimates derived. Hopefully some tasks will be completed within their estimates to compensate for the inevitable over runs. Realism and sometimes pessimism comes with age, experience and suffering, especially from those who have been associated with projects that have overrun. The recent examples of UK government software project disasters show how difficult it is to get the estimates correct, e.g. the new Air Traffic Control Centre at Swanwick where the hardware has been in place for some years and yet the software is still not finalised. An important by-product of the problems at Swanwick, which is typical of high tech projects, is that the system is technologically obsolete before it goes into operational service. Nothing is gained by ignoring new factors and slavishly following a partially irrelevant plan. Acceptance of the need to review and revise the system systematically is part of project planning.

The next stage is to assess if any new development work is required, or whether the requirement can be met from a stock or bought-in items. With this more detailed breakdown of the content, it should then be possible to estimate the effort required in man-hours or days, for each item to be developed.

There are two broad categories of cost, non-recurring costs and fixed costs. Non-recurring costs are one-off costs. They can be further divided into capital costs, which include depreciable costs such as buildings and machinery and non-depreciated costs such as land. Recurring costs, also called operating or manufacturing costs, are a direct function of the manufacturing activity. Each item machined will result in a recurring cost. Another classification system is the use of fixed costs, which are independent of the rate of production and variable costs, which change with the production rate. Fixed costs include: indirect plant (investment and overhead), management and administration, selling.

Methods for developing cost estimates can be broadly classified into three categories: methods engineering, costs by analogy, parametric analysis of past data. In methods engineering costing, the separate elements of work for a component are identified and analysed in detail and summed to produce a total cost for the part. Cost estimation by analogy involves basing the prediction of cost for a proposed design on the costs of a previous project. Allowance must be made for cost scaling due to the size and complexity of the project. For example it might be possible to base the cost estimate for the Airbus Industrie A340 four engine wide-bodied transport aircraft on the A310 two engine aircraft. This method requires a database of experience or published data. Cost estimation by analogy can normally only be used for similar products. It would not be sensible to base the cost estimate for a supersonic aircraft on the costs of the A310 aircraft due to the difference in technologies between subsonic and supersonic flight such as wingforms, powerplants and materials. In the parametric approach to cost estimation, analysis is used to establish relations between system costs and initial specifications (see Roskam (1986), Mileham (1993)). For example it might be possible to model the cost of developing a turbofan engine by $C = 0.14T^{0.74}N^{0.08}$, where C = cost (£ millions), T = thrust (N), N = number of engines produced (after Dieter (2000)). This method is particularly useful at the conceptual design phase.

The greater the technological advance involved, the larger the extent of underestimation. The well publicised figures for the escalation of forecasts of the development costs of the Concorde airliner show that these costs escalated almost linearly from £150 million in November 1962 to £1065 million in June 1973 (Twiss (1986)).

All costs in a manufacturing company are initiated in the design. It is in the manufacturing stage that the major costs are incurred. Examination of these costs by careful analysis, on the assumption that cost patterns do exist, almost invariably yields results, sometimes surprising, sometimes confirming what always has been suspected but never actually questioned.

For inexperienced or student engineers estimating is a particular problem. One method to overcome this, used at the University of Sussex, is based on the calculation of processing costs following the work of Swift and Booker (1997). Students, especially at the start of their education are often not even aware of the scope of material choices, let alone manufacturing techniques. In order to overcome this, most engineering degrees include courses in materials and manufacture. The work of Swift and Booker, if used in the educational context allows these areas to be pulled together with the added benefit of cost evaluation at the concept and detailed design phases. The process capability of each manufacturing technique can be described to students in summary form considering the physical process, material capability, economic factors, typical applications, design aspects and quality issues. This information can then be utilised in the estimation of the cost of manufacture of a proposed concept using:

$$C_m = VC_v + RP \tag{1}$$

where C_m = cost of manufacture, V = volume of material required, C_v = cost of material per unit volume, R = relative cost coefficient taking into account complexity of shape etc., P = basic processing cost for an 'ideal' design of component.

A given company will likely have its own data for the costs of materials and particular machining processes. In the absence of this information, Swift and Booker present an estimate of basic processing costs against quantity based on an ideal design along with material costs. The factor R in Equation 1 can be modified to account for complexity and other factors where a design departs from a simplistic 'ideal'. Equation 1 accompanied by a knowledge of processing capability therefore empowers a student to provide a cost estimate for a number of competing concepts. Based on the comparison a quantitative decision on cost can be taken as to which design is favourable or an estimate made for a contract bid.

It is here that the matter of contingencies arises. The estimates should include a percentage to account for future cost changes during the period of manufacture, installation, or construction. There are also other contingencies, which relate to errors in design and manufacture of scrap, employee absenteeism, lack of resources and delay in supply of materials. Experience will give an indication of how large a percentage should be added. With a significant development content it may be as high as 30%. Possible changes in the value of currency during the period of the business also needs consideration in the final bid.

With every contract bid there are commercial costs to be considered. The cost of negotiating the deal, travelling to meet with the client and also the overheads of running the organisation that need to be recovered. All require estimating in terms of the amount of effort required for the particular tasks. An estimate of the delivery date, which takes account of the client's expectations, is essential. Having collected all the data and consolidated the estimates into a business plan, the management must then decide on the goodness of the estimates and whether the business is viable. In addition the decision whether to submit a bid or not and the actual price offered for consideration, must take into account many factors. Intelligence obtained on the client's budget for the project, on the technical compliance of the bid, and the knowledge of the ability of competitors to meet the present requirement. This will be based on the systems offered by them to other clients where they have been successful. The decision does not alter the estimates but allows the amount of profit to be adjusted.

Control is not possible without a project or task being broken down into work packages and the estimates of cost and time, which relate to each package. On the assumption that it is viable, the implementation should be controlled using either bar charts or some form of critical path analysis. For each activity the earliest and latest time should be established. This will give the limits on the final delivery date, which must of course conform to the contractual obligation. If this is not so, some rearrangement of the logic has to be made, or extra effort or resources brought in, to endure that the delivery date is met. An analysis of 475 small British projects showed that these projects had durations ranging between 1.39 to 3.04 times the estimated time, these errors being greater than the range of cost estimates of between 0.97 and 1.51. For large projects, cost escalation is more significant than slippage in the development time. In some cases 'crash' action can save a slipping programme. However beyond a certain point slippage in time results in substantial extra costs and loss of reputation and possibly a claim for damages by the client.

All these problems make bidding and implementing projects a risky activity, but many organisations make a great success of their businesses and show considerable profits, but there are unfortunately many that do not succeed and their failure can in some cases be blamed on their inability to properly estimate costs.

3. ESTIMATES APPLIED TO UNIVERSITY PROJECTS.

It is a regular feature in education, particularly in university engineering departments that projects are almost invariably set with fixed end dates. Students should be encouraged to plan their programme of work in order to ensure delivery on time. It is not acceptable to suggest that the time taking to plan is a waste of time. As already seen, a plan even if quite simple, is essential to monitoring progress and will draw the attention of the student to any potential slippage and avoid many sleepless nights as the final date approaches.

A typical project will have a number of benchmarks or milestones, where submissions have to be made for assessment at intermediate points. It is a very useful discipline for the student to plan and estimate the amount of time and therefore the effort that has to be put in at each stage. Planning should not be considered as just another exercise to be done initially and not looked at again. Planning and estimating the times for each part of a project encourages the student to manage the work and distinguish between those tasks, which are urgent, and those, which are important for the future. Each milestone is a goal and these goals can be broken down into simple activities that can be done week-by-week, to make progress towards the goal. Over a typical week it is important to compare what has been done, with what was planned and to work out the reasons for any mismatch. Typical causes might be interruptions due to too much time spent answering the telephone, or getting sucked into peripheral tasks, or spending too much talking to others and allowing them to deflect the student from the line of though required. An analysis of a student's chart will allow things to be rescheduled, to do the unpleasant jobs as quickly as possible and separate thinking from doing.

An example of a typical question set to develop the student's knowledge of estimating, monitoring and control of projects, originally set at Manchester University in 1985 for the Management Module of the B.Sc. Electrical and Electronics course, is outlined in the table below.

The design must be completed before any other activity can start and all other activities must be completed before the testing can start. It may be assumed that each subassembly may proceed independently of the others, unless otherwise indicated.

Students are asked to draw a precedence diagram (see Figure 1) to indicate: the earliest completion date (answer 29 days), the critical path (answer ABEFGN), the float of activity H (answer 6 days). In addition they are asked if activity I were to be delayed by 8 days, what would be the effect on the earliest completion date (answer 2 days delay).

Table 1 Estimation of task durations.

	Activity	Days	Comments
Design	A	10	
Subassembly 1	B	5	
	C	3	
Subassembly 2	D	6	
	E	2	Cannot start until 2 days after B completed
	F	1	Cannot start until H completed
	G	4	Cannot start until C and I completed
Subassembly 3	H	2	
	I	2	
	J	3	
Subassembly 4	K	5	
	L	7	
	M	1	Cannot start until I completed
Test	N	5	

A bar chart is useful for simple tasks but as soon as there is a change of plan or a delay occurring after the chart has been prepared a major redrawing is required. With the Critical Path diagram, the effect of changes can be quickly determined and with complex diagrams analysis can be achieved with computer programs. The CPM allows precedence to be seen clearly by the introduction of 'Dummy' activities with zero time e.g. F not being permitted to start until H is complete.

In controlling a project using a chart or networks it is important to recognise how they can be adapted to various situations. Often the simplest conclusion arising from a delay in an activity, which would otherwise cause an overall delay, is to suggest that extra staff would be able to overcome the problem. This is not often a realistic solution. The better way is to examine the logic of the diagram and to re-arrange the various activities so that the items on the original critical path are no longer the determining factors. With complex projects where the number of activities runs into many hundreds, the use of Critical Path Analysis is the only viable way of monitoring and controlling the work, checking on the continuing validity of time and cost estimates and allowing revisions to be made in the easiest possible manner.

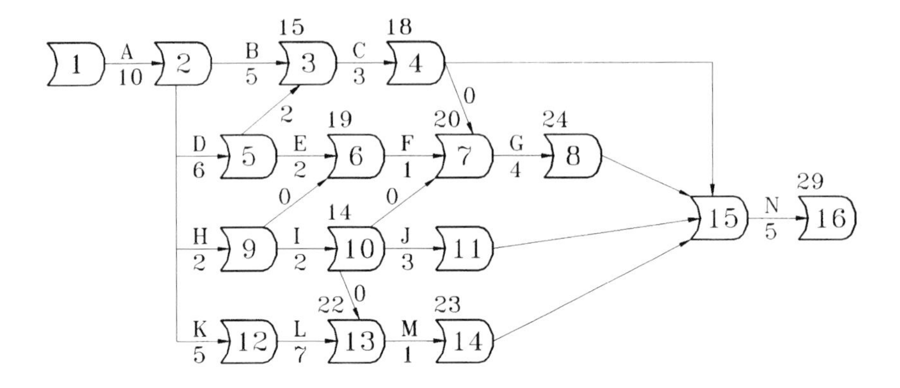

Fig. 1 Precedents Diagram

4. CONCLUSIONS

Without detailed planning and the associated estimates of time and cost, any project is effectively out of control as there are no reference points against which it can be monitored. Every student should be given sufficient guidance and encouragement to appreciate the value of an organised approach to all activities whether it is a school or university project, an investigation, or an experiment, or in a business environment. The approach described here provides a basis for estimation and can be used to develop core skills prior to the use of advanced database-software methods in agile engineering approaches.

REFERENCES

(1) Dieter, G.E. Engineering design, 3rd Edition. McGraw Hill, 2000.

(2) Mileham, A.R., Currie, G.C., Miles, A.W., Bradford, D.T. Journal of Engineering Design, Vol. 4, pp. 117-125, 1993.

(3) Payne, A.C., Chelsom, J.V., Reavill, L.R. Management for engineers. Wiley, 1996.

(4) Roskam, J. Rapid sizing method for airplanes. J Aircraft, Vol. 23, pp. 554-560, 1986.

(5) Swift, K.G., Booker, J.D. Process selection. Arnold, 1997.

(6) Twiss, B.C. Managing Technological Innovation. Longman, 1986.

Machine design to accommodate paraplegic operators

G T POWER and **W F GAUGHRAN**
Department of Manufacturing and Operations Engineering, University of Limerick, Ireland

ABSTRACT

It is generally accepted that society has a responsibility to allow all of its members enjoy a reasonable standard of living, and to lead their lives with self-respect and dignity as active members of that society. This dignity can be attained through a positive input into their chosen profession. However so many barriers stand between the paraplegic operator in an industrial/educational setting that this self-respect and dignity are merely aspirations. By acknowledging this fact one may endeavour to adapt and modify certain manufacturing tools, thus providing paraplegic operators with an opportunity to become equals in the industrial and educational arenas. As this modification is basically a design problem the investigation into the machines' functionality provides an insight into the possible solutions and their implementation, this coupled with the design analysis provides many possibilities into the future of machine design with respect to paraplegic operators.

Key Words: *paraplegia, manufacture, education, and machine design.*

1. INTRODUCTION

"There can be no joy of life without the joy of work"
Thomas Aquinas

For many paraplegic employees the technical sectors are becoming even more distant to them, even though as humans we are at our most technically advanced. For this reason it is apparent that we should be utilising state-of-the-art technology to assist paraplegic operators to enter and survive in the technical workplace. Work provides us with personal satisfaction and helps us to develop our social skills by interacting and co-operating with others. A person's work is undoubtedly one of the most formative influences upon one's character and has a significant role in the formation of our personality. Work is crucial to our lives and should not be denied to anyone, physical disability or not. The question which needs to be answered is how can paraplegics be integrated into the technical workforce to provide meaningful labour? This fundamental question can be answered through the area of ergonomics the study of human work and the design of various working conditions. Then by applying this knowledge to practical situations possible solutions may be uncovered. These solutions can then be assessed and viable possibilities are then explored. Through this paper it is intended that the importance of the initial stages of the design process be examined and the specific problems that face wheelchair users be looked at and discussed.

2. ERGONOMIC EVALUATION

Ergonomics draws from a wide range of specialised areas such as industry, medicine, occupational physiology etc. to gain an understanding of the rules that govern human work. Of course when the needs of a paraplegic wheelchair user are taken into account this becomes an even more complex problem for the workplace designer. It is accepted that all industrial machinery is designed for able-bodied operators and when a paraplegic operator is faced with these workstations obvious problems arise. Ergonomics examines, collects and classifies information and uses this information to formulate rules regarding the design of human work. However in the case of the pioneering paraplegic operator, as there is no previous standard to achieve or no fountain of information to draw upon. This makes the problem all the more complex as now the workplace designer is faced with a problem that wasn't seen to exist in the past which meant that all machinery was designed for one specific group of people, the able bodied. Now the designer must scientifically evaluate the problem through research and practice in order to achieve a suitable and worthwhile design.

2.1 Ergonomics of the disabled

Ergonomics in general is an extremely diverse topic this is due to the fact that as humans the only thing that we truly have in common is our individuality! This makes us varied and interesting, however this only complicates the job of the workplace designer, who would prefer a standard human specimen to apply to each workplace. Obviously this will never be the case and is even truer when the paraplegic operator is taken into consideration. Ergonomic work design needs to take variables like height, weight, strength, reach etc. into account, add the problems associated with wheelchair accessibility and the jigsaw puzzle may begin. When designing a workplace for a paraplegic operator there are three duties that must be carried out from an ergonomics perspective:

2.1.1 Workplace selection

This involves the correct selection of a suitable workplace. This is simply the selection of a place where the operator may carry out the specific task without discomfort. The decision on the correct workplace should be made by assessing the abilities of the operator and the demands of the task required, and then deciding whether the disability could be overcome by compensating it with ergonomic workplace design.

2.1.2 Workplace design

In designing this workplace to accommodate the operator it is necessary to take full consideration of the limitations that hinder the operator's functionality. This area is where the ergonomic studies will bear fruition as the new found knowledge will provide the workplace designer with the required know-how to prepare a suitable workplace that will accommodate the operators disabilities. This workplace should now be so suitable that the operator could conceivably develop his /her skills further and therefore achieve career development, which at the beginning of the study would have seemed unattainable.

2.1.3 Correct training

Adequate training of the paraplegic operator should result in at least average operational performance. This performance provides the operator with the dignity and integration into society that was the original goal, providing equality and dignity for the operator.

2.2 Work design for the disabled

The next aim of work design for disabled people must be to ensure that disabled people can carry out their occupation in their original workplace or be provided with a similar adapted workplace. Therefore the worker who was injured through an accident or illness should ideally be able to use all of his/her previously acquired skills and knowledge. Therefore work design will not only improve job satisfaction motivation and performance but also improve the workers self respect. This work design should strive to compensate for any disability by means of simple adaptation of the work, technical aids, the disabled person themselves and finally of the working environment. The adaptation of these four critical aspects would help to assist the operator realise that their "reduced functionality" can be overcome with the required adaptation.

2.2.1 Adaptation of the work

This is simply making the work being carried out more ergonomically friendly, for example footrests and armrests reduce the weight being carried by the body therefore allowing the operator to work for sustained periods. Also by providing tools within easy reach of the operator means that the worker is not straining to reach various implements, dials, handles etc. This would be an improvement for any worker disabled or not, but for the disabled operator it is one less hurdle for them to overcome and a very welcome modification.

2.2.2 Adaptation through use of technical aids

This method provides the disabled worker with various assitive technologies, which would be utilised to aid the worker in the completion of various tasks. The area of remote control units could be explored to enable wheelchair users to gain access to various parts of machinery therefore allowing them to gain complete control over a machine from a seated position, albeit at a greater distance away than a non-disabled operator. The disabled operator could still carry out the tasks and quite possibly at a greater degree of accuracy as many electronically operated machines operate at much higher levels of accuracy than manually operated machinery, due to the available technologies in this area i.e. limits, encoders and various other feedback units.

2.2.3 Adaptation of the disabled person

The disabled person must also strive to adapt to the new work area. As for many paraplegic operators they may have been trained initially when they were able bodied and would have therefore utilised "normal" work areas. This would then be a challenge to adapt to their new surroundings and to become proficient in their work.

2.2.4 Adaptation of the working environment
The organisation of the working environment is also of paramount importance, as disabled people have different needs to those who are not, so management need to recognise this and allow for certain alterations in the working conditions. It is when these measures have been carried out and no improvement has been made that other options must be examined.

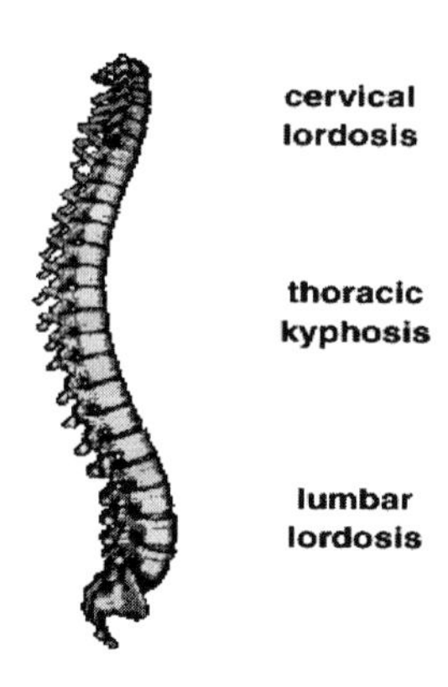

FIGURE 1

3. STUDY OF THE SEATED POSTURE

As wheelchair users spend approximately two thirds of their day seated, this is obviously of paramount importance and should be understood fully before attempting to design any device that would be utilised in a working environment. The idea that as far as the human back is concerned sitting is a high-risk activity is not a new one. However the harmful effects of sitting incorrectly are concealed because the damage is cumulative and builds up over time. And as the spine was not designed for long periods in this position it is possible to cause long-term spinal damage due to the lack of spinal movement. This lack of movement causes back muscles and ligaments to degenerate and also the backs' shock absorbers the inter-vertebral discs to degenerate as they are fed their required nutrients by a process known as osmosis which requires a fluctuation in pressure brought about by the movements of the vertebrae. However according to C Campbell "Suffering office back-ache" (1997) it is of extreme importance that any pressure exerted on the spine is done so in a manner that maintains the three natural curves of the spine: the cervical lordosis, thoracic kyphosis and the lumbar lordosis. If this natural curve is maintained, the load, which is exerted upon the discs of the spine changes drastically as can be seen from figure 2.

FIGURE 2

However controversy does occur when overall posture is discussed. A.C.Mandel discussed the idea of a person in the seated position, with their feet flat on the floor, their lower legs vertical, thighs parallel to the ground and their back upright (albeit maintaining the three natural curves) and then dismissed this posture as rubbish and a "man-made aberration". Mandel also puts forward the idea of supporting the lumbar lordosis by increasing the thigh-torso angle to approximately 110°. This would cause a problem as regards design of a device intended to be utilised in a industrial setting as this angle would cause the operator to lean back and away from the controls of the machine thus decreasing the reach of the operator. Mandel also dismisses the idea that lumbar supports solve a multitude of problems as they can be seriously misused as people can actually adopt postures that reverse the natural curve of the lumbar lordosis. However Mandel does accept that lumbar supports are generally worthwhile once they are used correctly.

4. PRODUCT DESIGN

There are many models of correct product design and depending on the source, it can take from five to twenty five steps in order to achieve a workable design. From an educational perspective it is important that no matter how many stages are employed to carry out your design, each stage is given enough attention. This is an extremely important concept to recognise. If one takes the problem definition stage into account it can be seen that the initial problem may not always be apparent at the first glance. Due to the fact that this stage may take up only a small proportion of the entire design cycle this extremely important step is often overlooked, or not given enough attention. As can be seen in figure 3 the final solution to a design problem can differ greatly depending upon the initial definition of the problem. It is apparent that the final design depends greatly upon the individual who actually defines the problem. This study ideally requires a paraplegic operator to assess and define the initial problem as by doing so the most suitable design may be produced.

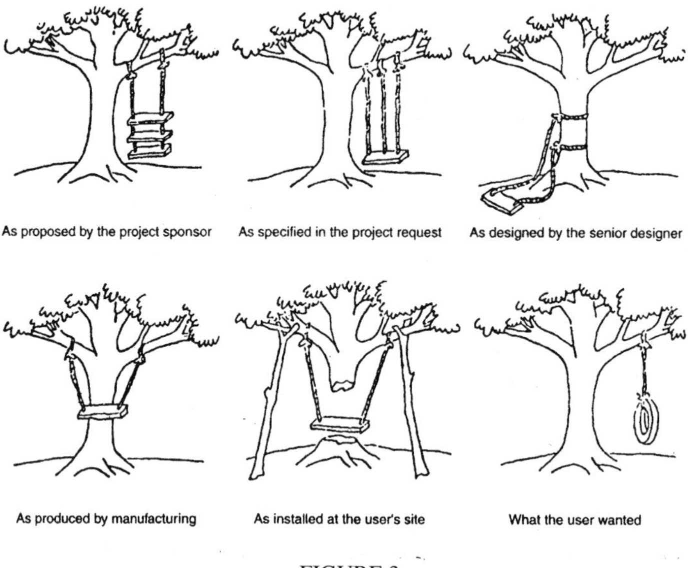

FIGURE 3

5. STUDY OF MACHINE ADAPTABILITY

After much deliberation the study then progressed to the area where it is at the moment, that being the adaptation of the machine to suit any paraplegic operator and not just one individual case. This scenario would determine whether the project was viable or not as if a solution could be found any paraplegic operator could conceivably operate whatever machine was chosen for adaptation.

5.1 Choice of Machine

The first choice encountered was what type of machine would be appropriate? This decision was made under the heading of use, functionality and the ease of which it could be adapted. The decision was made to try to adapt a vertical milling machine, as vertical milling machines are an extremely popular machine with a variety of uses and do not require the operator to be extremely mobile. However the main obstacle to the paraplegic operator is the height of such machines in some cases the tooling is at an approximate height of 1200mm above the floor, this has obvious implications for a wheelchair user. So to combat this problem alternative vertical milling machines were investigated with the Emco FB-2 proving to be the most easily adaptable as it could be positioned on any workbench regardless of height and most of the controls are within reach of the operator. This became the initial adaptation and allowed many of the controls to be quite easily accessed by the paraplegic operator.

5.2 Further Alterations to the Machine

The main problem faced with this type of machine was the accessibility of the hand wheels that are used to traverse the tables in the X, Y and Z directions. This problem is highlighted in fig 4 demonstrating an operator standing at a Bridgeport milling machine and operating it at a height far out of reach of a seated paraplegic operator.

FIGURE 4

To help the paraplegic operator gain control of these table feeds D.C. motors were attached to the leadscrews of each axis and this allowed the operator to control each axis by means of a pushbutton mounted on a remote control panel. The feed-rate of each axis was also controlled by potentiometers thereby ensuring that the machine was as easily controlled by motors as by the human hand. However the removal of the hand wheels meant that the operator had no control over the distance traveled by each axis. This was a fundamental problem and one, which needed to be solved in an acceptable way i.e. not losing any accuracy. To achieve this three linear encoders (one to each axis) were attached to each axis, these gave the operator the required feedback and allowed for machining with an accuracy of up to 5 thousandths of a millimeter. Other adaptations to the machine include attaching tool holders and other devices that made the operation of the machine that little bit easier by storing all the necessary implements in an accessible manner.

6. CONCLUSIONS

As this study is ongoing there are many areas that need to be examined including the implications in industrial and educational settings. As regards industrial use, this type of design problem can only evolve into a viable product if there is adequate research into employment rates of paraplegics and the disabled in general. By verifying the huge shortfalls in this sector industry in general may be faced with no option but to conduct research and development of the existing machines, more importantly with the disabled in mind. Due to the fact that most industrial milling machines need to accommodate much larger tooling and work-pieces than the prototype could accommodate this research could begin by tackling the larger vertical milling machines and assessing ways to adapt or design new machines for this purpose. From the educational perspective this type of machinery would suffice in the training of paraplegic operators however if the machine was a dual purpose machine (operable by means of powered lead-screw or hand-wheels) then it would also break down a number of barriers between the able bodied and the disabled as it would not be a "special" machine but a machine that was designed with all operators in mind.

Another option for educational institutions is the area of virtual training, whereby operators learn about the functions of machinery by utilising software programs to gain their initial experience. This software currently under research at the University of Limerick would help the operator gain valuable knowledge before ever having to operate a machine.

As this design is simply a prototype and is in its very early stages there are many areas that require fine-tuning in order for it to be applicable in both the industrial and the educational arenas much work needs to be completed and analysed before any concrete progress will be made. However every journey begins with one step and hopefully this step will be the first in a productive and worthwhile journey.

REFERENCES

(1) Dieter, G., 2000 "*Engineering Design: A Materials and Processing Approach*", McGraw-Hill, Singapore.
(2) Campbell C., 1997 "*System S An Engineering Solution to a Medical Problem*",
(3) Campbell C., 1998 "*Suffering Office Backache*"
(4) Mandel A.C., 1990 "*The Seated Man*"
(5) Carver *et al.*, 1978 "*Disability and the environment*". Elek Books.
(6) Chigier, E., 1987 "*Design for Disabled People*". Freund Publishers.

The design-education continuum – a moving field

M EASON
School of Engineering and the Built Environment, University of Wolverhampton, UK

ABSTRACT

Design and Education are both fluid mediums that respond to a number of outside influences and are therefore always developing and changing.
In many cases education through the age ranges has been a stepwise function.
The question arises as to whether a continuity can be found in a particular subject area across these changes.
The paper examines the experience of students and staff on university based courses to ascertain the continuity of design education.
From this analysis we may be able to predict if design will become a continuum and what major influences the future may hold for tomorrow's developing designers.

1. PREAMBLE

At junior school I envied those who made models in the slow learners class, though I did get to paint the fish tank and the goal posts. At secondary I was allowed the bliss of woodwork and metalwork so breaking the confined limits of my fathers garage. At O' level for the first time technical drawing was allowed. By A' level it was the maths, physics, chemistry trio, but not O' level metalwork on the side "not an academic subject". After a materials degree I entered industry, and was lucky enough to do a 3 month FE based course including design, and was thus able to draw parts for the machine tool test rigs of my post graduate research.
My first work based design task in the development department was a headstock test rig for a CNC lathe.
Terror! ask a senior designer.." Oh I don't do test rigs" and thus I became a designer...

2. A BRIEF HISTORY OF TECHNICAL EDUCATION

The Samuelson Commission (1882-84) (1) recommended the introduction of workshop skills into schools, as it was concerned with the decline in Britain's industrial performance. The report was intended to be the basis of a kind of apprenticeship, but soon acquired educational objectives:
"The objective of the instruction is not to create carpenters and joiners, but to familiarise the pupils with the properties of such common substances as wood and iron, *to teach hand and eye to work in unison, to accustom the pupil to exact measurements, and to enable him by the use of tools to produce actual things from drawings that represent them."* (1)
Wood work, cheaper and easier to mount predominated, though there were recurring attempts to put metalwork on an equal footing. Technical drawing was made a separate subject in 1898, so that by the end of the century, a pattern had been established which survived more than seventy years.

English education was steeped in a classical tradition that saw intelligence as distinct from the lower faculties used in practical activities, and something to be cultivated through the pure and abstract. Craft skills never enjoyed high status. In fact, manual instruction, at the time it was introduced in schools, was mainly associated with work houses and prisons.
In schools, practical work was regarded as being mainly for the less able:
" For boys, who are dull in all 'brain' work, and who's only hope is in mechanical work-writing, drawing, colouring, measuring, in which alone they can be profitably instructed...woodwork would be a delight and a real benefit." (1)
The 1944 Education Act which should have given practical education a boost by establishing technical schools for those with the talent, was never fully implemented. Technical schools at their height only took four per cent of the age group and workshop teaching was mainly relegated to the secondary moderns, the schools for the 11+ failures.
Craft therefore tended to be the poor relation of English education and those connected with it, particularly the teachers, were always striving to make it something more.

If we look at a typical text book of the time (2) we can see what, if any change is seen to have occurred in expectations of the student.
"The successful teaching of engineering drawing........Let us consider at the outset what our aims are in this respect; they are both educational and vocational, and the best results are obtained when both points of view are kept in mind throughout the course. In the course of his training in engineering drawing the student should gain and develop:

A clear conception and appreciation of form, proportion, and purpose.
Speed and accuracy in the use of pencil and drawing instruments.
The ability to think in three dimensions.
The power of expressing his ideas in construction work quickly and clearly by freehand sketches.
Competency in original design."
".....and that the best way to gain their interest, without which no teaching is effective, is by constantly linking up work in class with their experience of or interest in engineering"
“Finally, since our aim is to produce competent designers in engineering work, the elements of design should be introduced as soon as opportunity arises, and encouragement given to initiative and originality as soon as the preliminary stages have been thoroughly completed.”

We can now see that wider skills and competency were being recognised with expression of ideas, originality and elements of design being considered.
When the breakthrough came in recognition of the importance of education in the areas discussed, it was by association, on the one hand with 'design', as indicated in the extract above, and on the other with technology, as indicated in the following.

"Design is the essential purpose of Engineering. It begins with the recognition of a need and the conception of an idea to meet this need. It proceeds with the definition of the problem, continues through a program of directed research and development, and leads to the construction and evaluation of a prototype. It concludes with the effective multiplication and distribution of a product or system so that the original need may be met wherever it exists.
What distinguishes the objects of engineering design from those of other design activities is the extent to which technological factors must contribute to their achievement." (3)

The book goes onto consider design by evolution and technological change, which in the earlier days had limited the development of design education, as opposed to design by innovation which can produce major change quickly.

The elevation of craft into technology was mainly associated with "Project Technology" launched by the Schools Council in the wake of Harold Wilson's, "white heat of the technology revolution speech."
Harrison, then head of the craft department at Loughborough College, appointed to run the project, successfully fended off the science lobby led by the Association for Science Education and Council of Engineering Institutions (4) and managed to establish craft as the route to Technology in schools. It is not clear how many craft departments took to teaching technology, rather less than 5% it is suggested, (5) but from Project Technology came, in 1970, the National Centre for School Technology at Trent Polytechnic that was: "soon pouring out new material...and developing a highly effective public relations operation which kept the teaching of technology high on the political agenda."
Yet design in many cases at school was still seen as technical drawing.
"This book aims to cover the whole range of subject matter relevant to GCE O-level and CSE examination syllabuses in technical drawing." (6)

"Design... to draw the preliminary out-line of, sketch for a model...." (7) a modern definition would be quite different, and in terms of an individuals concept dependant on background. I am aware in this paper that I am coming from a particular point in my discussions on design.

Perhaps consumerism and creativity of the sixties along with work being done by the Design Council widened the growing perception of design as identifying needs, thinking creatively and communicating ideas in a broader sense.
John Eggleston at Keele University was trying to turn Handicraft into a new and exciting subject. He seized this new thinking on design and said; (8) "Problem solving strategies became the order of the day. Teachers acquired new vocabulary. Design methodologies using analytical and synthetical criteria moved logically from need identification to optimised solutions and their evaluation. Ways of extending the material boundaries of wood and metal were explored.
It was around this period (1976) that James Callaghan commented on the school curriculum particularly on subjects that were "not ultimately useful"

Craft studies now began to encompass a wide variety of other subjects, including design, workshop technology and technical subjects. The term Craft, Design and Technology was applied by the Department of Education and Science to bring some sort of order in, for example, its document The School Curriculum (9) and so a new subject was born. From the mid 1980s CDT began to take over from woodwork, metalwork and technical drawing as the 16+ examination.

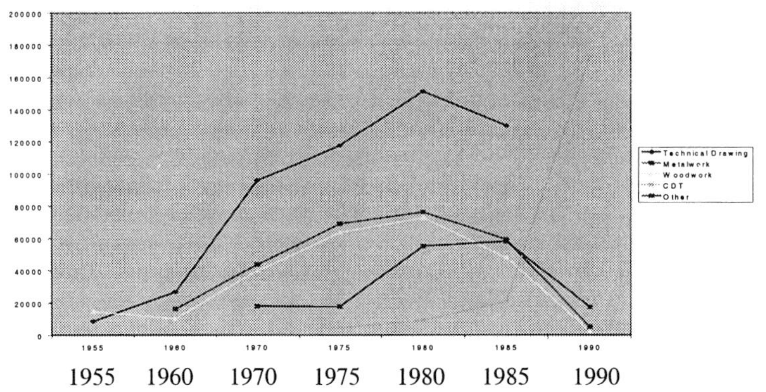

Fig. 1 Number of entrants to technical subjects at school by Year (GCE GCSE)

The trend for woodwork, and metal work are almost identical peaking in 1980 then falling away, with the 'other' combined technology areas peaking in 1985 and again falling away to 1990. Technical drawing was beginning to tail off from 1980 to 85, whereas CDT starting in 1975 had a very rapid rise from 1985 onwards

This was the period of the Red Books between 1977-83 which laid down areas of experience, but it was not until 1988 that the education act set out 10 subjects by law for the first time, until then technology was not classed as a National Curriculum subject.

The working party of 1988 was asked to view technology as; *"that area of the curriculum in which pupils design and make useful objects or systems, thus developing their ability to solve practical problems."* . Information Technology was also included: " Technological education should equip pupils with basic IT skills and develop awareness of the potential use of IT and computer technology whether in the business office or manufacturing or commerce.

So where have the changes brought us to? If we compare the text above with that on the first page the change is not immense.

"In 1950 our economic and social structures were dominated by mass industrial workplaces. There were clear distinctions and demarcations between occupations. Certification was designed for and provided to only a small elite.

All these certificates (other than those of the RSA) were based entirely on written examinations designed only to enable universities to select entrants. The examinations were norm- referenced - that is, performance was assessed in relation to the other candidates. At that time socio-economic structures had much in common with those of 20 years earlier. It was therefore legitimate to compare the qualifications of 1950 with those of 1930. It is illegitimate for our media and politicians to compare the situation now with the situation when they were at school in 1950 because the societies of 1997 and 1950 are worlds apart. In 1997 the mass industrial workplaces have largely gone. More diverse skills are needed than in the past, the key skills of: communication, application of number, use of IT, team working, problem solving, improving one's own learning." (10)

These are areas deeply entrenched in the philosophy of design.

Entries	1952	1995
GCE O' level	829,710	5,419,742 GCSE
GCE A' level	107,676	730,359
RSA component	130,132	1,470,890
First degrees	26,490	273,032

(The increase in Art and Design related first degree students over the past 30 years has risen from 4,500 to 72,000.) (11)

Foundation Target 1	1990	1996	2000
% of 19 year olds with 5 GCSEs at grade C or above or GNVQ 2,NVQ 2.	52	69	85
Foundation Target 2			
% of 21 year olds with 2 A' levels an Advanced GNVQ or an NVQ 3.	31	45	60

So how then is the relevance of increased education perceived and what is the role of design? "In the emerging global economy, human creativity is fast becoming the single most important arbiter of national prosperity and security.
The distinctions between arts, technology and science - which have become ever more rigid in the two hundred years since the industrial revolution - are just now beginning to dissolve"(12)

The number of pupils taking a range of subjects such as woodwork, metalwork and technical drawing shows a build up from 1955 then peaking around 1980 falling sharply to 1990. Whereas Craft Design Technology CDT, and derivatives, rapidly increased from 1985.
In our analysis we would therefore expect to see students below 25 years to have CDT backgrounds and 25 plus to mirror the previous range of subjects.

3. RESEARCH

The preliminary stages of the research have involved interviewing students who were predominantly on a Computer Aided Product Design course (85%) with a smaller percentage on Computer Aided Engineering Design (15%). Staff teaching in these areas were also provided with a similar questionnaire. The areas under question were the type of educational establishment and type of qualification studied. Design specific courses were then highlighted (Q1) and their contribution to the degree level course of study was sought (Q2) along with a question on whether design education had been seamless from early school onwards (Q4). A further question asked what other subjects would have helped with the present degree study (Q3) and finally whether the present course would interface into the role of a designer in the work environment. (Q5)
The research is ongoing at present to provide a fuller picture.

4. RESULTS

The percentages for post 16 education establishment and qualification type compared to age are as follows.

Age.	18+	21+	30+
FE	26	50	71
6th Form	74	50	29
A' level	77	55	17
GNVQ	17	21	8
BTEC	6	24	
HND,HNC			75

The data for the question related to the seamlessness of design education Q4. and the value of the present course to your future as a designer Q5. is as follows.

Q4.	Yes	77	63	37
Q5.	Yes	88	87	100

The questions related to which design related subjects had been studied at school Q1. which previously studied subjects assisted you in your present course Q2. and what subject would you have wished to study at school to help with your course Q3, produced the following.

	Q1	Q2	Q3
Design & Technology	39	33	8
Art and Design	15	12	
GNVQ . & Manuf.	13	16	
Design Communications	8		6
Design	8	12	
Art	6		12
CDT	6	2	2
Metalwork	1		
Computing	1		14
CAD		12	39
Wkshop/Manufacturing		14	18
Physics, Maths		9	

5. ANALYSIS

As we would have expected the majority of 18+ students have entered through the 6th. form A' level route whereas the 21+ are evenly balanced with the FE route.

The older students and staff are far more likely to have come through an FE route with certificates and diplomas.

In terms of the seamlessness of design education the older respondents tended to see little structured coherence to the extent that the 55+ stated that practical work and design were frowned upon at school (1955). The 40+ range (1970) tended to have undertaken woodwork, metalwork and technical drawing but generally not to a certificated level, 37%.
As we come to the 21-25 age range starting secondary in 1986-90 we see an increase in the agreement with the seamlessnes of design 63%. The 18-20 age range starting secondary 1991-1993 showed a greater agreement that design was present throughout.77%.
Most positive comments about seamlessness indicated the secondary school level onwards but few respondents commented on design at the infant or junior levels.
Other comments tended to indicate that some design work at the GCSE level was not particularly appropriate but this may reflect on resource issues and enthusiasm at the age.
The figures would tend to reflect that the introduction of technology into the curriculum brought more people into contact with design.

As to the present course of study there was agreement in the major student groupings at 88% considering the course would lead on to design in the working environment. The 12% comments tended to be specific to seeing a need for greater engineering input or graphics input.

In terms of the subjects studied at school level Design Technology, Art and Design, GNVQs and Design courses predominate. This would be expected in that students were asked to indicate design related courses taken. In virtually every case design had been studied at school and this had shown to be useful in their degree studies.
The previous study of CAD, workshop/Manufacturing and Maths/ Physics also came out to have been useful to the students though not necessarily studied as an individual subject.

In terms of shortfall on subject matter at school CAD came out predominantly at 39%. More computing and workshop/ model making was also indicated as was art and graphics. This may particularly reflect the course content but also reflects on the growth in technology. At school the computer resources would have been far more limited as would the software and the extent to which workshop space and time was available.

The present move in equipping schools with more powerful computers, software and IT facilities will mean that these 'needs' areas will decline. Whether the move to greater use of computers decreases manufacturing and model making thus making it a needs area remains to be seen. This may then impinge on the use of newer rapid prototyping or computer based only approaches to design and manufacture.

6. CONCLUSION

The work so far undertaken tends to indicate that design is becoming more seamless and that it is becoming far less rigidly defined. It is also apparent that it is being technology driven through the use of computers.

REFERENCES

(1) Maclure. J.S. (1973) Report of the Royal Commission Technical Instruction(The Samuelson Report), 1882-84. Educational Documents England and Wales:1816 to the present Day. London: Methuen , pp121-127.

(2) First Year Engineering Drawing, A.C.Parkinson, Sir Isaac Pitman and Sons, Ltd. 1939. p.V.
(3) Introduction to Design, Morris Asimow, Prentice Hall,1962, p1-2. Library of Congress Catalogue Card Number 62-10550.
(4) McCulloch. G.,Jenkins.E. and Layton.D. (1985). Technological Revolution? The Politics of School Science and Technology in England and Wales since 1945. Lewes: The Framer Press, pp 159-60.
(5) Penfold. J.(1988). Craft, Design and Technology: Past, Present and Future, Stoke: Trentham Books p121.
(6) Geometrical and Engineering Drawing, Introduction, K.Morling, Edward Arnold, 1974, 0 7131 3319 8.
(7) Routledge's New English Dictionary, Reprinted October 1926.
(8) Eggleston. S.J. (1976) Developments in Design Education. London: Open Books.
(9) DES (1981) The School Curriculum, London: HMSO,p17.
(10) A qualified success or licence for all, Martin Cross, RSA journal Nov/Dec 1997 pp89-90.
(11) www.design-council.org.uk/dib/facts/f-facts.html.
(12) Consulting on NESTA, RSA Journal Nov/Dec 1997 pg47,48.

International Comparisons

The engineering design centre of the MPEI – a history, experience, and development

V VZYATYSHEV, Y KANDYRIN, and **F POKROVSKIY**
Engineering Design Center, Moscow Power Engineering Institute, Russia

SYNOPSIS

The Engineering Design movement – an interesting societal phenomenon in the life of the Moscow Power Engineering Institute (MPEI) is described in this paper. Formed in 1984, when a team of the teachers and researchers became to develop the "human methodology" of design engineering activity. We were convinced that this activity (Engineering Design - ED) has specific features, different from learning and research activities. Therefore ED should be the base of engineering education, in view of this the Engineering Design Centre (EDC) was created. The EDC has done a lot of rather interesting work for the last 9 years – even for 2000.

1. INTRODUCTION: THE ORIGIN OF THE CONCEPT OF ENGINEERING DESIGN AND THE AIM ITS METHODOLOGY IS TO SERVE

The Engineering Design Centre (EDC) has been working at the Moscow Power Engineering Institute (MPEI) since March 1988. The initiative for establishing EDC came from members of the "working team on the ED problems " created by 21 people from 15 chairs within 8 faculties of MPEI - from various fields: radio-electronics, electromechanics, hydromechanics, automation, etc. Almost all of them were experienced professors. Many of them were inventors: the owners of more than 50 patents. They also carried out joint work with industrialists. Professors N.F. Iljinsky, V.K. Losenko, S.I. Baskakov, W.A. Delectorsky and instructors Yu.S. Sakharov, A.0. Gornov, I.I. Dzegelenok and I.M. Koronevsky have been active members of EDC.

The ED movement was begun in January 1984, when at the MPEI meeting in Firsanovka the problems of design-technological training were discussed. Approved at the meeting, the societal initiative which soon resulted in discussions at seminars and on pages of the MPEI newspaper "Energetic", began from a wide circle of engineering activity problems. But already by the end of the second year the movement has got to key headings:

- "Engineering Design and its role in training specialists";
- "Engineering Design - connecting science with production".

A session of Council was formed in 1987 to discuss the "initial accumulation of the methodological capital". As a result of four-hour discussion the Council recommended to include a discipline "Bases of Engineering Designing" into the curricula of all developing specialities of the MPEI.

The team was concerned about the situation in Russian engineering training. The main trends, gaining strength in education, were technocratic tendencies, formal logic and formalised procedures. This resulted in the dehumanisation of the educational process and a sharp decrease of creativity in the engineering training. According to a remark made by Professor Christo Butsev, during the course of training "enthusiasm dies out in the students eyes, and an active interest for engineering becomes lost" (January 1990). Meanwhile, methods based on the mobilisation of human factors and potential are, little by little, being forced out. The supporters of the tendency often state that those methods pertain to the arts and cannot even be taught.

This process has a long history. 70 years ago Prof. A. Sidorov (2) had written about the unnecessary formalisation and "theorisation" in the engineering educational process. He cited examples of blunders made because of insufficient accounts of human experiences, and lack of systematic analysis. Since that time the situation has become much worse.

Computers and CAD technology have not helped to remove the contradiction, but have made the problem more obvious. Some team members (Y Sakharov, Y Kandyrin, I Dzegelenok) should acknowledge that work in CAD and especially in its educational applications motivate the analysis of human activity. We have come to the conclusion that in order to explain the essence of ED one should carry out an analysis of contradictions:

- between formal analysis and creativity;
- between algorithmic and conceptual thinking;
- between a computer and a human;
- between research activity and design activity.

Every one of us has dealt with these contradictions for years and is looking for its removal from various special subjects (3-5). Our joint research in the team, and later in the EDC, resulted in raising our understanding of the problem to a new level. The MPEI seminar "Methodology of ED" involved volunteers from the MPEI and other organizations and presided over nearly 100 discussions during five years (6,7). We have also enlisted philosophers, psychologists, lecturers and methodologists to help the engineers. In 1989 a seminar on engineering training was set up attached to the Scientific Council of the USSR Academy of Pedagogical Sciences (8).

2. THE SOCIAL TECHNOLOGY FOR OUR COLLECTIVE ACTIVITY

The basic element of our activity was and is the Seminar (since November 28, 1986). From October 1991 (after the meeting and discussion with M. Osborne and P. Booker at the Institution of Engineering Designers, Westbury) it was named as "Design and Management in a Society, in Education and in Engineering". Just its shape in the large degree determines the style of EDC activity and later – the image of Social Technologies Center (STC) - was found in February 1997. These are some of the facts and hypotheses:

- There was a problems selecting the technology for discussion - not so much because of the subject but because of the personalities of the lecturers.
- There was a style of reports and discussions, when not so much in search of a single truth, but that the variety and originality of the approach to a problem is important.
- During discussions we asked less questions, and we act with "thinking correlation" more. The participants of the seminar quite often speak all without exception.
- Everyone receives a copy of the basic suggestions and arguments of the report.

- During a seminar there is always an audio and/or video recording.
- After the seminar participants quickly receive the documents with the basic results of the discussion.

3. ENGINEERING DESIGN IN THE SYSTEM OF CONTRADICTIONS: SCIENCE, ART, KNOWLEDGE, CULTURE, SOCIAL NEEDS

Named contradictions in the ED are reflections of contradictions in World culture. It is accepted that one should distinguish the culture of science and the culture of art (Fig. 1). Art and science imply different ways of cognition of the World and its laws. We consider ED to be a component of the culture of humankind, the methodology of the World transformation.

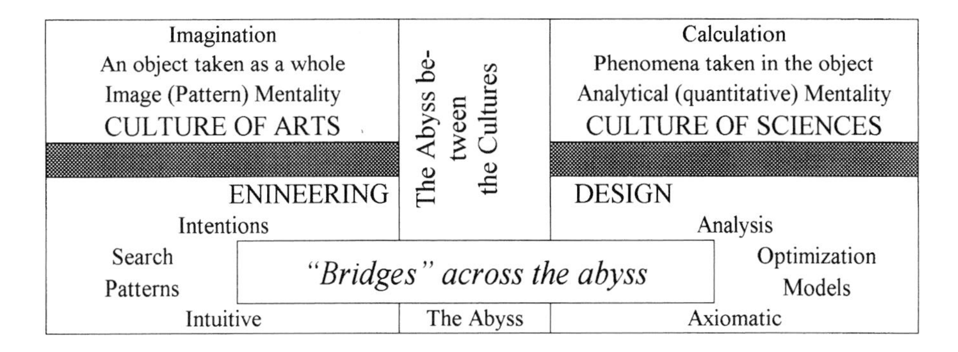

Fig. 1 Humankind and the two World cultures: arts and sciences

ED can bridge the abyss between those two cultures. The abyss is more serious for engineers than for the rest of humankind. Workers of art deal only with art, people of science do quite the reverse. They may not notice the abyss at all. Only the engineer-designer has to live and work on both sides of the abyss, and to cross it repeatedly. The methodology of ED is designated to help in building "bridges" and to provide an engineer with reliable maps, preferably in two languages. J. Dixon gives a visual comparison of the aims of various kinds of activity: of analysis, research and design. We have developed similar comparison over a number of quotients - see Table 1.

Table 1 Quotients of kinds of activity

Kind of activity: / Quotients	Research activity	Design activity
Aim of activity	Establishment of natural laws	Creation of a new object
Methods of achieving the aim	Axiomatic, formal, logical	Heuristic, intuitive
Degree of generality of the aim	As broad as possible	As concrete as possible
Approach based on the result	Doubt, skepticism	Subconscious belief

Character of the result	Unambiguous	Plenty of variants
Criterion of the result	Truth	Efficacy, economy
Dependence on the personality	Weak	Strong

4. ENGINEERING DESIGN: PROFESSIONAL METHODOLOGY FOR THE DEVELOPMENT OF CREATIVE ACTIVITY

Up to 1986 the concepts described had been pedagogically realized only in special courses given by teachers - members of the EDC - and in the course "Introduction to a Specialty". These initial experiences were carried out independently. After January 1987 (1) a systematic research on the basic course 'Foundations of Engineering Design', invariant to the MPEI specialties, was organized. One can find below a short description of one of these experiments.

4.1 Scientific hypotheses

This was to be a course which discloses the systematic character and inherent beauty of ED, which teaches ED methodology and gives a student an opportunity to formulate their own individuality as a professional. This course is intended to stimulate students' interest for:

- the self-realization of inherent creative inclinations;
- the methodology of activity and acquiring special knowledge;
- cooperation with a teacher in the field of knowledge and methodology.

4.2 General description of the course

One of the experimental courses made in 1988 was delivered to 159 four-year students in 18 engineering specialties. The course covered about 100 hours given within one month: lectures on the ED methodology (30 hours), computer training (12 hours), seminars on ED procedures (30 hours), consultations, involving 28 teachers. Besides these, participants were: reader E. Starovoytova, instructors N. Morgunova and I. Antonov, engineers S. Bankov, B. Ryabov and G. Raevsky and others. The course also included the core – the course project.

4.3 Student projects

The course project was assigned so that it required personal and individual approach from the student. Every student had his/her own problem to solve, an individual formulation of the problem, and a particular region of search; they were to their find own way to a solution. It is by the individuality of the assignments, as it turned out later, that one of the secrets of the success could be explained.

Four months before the course the students were offered a choice of design problem to develop in their projects. The problem should be connected with the provision one of six vital human needs (power supply, information, lighting, water, comfortable surroundings, tools and devices) in one of six situations (a city flat, a country cottage, a study room, a ward in a hospital, a tourist tent, a car). The student could also work beyond the matrix offered.

The students formulate their own design specifications, specify and substantiate quality quotients, conditions and restrictions. For this they have to think over the requirements, build and analyze the tree of aims, make the selection of aims, and interpret them into the language of technical terms. They build the tree of problem solutions and the morphological box of the variants of the solution. Every student formulates criteria, selects the most important of them, and substantiates his or her variant of the truncation of the solution tree. As a result, students find efficient solutions by themselves. Their number is usually from 4 to 15. The student finally chooses the best solution, makes a model or constructive development of the variant chosen further on.

The stages listed are typical procedures of ED (9). At every stage a specified method is required. The latter was given at the lectures, discussed during practical classes and was immediately made use of by students. Hence, the course was more oriented towards activity rather than to acquiring knowledge. As a result, the majority of students have acquired the foundations of systems thinking and structural syntheses. Note: all this has been achieved in four weeks.

Along with that, we have obtained another striking result. The interest displayed for a concrete problem, understood as an individual one, aroused an interest for factual information: students worked in libraries, consulted with specialists, and even studied the proceedings of the latest international conferences. The results of the experiment have surpassed all expectations. Of special interest was the realization of "codevelopment" of teachers and students. A student develops by making use of his or her existing knowledge, searching for insufficient knowledge and mobilizing all his abilities for the solution of a particular design problem. In this process he appreciates the teacher's methodological assistance very highly. In a questionnaire, when asked to "estimate the efficacy and usefulness of various kinds of classes" students of all groups gave the highest marks to "the personal contacts with a teacher during consultations on the project". As for the teacher, he develops when he gets to know students' papers, the examples of the methodology application in solving problems and by way of studying individual activity styles of various students. In conclusion, we shall cite several typical students' opinions.

4.4. Students' opinions

- "The course makes you think and work independently. During that month with the project I spent more time in the library than during all the previous years." (Student T.E.)
- "The course has been a great help for me. The number of variants in my project amounted to a six-digit number. It would be impossible to process them without the methods of the course." (Student A.J.)
- "The course is intended for a human and teaches him how to solve problems successfully irrespective of the situation at present and in future." (Student J.K.)
- "The course is worth taking because it develops students' engineering thinking." (J.K.)

Many students also point out that the course made them ponder over the meaning of engineering activity for the first time.

The successful results of the experiments permitted us to introduce the course "Engineering Design Foundations" into the curricula of 18 specialties at the MPEI.

5. CONCLUSION: PROSPECTS AND DEVELOPMENT TRENDS OF ENGINEERING TRAINING PROFESSIONAL PEDAGOGICS

ED methodology is in the process of development. For its improvement it needs interrelated computer support and psycho-pedagogical ensuring of the educational and training process.

Computer support is to be based on the psycho-pedagogical laws of guided development of creative personality. Meanwhile, the technology of the ED training will essentially depend both on the extent of the computer support and on the available resources of creative teachers.

In our opinion it is expedient to have the following kinds of the computer support:

1. Design problems of increasing complexity (involving different numbers of procedures). The main principle of the dialogue between a student and a computer can be described as follows: the student is to perform the greatest amount of solutions and choice operations, while a mass of calculation and logical operations are to be carried out by the computer.
2. The aim of the initial stage of work may be to master separate procedures and, probably, to bring simple procedures to automatics. The development of an individual style of student activity is desired at this stage. The analysis of typical mistakes and a proper recommendation and/or suggestion of the correct solution by the computer may contribute to the latter.
3. The next stages are to be devoted to mastering strategies - to the combination and sequence of procedures (6). Here for the development of an individual style some strategies should be provided. A student should be offered an opportunity for making a choice.
4. It is desirable for the complex of computer assignments to include a subsystem of the assessment and incentives of the decisions taken by a student. One can assess the individual style of a student's activity by the pattern analysis of decisions taken by him or her.

For the development of professional aspects of engineering pedagogy it is also expedient:

- Along with the methodology of ED, to develop a broader methodology of "productive engineering activity" (PEA). Prof. G. Morgunov, a member of EDC (10), discusses the category of PEA in detail.
- To perfect the methodology and to make it more concrete, for instance, in the process of postgraduate engineers' training. Here PEA of methodologists and engineers is of primary importance in the course of their joint solution of design training tasks.
- To formulate a branched structures of engineering activity and to look for the ways of adapting those structures to various psychological types of users

Finally, the widespread applications of the methodologies of ED and other kinds of engineering activity imply building the whole process of engineers' training on this foundation. Methodological experiments in this direction were started by EDC of the MPEI in 1988 (10). Described here concepts have been developed and put into action by dozens of people.

To summarize, the aim of this concept can be expressed as follows: to find ways and means of maintaining, disclosing and developing students' inherent creative abilities given to them by God, and keeping alive their interests.

REFERENCES

(1) "Inzhenernoe proektirovanie i sovremennyi spetsialist", *Vestnik Vysshei Shkoly*, 1986, **(6)**, pp. 7-19 (in Russian).

(2) Sidorov, A.: "Osnovnye printsipy proektirovaniya i konstruirovuniya mashin", Moscow, (MAKIZ, 1929) (in Russian).

(3) Vzyatyshev, V.: "Metody poiska proektno-konstruktorskikh resheniy", Moscow, (MPEI, 1983) (in Russian).

(4) Dzegelenok, I.: "Logika poiska projektnykh resheniy", Moscow (MPEI, 1984) (in Russian).
(5) Khoroshev, A.: "Syslemnoe proektirovanie mashin", Moscow, (MPEI, 1989) (in Russian).
(6) "Deyatel'noslnaya kontseptsiya professional'noi pedagogiki ingenernogo obrazovaniya". Sbornik, Moscow, (MPEI, 1989), (in Russian).
(7) Vzyatyshev, V.: "Struktura protsessa inzhenernogo proektirovaniya i problemy podgotovki radioinzhenerov", *Radiolekhnika,* 1989 **(3)**, pp. 91-95 (in Russian).
(8) Vzyatyshev, V., Orekhanova, G.: "Razvitie tvorcheskogo potentsiala lichnosti v protsesse podgotovki inzhenerov - sozdateley novoy tekhniki", *Izvestiya APN SSSR,* 1990 **(1)** (in Russian).
(9) Vzyatyshev, V.: - in this Conf. Proc.
(10) Morgunov, G., Zuev, Yu.: - in this Conf. Proc.

Experience of engineering design education at the MPEI

G MORGUNOV and **Y ZUEV**
Engineering Design Center, Moscow Power Engineering Institute, Russia

SYNOPSIS

In this report the essence of an eight-year educational project realised in the Moscow Power Engineering Institute (MPEI) is described. This project is especially interesting as an example of fruitful collective activity in Engineering Design training. It can be considered also as a good illustration to problems of social management in the field of engineering education.

1. TARGET SETTING

As is known (1), educational systems (ES) of the advanced countries, and especially USA, develop as liberal, ample to students opportunities of a choice of a training "trajectory" and give them a wide spectrum of studied disciplines.

The new requirements that showed by processes of entry of some the countries in the post-industrial period of development have revealed disturbing defects of this ES. In our opinion "silent crisis" of the ES is connected with "supporting" character of training. The orientation of the present ES as a whole on the last experience takes place; more details in the report (2). Besides that the easing of fundamental natural-science education, for example, in such countries of the Western world, as the USA (3,4) is observed. As the result of those phenomena the slogan about the necessity of the formation of a continuously training society was proposed (5).

This confirms, in our opinion, an urgency of development of more perfect models of education (ME) whose basis is harmonious and practically universal. At the construction of such models we use the methodology of Engineering Design Activity (6-8). Another possibility, we believe, is the use of the system-creative pedagogic and theories of management and self-organising - synergetic - in the applcation to sociological problems (9,10).

2. ESSENCE OF THE PROJECT – THE ACTIVITY APPROACH

An innovative movement, which key words are "Engineering Design" (ED), and one of which products the described project became, was created in MPEI in 1984 (11). After discussion in 1986 at the MPEI council the activity problems of engineering training (6) Prof. V. Vzyatyshev has been reading to MPEI teachers the lectures "Bases of engineering design".

Among his students there were a lot of teachers, including senior lecturers V. Levin and Y. Zuev. They have taken the ED training ideas to heart and have begun to develop them as though they were trained for this all their lives. In the spring 1987 they offered the idea of the described project: all the education and training of engineering designers should be impregnated with design methodology (7), all disciplines should be connected by system structure of design activity, as is a body of the human – by a skeleton and nervous system. As a result of the discussions which have followed this idea, the educational project described below was formed.

For its realisation the belief of a MPEI top management was necessary, but the initiative of many teachers was even more important. Fortunately, the MPEI rector Igor Orlov believed in the project and approved the idea. We went to the MPEI chairs and faculties and found about seventy teachers of all areas of engineering education from the English language and philosophy up to special disciplines and computer science. All these people including the project teachers team (PTT) - have expressed a sincere desire to use their time and forces for this new and interesting project.

Two new innovation chairs have been created: "The Hydromechanics and hydromashines (HHM)" of "Power machine building" MPEI faculty and "Designing and manufacture of radio equipment (DMRE)" of "Radioelectronics" MPEI faculty. The chair HHM initiated the undertaking. The chair DMRE (Prof. V. Vzyatyshev) joined later. Initiators there became Yu. Sacharov, E. Starovojtova and A. Nazarov. Statements of social-educational project and its preliminary results are described in more detail in (12,13).

PTT seminars began in autumn of 1988 at these chairs. Within one year at 1-2 times per week for 3-4 hours these remarkable people worked free-of-charge and with enthusiasm. Participants of PTT seminars collectively, in an atmosphere of mutual understanding, tolerance and search of compromises have made new educational curriculum (EC), have co-ordinated programs of key disciplines and have agreed upon interactions. Looking back on seven years, we count the work of these seminars by a rather successful increase in qualification, especially in interdisciplinary spheres of educational problems.

Having started with the structure and methodology of ED, taking into account the features of a professional subject domain, the PTT have begun to compare and in common to discuss the role each of their disciplines. They put on the two-faced problem: 1) the formation of experts and 2) the development of humane highly educated persons.

The activity of this remarkable collective is a striking illustration of that enthusiasm, which was observed in our country in those years. It is impossible to list all participants of it, but mechanics A. Khoroshev and D.Korj (14), mathematics A. Kirillov, philosopher L. Voronkova should be mentioned. The experience spent within the framework of a speciality "Hydraulic machines, hydrodrives and hydro-pneumo-automatics means" is described below.

3. SYSTEM DESIGN OF DEVELOPING EDUCATIONAL PROCESS

System-creative combinations of the teaching and educational interactions of teachers and students are base cells of open synergetic system of education process (SSEP), having property of generation of new qualities. The scientific-methodical and educational procedures make a hierarchy of educational bodies. All of them have pedagogical, developing and cognitive functions. Structure SSEP and also cumulative results of the above-stated interactions should provide consecutive progress with achievement to the greatest degree desirable purposes and qualities for every trained person.

The body forming managing functions was the creative PTT of highly skilled teachers (about 60 persons) informally interested in the success of the project (the head - Prof. G. Morgunov). The basic contents of key educational procedures (EP) is reflected in reports on scientific-methodical works, in curricula, programs and also in the publications – see (12, 15).

For the first time in our country in the curricula of a technical college the compulsory cycles of the disciplines containing the beginnings of knowledge about achievements of human culture interconnected with disciplines from physical, mathematical and technical cycles were icluded. These cycles were constructed by a hierarchical principle.

Let's specify base procedures of designing process for SSEP:

- Target setting - the formulation of the training purposes of engineers as harmoniously advanced and professionally highly skilled persons.
- Conceptual synthesis - the formulation of necessary and sufficient conditions for achievement of the result.
- Abstract synthesis of the EC - balance of knowledge of the spiritual and rational contents.
- Quality standardisation of the graduate - formation of the concrete contents for the elected spheres of knowledge and technologies.
- Structural and parametrical synthesis of the working EC.

As a result of the realisation of the above set procedures there was received harmonious hierarchical structure with logical transitions from discipline "Introduction in system of knowledge, the world of creativity and a speciality" to fundamental disciplines on humanitarian, natural-science and all technical disciplines. And then to base subjects of engineering power education and, at last, to special disciplines with the profound subject of training.

4. ESTIMATION OF THE RESULTS

On responses of the students, their training activity has turned them from rather unrelated and formal procedures to the natural and proved sequence of intelligent procedures. Understanding of basic ED methodology by students:

- has allowed them independently to find and confidently to defend design solutions,
- has removed usually noticed blanks in the logic of systems designing,
- initiated works on application of new physical effects for solutions improvement.

After the described methodological training the long construction works (during several semester of training) were organised. The last has allowed the students to realise their creative potential. As a result at the basic part of graduates the heightened interest to productive practical activities, and at oversystem - the raised demand for such experts was observed.

Within the framework of the speciality "The Power Machine Building" the direction of engineers training by means of system-creative process was realised. In this process the mutual interactions between training and trained persons were realised. Such interactions, we believe, determine the synergetic system of education process (SSEP), which have the raised (increased) ability to educate and train high quality engineers even in conditions of the modern crisis conditions of the Russian society.

The openness of educational process achieved in the project realisation has caused its operative to react to changes of oversystem requirements. So, the answer to qualitative changes of the markets of the offer and demand was established without narrowing the main the speciality, three specialisations: "An Engineering Design and researches in hydromechanics engineering"; "An installation, adjustment and operation of the hydraulic equipment"; "Engineering business and management in hydromechanics engineering".

5. THE CONCLUSION

In February 1995 the first experimental groups of engineers–hydromechanics and the engineers radiotechnicians graduated from MPEI and franchised degree projects. Trying to judge them, we have seen in their eyes, in depth of engineering studies, in confidence and ability to assert the ideas much that we planned more than seven years before. Last three releases of engineers on the speciality have shown: our graduates besides qualitatively increased vocational training, have better adaptability to new social and economic conditions in the country, more advanced humanitarian and social-ecological outlook, active skills of acceptance of quite proved and most preferable decisions, considerably increased freedom and culture of professional dialogue. This influence was especially obvious in the group of the best and strongest graduates.

REFERENCES

(1) Galagan, A.: "Universities in regional economic and administrative structures of USA, the countries of the Western Europe and Japan", Moscow, *RIHE, 1994* (in Russian).
(2) Romankova L., Vzyatyshev, V. - in this Conf. Proc.
(3) "A Nation at Risk: The Imperative for Educational Reform": A Report to the National and Secretary of Education, US Department of Education, Washington D.C., Apr. 1983.
(4) Matjuhin, V. "Overcoming of the crisis phenomena in an education system of USA (1980 - 1990): methodology and results", Moscow, *RIHE, 1994 (1)*, 128 pp. (in Russian).
(5) "By the Year 2000. First in the Word": Report of the FCCSET Committee on Education and Human Resources, FY 1992 Budget Summary, Washington D.C., Feb. 1991.
(6) "Engineering Design and the modern expert: *The Bulletin of the Higher School, 1986 (6)*, p. 7-19 (in Russian).
(7) Vzyatyshev, V. "Engineering design and creative abilities", Scientific achievements and the best practices in the higher education, Moscow, RIHE, 1992 (**11**), 28pp. (in Russian).
(8) Vzyatyshev, V. - in this Conf. Proc.
(9) Haken, G.: "The Synergy", Moscow, *The World, 1980, 198*, 240 pp. (in Russian).
(10) Prigojin, I., Stengers, A.: "The order from chaos", *Progress, 1986*, 431 pp. (in Russian).
(11) Vzyatyshev, V., Kandyrin, Yu., Pokrovskiy, F.: - in this Conf. Proc.
(12) Vzyatyshev, V., Morgunov, G., Zuev, Yu., Levin V.: "Experience of innovative education in the MPEI", The International conference "Problems of the Academic and Professional Mobility on the eve of the XXI century", Almaty, IHEAS, 1996, p. 73-75.

(13) Morgunov, G.: "Preparing specialists for Productive Engineering Activity", *Int. J. Continuing Engineering Education*, 1991, **1**, (3), pp. 235-248.
(14) Vzyatyshev, V.: "Engineering Design and Creativity", *Int. J. Continuing Engineering Education*, 1991, **1**, (3), pp. 219-234.
(15) Korj, D., Khoroshev, A.. - in this Conf. Proc.
(16) Morgunov G., Golubev, V., Zuev, Yu.: "Modern innovations in the higher technical education and their applications at the MPEI HHM chair", Moscow, MPEI, 1996 (in Russian).

Methods for machines engineering design

D KORHZ and **A KHOROSHEV**
The Faculty of the Principles of Machine Design, Moscow Power Engineering Institute, Russia

SYNOPSIS

A new system of general technical education and training is suggested and introduced into the educational process. This system is intended to raise the creative abilities of the students, the independence and responsibility of their opinions and to help them rapidly adapt to variable conditions and new tasks. The results of the analysis of its practical application are given. In 1999 on the basis of teaching experience the textbook "Introduction into management of mechanical systems designing" was published, covering all basic aspects of design.

It is impossible for Russia to join the developed countries without radically improving its educational system. It is necessary to take into account that the life and work conditions in our country differ from western and eastern ones. Therefore formal copying of other educational systems will not give positive results, as each of them is characterised by its own culture, traditions, socio-economic features and historical experience.

1. ENGINEERING DESIGN CENTRE AND TECHNICAL EDUCATION

Technical education is a component of the educational system. Thus a search for ways of increasing its efficiency will help in perfecting the education system as a whole. On the other hand this task is slightly facilitated, as the technical education unlike humanitarian and economic ones has precise criteria for results estimation and is based on objective laws.

In 1986 an Engineering Design Centre (EDC) was created at Moscow Power Engineering Institute (MPEI). It was aimed at a search for ways of perfecting engineering education and co-ordination of related activities (details in the reports written by V. Vzyatyshev). "The Faculty of the Principles of Machine Design" of MPEI has taken part in work of the Centre. The logical system of general technical training was created and introduced into the educational process. The analysis of accumulated and practice-proven pedagogical experience (more than 10 years) has shown an importance of the following aspects listed below.

1. Necessity of intensifying general educational training. It enables orientation in various spheres of activity and quick adaptation to variable conditions and new tasks.

2. The connections between disciplines and the logically defined sequence of training must be established. Content and kind of knowledge on each discipline should be determined by the needs of real life and the importance for study of other disciplines. Each following discipline should develop and consolidate the knowledge gained earlier.

3. Orientation of education towards the solution of real tasks: teach an ability to choose the purpose, place priorities, offer variants of solutions and select best of them, offer and choose solution methods, co-ordinate technical decisions with social, economic and ecological requirements and give practical recommendations. Tasks should be made less formalised and more typical. Attention should be paid to aspects of activity that graduates will face in life. It is necessary to use combined tasks (unified for several disciplines) that join the separate parts of the knowledge in a uniform system.

4. Training computer skills should be combined with the development of human abilities: to set targets, choose a method of solution and make a decision. Teaching an ability to solve the tasks in a qualitative way and to estimate the possible results without a computer.

5. Construction of the educational system in the form of a chain: concepts and terms of the discipline – analysis methods - synthesis methods. It is the possession of synthesis methods that allows competitive solutions to be found. The lack of a synthesis method in a discipline sharply reduces its practical importance.

6. Training in alternative thinking; make the student try to suggest several variants in any situation (analytical models, methods and ways of solutions etc.). It develops the initiative, allows student to feel sure in uncertain situations, and raises the validity of accepted decisions.

2. SITUATION ANALYSIS

The detailed explanation of above-stated information is given below.

The process of general technical education, like any other, should have a definite structure. It is convenient to have a structure where the knowledge gained at each stage is sufficient for the fulfilment of certain kinds of activity. Such a structure includes the following stages:

1. Nomenclature and basic elements of discipline (basic structural elements)
2. Laws of discipline
3. Principles of discipline:
    - 3.1. Analysis methods
    - 3.2. Synthesis methods
4. Productive activity

Stages 1 to 3.1 provide knowledge in a form of data set, but are unproductive. Discipline, which includes only stages 1 and 2, is of descriptive type and therefore is inadmissible for university. Today there is a demand for disciplines and teaching synthesis methods (stages from 1 to 3.2 inclusive). Stage 4 provides skills in putting received knowledge into practice - it provides efficiency of education.

The main problem in education is whether to teach knowledge or skills. Unfortunately, it is often decided in favour of knowledge. Sometimes, as a result the student turns out to know everything, but cannot apply it in practice (cannot really do anything). This lack of skills for productive activity causes a reluctance and fear to be engaged in real work in the future. This is nobody's fault, as the student has been carefully taught - but what subjects were set, so that such results are received? Only the sum of the skills makes the monitoring system of education more exacting and should become the assessment criterion for education quality. Of course, the skills are based on knowledge.

Requirements on training specialists in any discipline are proposed to satisfy the list of skills. The complexity of a task and the list of skills necessary for its performance, determine the disciplines list and content. A faculty establishes rules, concepts and composition of elements base, which are given to students in order to gain the skills listed above. An estimation of faculty work should be based on the sum of the skills which are required during the education and later. It is clear, that in this case an educational plan of speciality should be created in the following way: industry requirement – special disciplines – general technical disciplines – general educational disciplines – fundamental disciplines.

It is possible that the content of some disciplines will be sharply narrowed (some questions will disappear as useless or because of duplication) or they will be found needless. We should try to find the connection between disciplines, especially when experience implies their existence, but they can not be visualised.

Life is known to differ from the model in insufficient distinctness and responsibility for taking decisions. If the training of students is directed towards future real work, then educational tasks should be presented in a realistic view. First of all this includes:

- the ability to set a problem,
- the ability to choose a method of solution in a justified manner,
- the ability to make responsible decisions.

3. SOME RECOMMENDATIONS

It is necessary to teach students from the first year, starting with simple tasks and finishing with course work and projects. Such tasks should include incomplete problem situations, without indication of the scheme or the solution method. The teacher is supposed to be an adviser rather than a supervising instructor or authoritarian person, who utters absolute truths and knows absolutely everything. This produces too high demands on the teacher, and a reluctance to appear unaware badly affects the character of some teachers. On the other hand, the teacher-adviser will not allow the student to shift the responsibility to another (if you made it, then you are responsible for it), developing the initiative and creating comfortable and unconstrained study conditions. An ability to make a choice (to suggest variants of solutions and to choose the best one) builds up the following features in student character, such as tolerance, respectfulness, and a tendency to co-operation and mutual understanding.

There are a lot of possible ways for solutions of real problems. The choice of a final solution depends on specific conditions and the given quality parameters. Each situation has its own optimal solution. The understanding of this fact brings up the tolerance in the perception of opinions of a companion, the desire to understand him and to investigate fairly all details. On the other hand all real problems have a lot of criterions. There are no common objective methods of their solution yet, in practice the choice is somewhat subjective. This emphasises once again a lack of absolute truth and forms in character a feature of respectfulness. Also, experience shows that compromises are most suitable for multi-criterion problems. Thus, the student is accustomed to search for points of contact with his companion and to find the best sensible variant by common efforts.

In connection with continuous development of computer science and its intellectual progress the following question becomes more and more urgent: what differs computer from a man? If we do not want to train specialists, whose functions the machine performs more effectively, then we should precisely know what a man can do and a machine cannot. From this point of view it is necessary to teach students human activity rather than computer-assisted, automated procedures. This includes the ability to set a problem, choose a method of solution, make a decision on the basis of work results and skills of "qualitative" (without calculation) problem solution and estimate possible results without computer.

The quality of education is backed-up by constantly using knowledge and skills learnt before. One of the effective ways to implement this is to make student's work complex and to include knowledge of several disciplines. For example, a work consists of several logically connected parts, one faculty gives and controls the work as a whole, and another one monitors performance of their parts. The complex work helps students once again to recollect and involve earlier gained knowledge, to integrate this with other courses in a single system and to understand its practical use.

The faculty of the principles of machine has carried out the practical implementation of the new approaches in training specialists within three-term courses.

Since 1988 first year students have been taught a one-term course, named "The principles of design". This states the general methodology of the design activity of a specialist. There are no rigid connections between design methodology and the particular development object, as first year students have no knowledge of the specific object domain.

At the same period of time the second and third year students carried out a complex course work (CCW), dedicated to design the objects from a field of general mechanical engineering. The work was conducted as the educational experiment (detail in the report of G.Morgunov, Yu.Zuev, V.Vzyatyshev). An educational course contains two basic parts. The first part of the course was devoted to study the elements of machines, typical or standard methods of the analysis and design techniques. The second part of the course considers problems of element synthesis for a mechanical device as applied to all stages of designing. These stages include formation of a technical project, search for the optimum decisions at the stages of choice of structure and parameters, and the final product release as a complete set of inclusive engineering documents. A feature of CCW was the necessary use of consultants from other university faculties. This strengthened the connections between disciplines, thus students consulted with the faculty of electrical engineering in order to choose the electrical engine. Processes of the manufacture of some elements were studied at the same faculty, the teachers of faculty of engineering graphics advised the students on rules of design documentation execution. The estimation of CCW was given by a commission, which comprised teachers from various faculties. This was the final stage of serious qualifying work of a student in a sphere of general technical education.

4. ABOUT A NEW DISCIPLINE "THE PRINCIPLES OF ENGINEERING DESIGNING"

The accumulated ideas and workings have been used in the discipline "The principles of engineering designing". The second and third year students within a three term course study the discipline. The curriculum covers the following: attending lectures, carrying out laboratory test, practical training and independent research. We could emphasise three basic targets, which underlie this discipline.

1. Study of elements of machines. The students should know functionality and design of machines, criterion and methods of designing and techniques of verifying calculations for typical elements of machines.

2. Study of designing methods for technical systems. The following design stages are studied: requirement specification set up, choice of physical principles of operation, structural and parametrical synthesis. Emphasis is placed on independence of design principles and methods, as well as on necessary consideration of a number of decision variants and choosing an optimum one.

3. Securing received knowledge and the ability to use them in practice by means of designing a technical system. It is necessary to have skills in reading drawings and the creation of graphic images both by hand and by computer. Special attention is placed on the defence of the chosen decisions.

Some difficulties which the faculty faced while realising those tasks should be noted.

The overwhelming majority of tutorials are devoted to parametrical synthesis of specific objects. There is not enough educational literature which completely and consecutively considers problems of designing as a whole. Experienced teachers often feel that there is no stage of refinement in a technical project and the customer completely accounts for definition of the technical project without the designer. This is possibly the reason why problems of definition of technical projects are not considered in the textbooks. Our faculty developed a sequence of procedures for the definition of technical projects. For example, we have offered to carry out an analysis of all external connections of an object for designing. To do this a hierarchical structure of technical systems must be constructed. In order to find out the parameters of quality it is recommended that a tree of objectives is created. As a result it is possible to create a concrete definition of the problem and to find out the conditions unknown beforehand, being capable of producing significant influence on the variants of the decision.

Usually the structural synthesis is absent in training to design. This means that it is known the scheme in the task and it is necessary to solve a problem on definition of object parameters. In reality the designer is forced to carry out a transition from given parameters of the object to the decisions at a scheme level, during which the given function is realised with the required quality. Traditionally, this transition is accomplished by using prototypes of the projected device. However the choice of the prototypes is carried out randomly, without deliberate and reasonable analysis of the possible variants, and such choice is beyond the strength of a beginner. The faculty has suggested an approximate algorithm for the formation of the possible scheme as applied to mechanical devices. It allows any student, even the ones with no design experience, to generate a set of variants of the scheme of mechanisms, carry out a comparative analysis and choose the preferred decisions. There is the problem of quantitative comparative estimation of scheme variants, as at a stage of structural synthesis the exact values of parameters of technical system are unknown.

5. CONCLUSION

The experience of teaching the discipline " The principles of engineering designing", based on the methodology of training alternative thinking, as well as the search and substantiation of optimum decisions, has been shown to prepare the students to solve problems in indefinite conditions.

A comparison of student behaviour while defending the course projects has shown the advantage of the new system. Earlier, when asked the question why was one or the other solution accepted, it was possible to hear the following: the data is given in the book, or the consultant advised me to do this, or I do not know the answer. Now at defending of the course work the obligatory substantiation of taken decisions became a standard. The students cope with it, even if they are asked to answer any question, connected with a theme of their project. Thus, the offered approach to design training allows young people to feel more confident and to adapt easily for rapidly changing external life conditions.

In 1999 on the basis of teaching experience on this discipline, the faculty prepared and published the textbook "Introduction into management of mechanical systems designing". It takes into account modern requirements and fully considers all basic aspects of designing.

The theory of decision-making in engineering design

V KADEL and **Y KANDYRIN**
Science and Education Department, Moscow Power Engineering Institute, Russia

ABSTRACT

Human relations, system of references and stereotypes of perception in the areas of creativity, business and sciences have become more complicated. The deficiency of strict decision-making methods valid under conditions of multiple criteria, fuzzy restrictions, and real-time change of the choice-influencing factors has become evident. The authors do not separate study and training from the design and social processes. The process of autonomous object design is considered as a method of both helping an engineer to perfect his qualification and a technical high school student to get the necessary working knowledge and skills. This process is based on the decision-making theory.

1. SEVENTH DEGREE OF FREEDOM

For thousands of years, living beings have used various mechanisms, such as the vestibular system, allowing them to function preserving their stability in the world giving them six degrees of freedom. Man has extended these opportunities using his knowledge of the laws of mechanics, physics and chemistry. With the formation and development of the social environment man began to feel the necessity of the seventh degree of freedom, i.e. of the orientation and self-development in this social environment, but he lacked the mechanisms necessary for it. However, life constantly requires that he shall do the decision-making (DM): during profession choice, house purchase, project implementation, business organization, etc.

It is known that the DM efficiency abruptly decreases with the increasing duration of the process of choice, and expenses for its realization grow simultaneously (Fig. 1). It is reasonable to treat any DM's procedure and stage, including design, in terms of information movement and treatment, i.e. taking the time factor into account. It requires the task features formalisation, which is very difficult for man as he shall deal with a "mish-mash" made of purposes, criteria, choice-influencing factors, alternatives and their estimations, and all this "mish-mash" must be analysed and estimated simultaneously. So, professionals began to feel the necessity for knowledge in economy, marketing, management, psychology, etc, besides their special disciplines. However each discipline uses its own definitions, concepts, terms, i.e. different languages, that drastically reduces the efficiency of acquired knowledge application.

Decision-making theory (DMT) helps to overcome all these difficulties. It allows man to structure his environment, to position himself in the "problem - situation" field, to state a problem correctly on a typical approach basis (Fig. 2) and to build an algorithm of the decision, first in the terms, definitions and concepts of the theory itself, and then in those claimed by the problem.

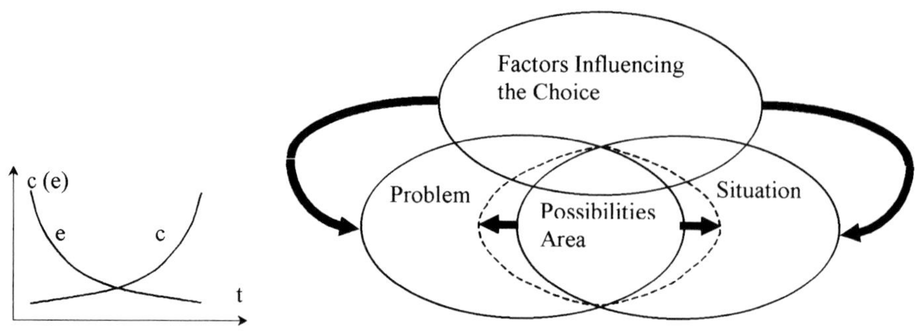

Fig. 1. Making Decision Efficiencies & Costs varying in time

Fig. 2 External Design Stages

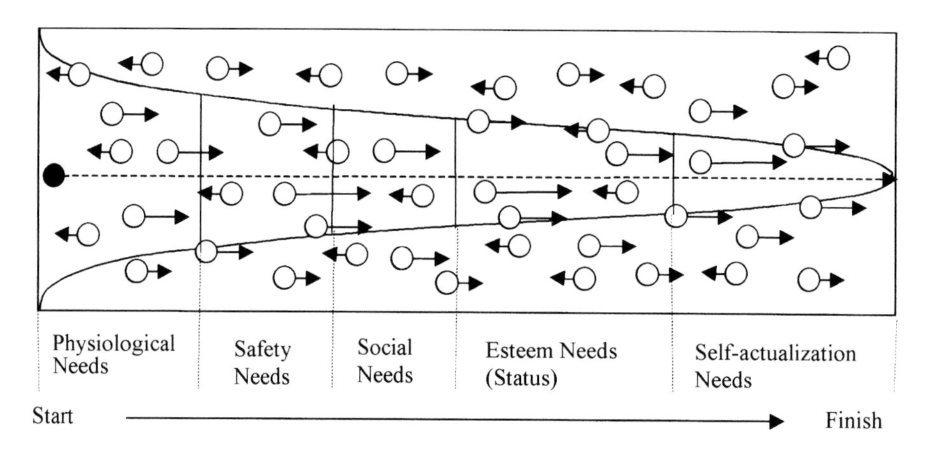

Fig. 3. Social Success Model

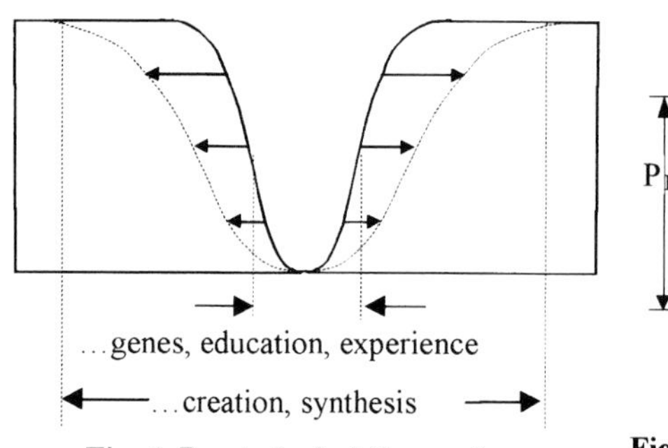

Fig. 4. Psychological Perception

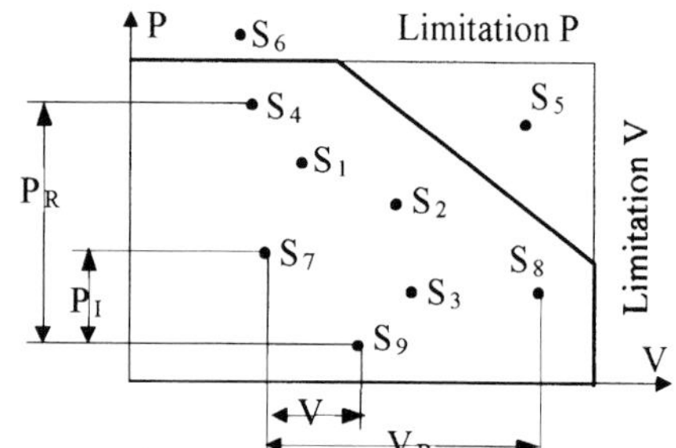

Fig. 5. Determination of Making Decision Strata in Euclidean Space

The higher the positioning level and wider the problem environment (choice-influencing factors) taken into account, the more effective the final decisions. Man must understand that any success of his is not possible without the success of the problem environment - contractors, customers, firms in general, and even the nation. That's why at the very beginning of the study I offered the students (employees) the Social Success Model. The equation of a perfect-liquid laminar flow describes the ideal utopian society, the one of a viscous-liquid laminar flow is representative of the real society (Fig. 3) and the turbulent flow equation is for revolutions, wars etc. Liquid molecules (individuals) are inseparable in the laminar flow, but their moving speeds differ. That's why an individual who will be able to reach, through some positive effort, a more favorable environment from his current position will achieve success. An individual who will break the rules of the game valid in his environment will be transferred to a lower strata of the flow and will never achieve any success.

The greatest difficulty consists in the discrepancy of experience and knowledge, received during training on the one hand, and innovation on the other hand. Man is compelled to lean on his own experience and on training, i.e. another man's experience, which are always of a retrospective nature and do not allow him to effectively solve new tasks. So the person acquires only one thing - stability given by the "filter of perception" (Fig. 4). Indeed, the more indeterminate is the environment (criteria, conditions, choice-influencing factors) the higher the system level of the decisions made, and the greater the need for tradeoff decisions and, hence, the greater the responsibility – the less certain and comfortable man is, the more suppressed his creative capabilities and this reduces the probability of effective DM. Wishing to avoid this discomfort, the man moves farther and farther into the safe area of the "filter of perception" where everything is familiar to him, but where the conditions, outcomes, alternatives are estimated unconsciously and inadequately. Unfortunately, the skill and desire to make tradeoff decisions cannot be taught, but need to be carefully learnt as more an art than science.

The lack of skill and willingness to make a choice from several alternatives is aggravated in that there is a fashion on technical decisions, ways, structures, circuits, elements and on the axiomatic rules describing them, and the fashion varies from time to time. Like any fashion, it is associated with excessive enthusiasm for any one approach to the detriment of any alternative decisions. The same applies to various schools and directions dominant in various educational and design organizations. In other words, the wider the experience, the smaller the flexibility. The majority of the inventions have been made by the people with average IQs, and just a few have been made by those whose IQs is greater than 90, because the common sense (the realized) prevails over perceptibility (the intuitive), and the correct idea "to believe means to see" is substituted by the wrong idea "to see means to believe". We propose to make a wider use of the devices of invention: empathy, inversion, analogy, imagination. A detailed description of the above-stated approach can be found in (1) and partially in (2).

2. ALGORITHMS OF MULTIVARIATE OPTIMIZATION

When choosing from the optimum decisions, the customer, i.e. the whole autonomous object (AO) designer, and performer use new characteristics, which have not been taken into account before and the relationships between the quality parameters (QPs) found in the course of designing, thus changing the system of preference. In other words, the real optimization differs from abstract (theoretical) optimization in that it is dynamic and is made on a basis of the "floating" structure of the QPs, varying in view of intermediate results. To each stage there generally corresponds its own set of initial data, QPs estimation criteria, optimization problem content and problem solution method, as well as models. All of the QPs and characteristics of the power electronic system (PES) to be determined are found as follows. In the first layer, the ranges of their probable values are determined. For Fig. 5, they are the volume V_r and the loss power P_r. In the second layer, the ranges of their optimum values - V_i, P_i – are found by selecting the Pareto-optimum solutions in the affine space of all the parameters. And, at last, in the third layer of the QPs, their fixed values are determined, for which purpose use is made of the finalized models allowing the parameters to be determined on absolute scales.

The use of two kinds of hierarchy – by the design stages and by the decision-making layer - allows man to pool all of the sought-for characteristics, QPs and the initial data into groups by the degree of reliability of determination of their values at various stages. This aggregation permits relating of these characteristics to development stages. It is carried out by the following groups: W_1 to W_{26} are the quality parameters (efficiency, volume, mass, etc.), R_1 to R_{15} are the output characteristics (configuration, connection circuits, equivalent resistances, etc.), Q_1 to Q_{10} are the design, technology, etc H_{11} to H_{16} are the characteristics of energy generators, D_1 to D_4 are the operating conditions, G_1 to G_{17} are the characteristics of the loads. Then a stage-by-stage designing table is made, which specifies at what stage what solution layer shall be determined for each of the characteristics. For each stage the models satisfying the requirements of consistency and reliability have been developed: I - block, structural diagrams; II - functional models of transformation (FMT); III - basic electronic circuits; IV - equivalent circuits. The FMT is a model specially developed for computer-aided calculation, making it possible to calculate relationships and cross-influences for all of the electrical circuit characteristics. Not devices but operations — gating, transformation, averaging (filtration), regulation etc. - are the nodes of this model. Thus, all of the model connections become sequential (there are no cross-connections) and all of the control circuits are replaced by an equivalent resistor grounded via a current generator. The FMT also makes it possible to calculate bar charts of the relative input of each operation into the quality parameters of the entire device, thus facilitating the quest for their improvement. The algorithm of work at the stage is given in Fig. 6.

To find the set of Pareto-optimum solutions πS, the equation $\pi S=S/S^*$ can be used where S^* is the set of the nonoptimal solutions. The system is $s\in S^*$, if there exists such a variant of $x \in S$, where $k_i(s) \geq k_i(x)$, $i= 1, m$, and $k_j(s) > k_j(x)$ at least for one of j, where $k_j(s)$ is the value of the ith criterion; m is the number of criteria space dimensions. The search of set S^* proves to be easier than that of πS since it is enough to find at least one condition-satisfying variant of $x \in S$ so as to ascertain the variant's belonging to the S^* set. To formalise the decision rule, we shall introduce a nonlinear function of preference between PESs S_n and S_m

$$F(S_m,S_n)=\begin{cases}1 \text{ if } k_i(S_m)\geq k_i(S_n); i=\overline{1,N}, \exists j: k_j(S_m)>k_j(S_n)\\ -1 \text{ if } k_i(S_m)\leq k_i(S_n); i=\overline{1,N}, \exists j: k_j(S_m)<k_j(S_n)\\ 0 \text{ if } k_i(S_m)=k_i(S_n)\end{cases}$$

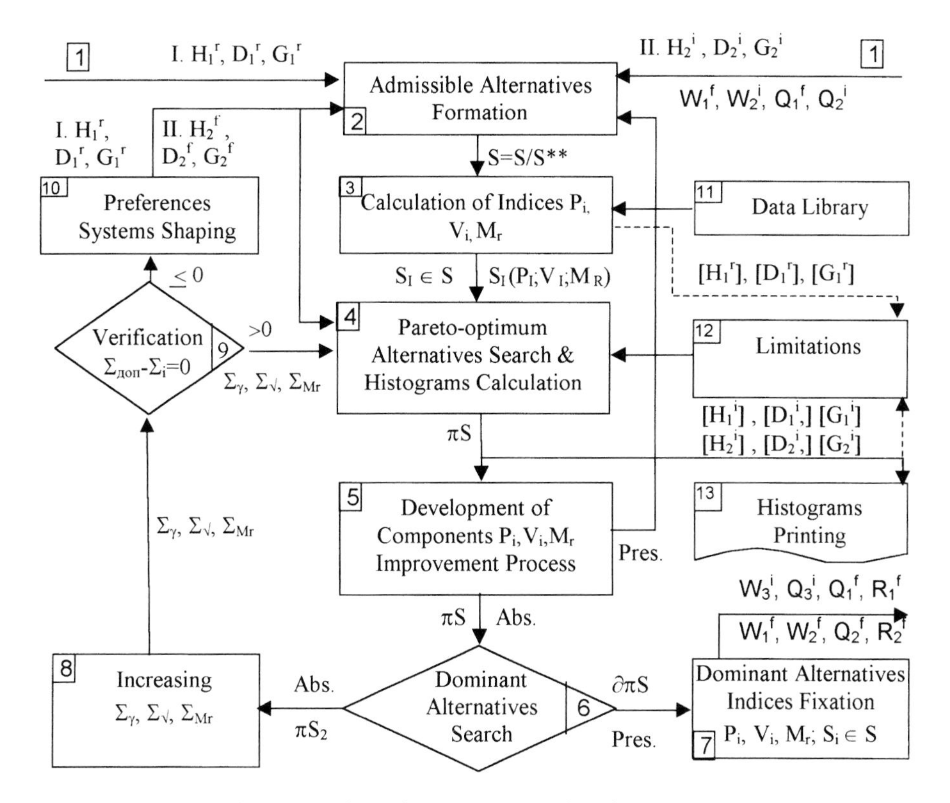

Fig. 6. Algorithm for New Alternatives Search

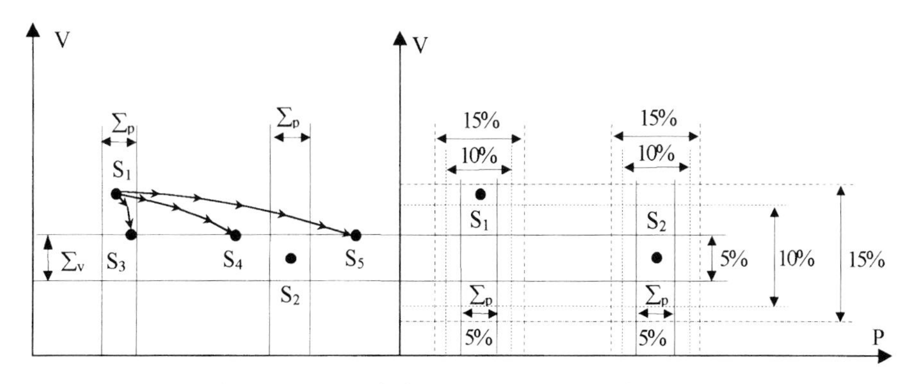

Fig. 7. Methods of Fixed Alternatives Finding

At F (S_m, S_n) = 1 the variant S_n shall be preferred to S_m; at F (S_m, S_n) =-1 S_m shall be preferred to S_n; at F (S_m, S_n) = 0 S_m and S_n are not comparable. So an algorithm for computer-aided segregation of subset S' from set S is constructed. Generally, the function F (S_m, S_n) can be generated for space of criteria, characterized by any number of dimensions. In the three-dimensional case F (S_m, S_n), a_i being preference index by i-th criterion:

$$a_j = \begin{cases} 0, [k_i(S_m) - k_i(S_n)]/\min[k_i(S_m), k_i(S_n)] < \varepsilon \\ 1, k_i(S_m) > k_i(S_n), [k_i(S_m) - k_i(S_n)]/\min[k_i(S_m), k_i(S_n)] \geq \varepsilon \\ -1, k_i(S_m) < k_i(S_n), [k_i(S_m) - k_i(S_n)]/\min[k_i(S_m), k_i(S_n)] \geq \varepsilon, \end{cases}$$

$$F(S_m, S_n) = \left\{ \text{sign}(\sum_{i=1}^{m} a_i), \text{ if } \text{sign}(\sum_{i=1}^{m} a_i \prod_{i=1}^{m} a_i) \geq 0; 0 \text{ if } \text{sign}(\sum_{i=1}^{m} a_i \prod_{i=1}^{m} a_i) < 0 \right\}$$

The research of this set πS and comparison of alternatives for individual QPs frequently give nontrivial results since they help find technical contradictions, thus facilitating the heuristic engineering activity. For example, the research of estimates of the same FMT operations on the QPs used to estimate the entire variant reveals the operations which make the greatest input into deterioration of the QP under consideration.

It is difficult to make a verbal description or to present a graphic interpretation of the work of the Fig. 6 algorithm (its cycles) because virtually all of the operations are performed by the computer and at each of the cycles the QP is represented by a space having no fewer than three dimensions. Fig. 7 provides, on the left, an example of the analysis of systems on a basis of two quality parameters - P and V - only. As can be seen, the systems S_1 and S_2 do not have a dominant kernel. The bar chart hard copy (Module 13) has shown, that the volume of S_1 basically depends on the frequency of transformation. By increasing the frequency of transformation (Module 6 work), we can synthesize variants of systems with other Vs, but it is invariably associated with worsening of Ps. If the P and V calculation gives the path S_1 - S_3, the system S_3 thus obtained is identical with S_1 within $\Delta_{\Sigma p}$, but it is more preferable than S_2. Thus, S_3 represents the dominant kernel dπS, and the problem of the third layer is thus solved in Module 6. If the calculation gives the path S_1 - S_4, the new system S_4 is more preferable than S_2 in respect to the parameter P, however it is not comparable with S_1. Thus, the domain has changed: systems S_1и S_4 are compared there. In itself, the result is good: a new system has been added to the library (Module 11) to be used in future system for the following calculations. However the dominant kernel is not present in the domain πS, and there is no third-layer solution. And with the path S_1 - S_5 all of the new systems, including S_5 are worse than the earlier systems S_1 and S_2. The Module 8 work is illustrated by the right part of Fig. 7. By consistently decreasing QPs determination accuracy at $\Sigma \leq 15\%$, we have that S_1 becomes comparable with S_2 Σ_v, but it is more preferable than the latter system in respect to P. If Σ_v is within the specified and actual accuracy of calculation with the use of the appropriate model, the third-layer problem can be regarded as solved. If no, the system of preference change is necessary. The calculation algorithms and software have been developed and examples have been provided for all of the stages (2).

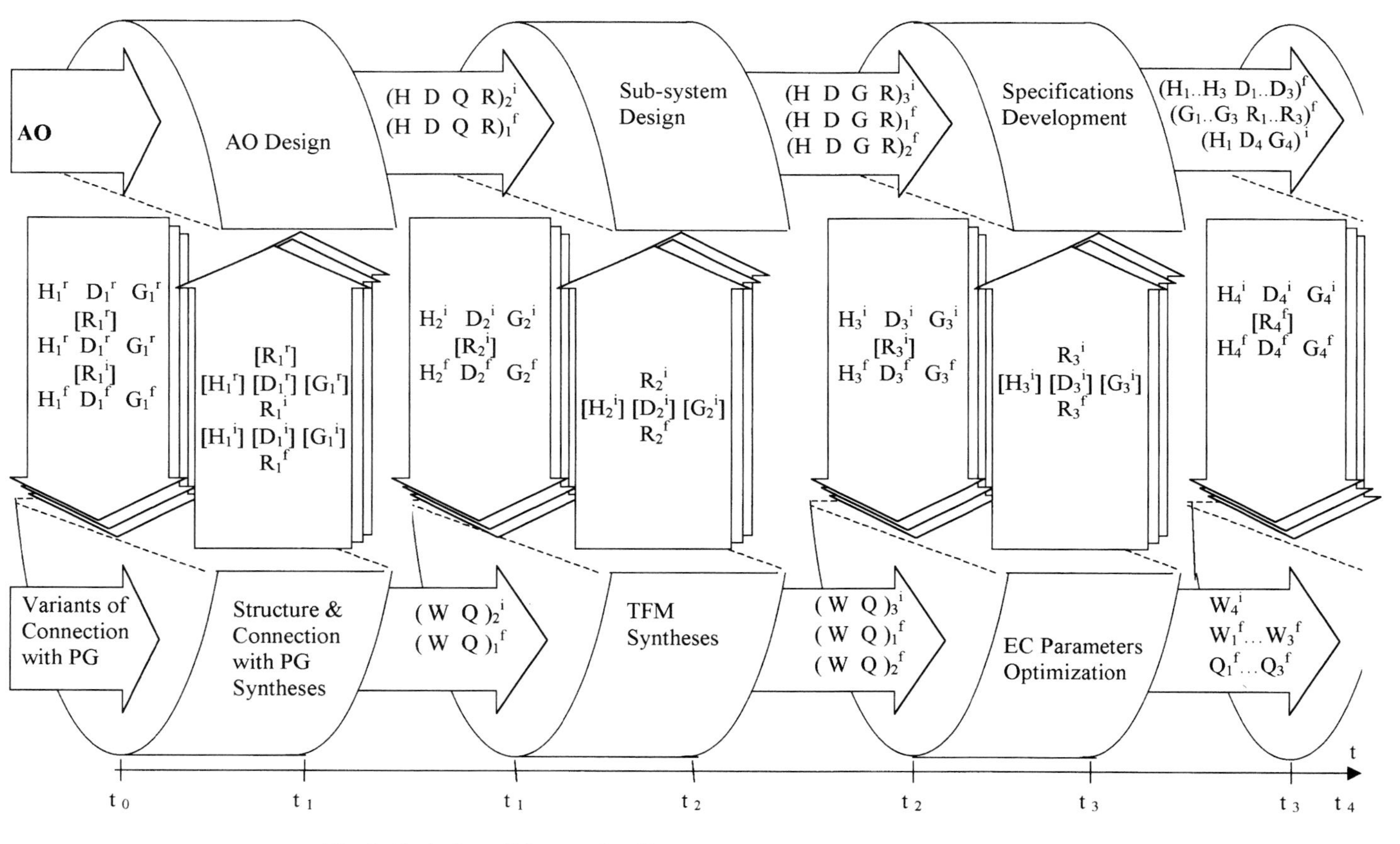

Fig. 8. Technique Scheme of AO's Power Electronic Systems Real Design

3. FINAL OPTIMIZATION OF QPS

The number of equivalent circuits dimensions is great because of the large number at the 3rd stage. So the convolution of the vector criterion into a scalar one is necessary. The least biased and strongest technique is the application of the maximin criterion: $\varepsilon_{ij}^{min} = \min_j \varepsilon_{ij}$, the worst parameter of every PES; $\varepsilon^*_{mn} = \max_i \{\varepsilon_j^{min}\}$, the best of the worst parameters of all PESs, where i = 1, ..., 2m.

The subsequent amplification of the decision rule can be based on the inputted information about relative importance of criteria. Then, it is expedient to take advantage of the consecutive concessions method: the QPs are arranged in a decreasing order of importance, the most important parameter ε_1 is maximized and its greatest value ε_{1max} is determined. Then the allowable decrease (concession) value $\Delta_1 \geq 0$ is specified for the parameter ε_2 on the condition that the value of the first criterion shall be not smaller than ε_{1max}-Δ_1. Thus, the initial alternative set Z_0 is narrowed down to Z_1. The concession value $\Delta_2 \geq 0$ is specified again (for the second parameter) and so on:

$$\varepsilon_{1\max} = \max_{Z_0} \varepsilon_{1j};\ Z_0 \supset Z_1 \overset{def}{=} \left\{ \varepsilon_{1j} : \varepsilon_{1j} > \varepsilon_{\underset{Z_0}{1\max}} - \Delta_1 \varepsilon_1 \right\};$$

....

$$\varepsilon_{n\max} = \max_{Z_{n-1}} \varepsilon_{1j};\ Z_{n-1} \supset Z_n \overset{def}{=} \left\{ \varepsilon_{nj} : \varepsilon_{nj} > \varepsilon_{\underset{Z_{n-1}}{n\max}} - \Delta_n \varepsilon_n \right\};$$

The procedure of designing PESs which constitute parts of more complex radio engineering systems shall of necessity take account of simultaneous development of other AO systems (the top half of Fig. 8). Therefore it shall be based on a flexible process of multistage optimization consisting of no fewer than four stages and using the criteria which structure and number change from stage to stage, reflecting the increase of information in the time elapsed from the beginning of the design work. The Fig. 8 diagram shows nothing but the beginning of the fourth stage - updating of the documentation by the results of testing of an AO as a whole. It can be repeated during subsequent testing and operation. Calculation algorithms and software, techniques of heuristic search for new decisions, some dozens of examples of the PES and its components and the results of their modeling have been developed for all of the procedures of section (2). The above-described process of designing PESs for an independent object (plane, robot, planet research vehicle) makes it possible to solve educational tasks too, so it was used as a basis for a manual for the students of special disciplines and post-college education. Using it, the engineer improves his qualification, and the student of a technical college gets the knowledge and skills required for work.

The proposed methods have been proven in the educational processes of several colleges, including the MPEI. Another sequence of the increasing force criteria (Pareto, Δ, L) has been proved during development of the Microstory R-C software and training system.

REFERENCES

(1) V.Kadel. The Decision-Making Theory as a Special Courses-Integrating Theory In: Collected Proceedings of the First International Conference on Mechatronics and Robotics. - Education Section. St. Petersburg, April - May 2000.

(2) V.Kadel. Power Electronic Systems of Autonomous Objects. Theory and Practice of Computer-Aided Dynamic Optimization. - Moscow, “Radio and Svyaz”, 1990, 224 pp.

(3) Yu.Kandyrin. Architectural Principles of Information Systems for Computer-Aided Multicriterial Choice. - Moscow, “Radiotekhnika”, May 1999.

Engineering design and innovative higher education – methodological aspects

L ROMANKOVA and **V VZYATYSHEV**
Engineering Design Centre, Moscow Power Engineering Institute, Russia

SYNOPSIS

The sense, which we put in concept of innovative education (IE), opposing its supporting education (SE), is brightly displayed in the approaches to such types of engineering activity as engineering design (ED) and management. The methodologies for IE and ED are compared in this report. It is established that close and multidimensional connections between these two concepts exist. In particular, it is shown that the structure of activity for the solving of vital innovative tasks (innovative technology) has the same shape as the structure of activity for solving the ED tasks considered in the report (11).

1. INTRODUCTION - HOW ENGINEERING DESIGN IN RUSSIA HAS AFFECTED FORMATION INNOVATIVE HIGHER EDUCATION

Innovative Education (IE) – the concept of reforming Higher Education in Russia, which is offered by the International Higher Education Academy of Sciences (IHEAS). The authors took an active part in the formation of its basic principles (3-5), and also initiated the discussion of the concept on IHEAS Academic Readings (6) in 1994-99. Today it is possible to approve that the basic IE principles are accepted by a scientific community of the Russian engineering higher schools. The question is: "Are there any connections between IE and ED?". This paper tries to discuss the answer.

The people supporting the IE concept frequently asked us: "How did you managed to formulate the basic IE principles and societal technologies of innovative activity? How could you generate its structure and describe many of the procedures?"

By reflecting answers to similar the questions, one of the authors (VV) recollected the discussions, which occurred after his lectures on the ED methodology at some universities and research institutes during the 1980s. Some people (students, teachers and engineers) were surprised: "You teach about the methodology of engineering designing, but it seems that this methodology without any changes can be used for the solving of a much wider class of tasks. Do you agree with this?" Thus, many people perceive the structure of the design activity as the structure of any innovative activity – such as the solving of new tasks.

This is why the authors have decided to write and present this report. It is seemed to us, that the training of the engineers in the field of ED must contain IE elements. The modern understandings about IE and about social technologies of activity in the generalised kind are submitted in the monograph (7) by other authors of the present report (LR).

2. DESIGN AND MANAGEMENT - BASIC TECHNOLOGIES FOR INNOVATIONS

2.1. Some definitions

The design (particularly, ED) is a drawing up of the ideal description for the future (device, technology, process, human activity etc.). In other words, the designing is "future-creating", but made before its realization, in an ideal (in the head, on the computer screen, on a sheet of paper - in ideas, in drawings, in words, in symbols). The management - planning, organization and control of the people, means, materials and time for optimal achievement of the given purposes. Realization of the project is also management. Managerial activity like design activity is also multivariate and multicriterial, and frequently iterative.

The interaction of design and management could be name as "the theory of practice". But, if the practice was always considered as alternative to a science, and "the practical experience" was accepted for mastering "during real life", methodology of design (MD) and methodology of management (MM) cover today rather capacious fields of knowledge, which can be professionally accustomed during educational process, and namely the IE can do it. It is necessary to emphasize that the division of the creative activity methodologies of the human being on MD and MM is largely conditional. A role and place of MD and MM in the IE are multifaceted and multidimensional. Let's name only some of them.

2.2. Some aspects of MD and MM in the IE

Activity aspect. SE concept deals with cognitive activity – to know, that already exists. IE concept deals with creative either transforming activity - or to create new, or to transform, that which already exists. MD and MM serve just to create and transform, and they differ from methodology of knowledge by many factors - see Table 1 in (8).

Humanistic and personality aspects. MD and MM correspond much more to the creative applicability of the human being and humanistic paradigm of education. Due to the inevitable solutions multitude they give to a person in his aspiration to the activity of self-realization and self-determination. On the contrary, in the world of the formalized sciences with their unequivocal answers for all correctly put questions the person feels uncomfortable. The scientific result does not depend on the person and his (her) subjective features. Accordingly, SE system is inclined to level the person. It was rather good when higher education trained engineers for the industrial "megamachine". On the contrary, the result obtained with MD and MM in the IE system depends on the person - the subject of the activity.

World outlook aspect. The gap between the philosophers professing different World pictures is traced through thousand years. The difficulties of IE introduction are connected, in particular, with that overwhelming part of the philosophers already explained in creating the World. Only a small part of the philosophers considered the process of creating activity of the Human in the World, among them there are Socrates, I. Cant and K. Popper.

Value and motivation aspects. MD and MM activity forces the student to reflect on system of his (her) values. It is well known that scientific results cannot be good or bad - they are true or false. And the results received with the MD and MM, are good or bad in senses of different values. Our experience shows (8, 12) that the educational tasks solved with MD and MM technologies cause large interest in the students. Solving a design task and, moreover, working under own pressure, distinct from any other variants, the student as though in passing and between times masters sometimes such volume of cognitive information, the study of which without activity motivation, would cause significant difficulties.

Aspect of knowledge, information and technologies management. A structure - divided into a system of procedures and operations - is characteristic for MD and MM as knowledge about activity. Such division enables the systematic reorganisation of the activity knowledge and, thus to make them subject of theoretical study, and then learning and training. The SE system transfers mainly knowledge, which is disciplined and multi-faceted. The IE system, working with MD and MM, perceives a natural way of integrative management of knowledge - around a structure of activity. Introducing MD and MM in educational process makes it more completed and integrated. The point is that MD and MM will fill in a sphere of rational substantiation of human activity. Already whole world has understood that the slogan "a science becomes direct productive force" is hopelessly antiquated. The place of a science in this phrase should be occupied by scientific rationality uniting a science and methodology of activity.

3. ED PLACE IN THE INNOVATION PROCESS

For an overview of the ED place let us look at the picture of the scientific-technological innovative process (see Fig. 1 in the end). ED transforms the world by way of producing objects satisfying new social needs. Doing this, ED makes use of the data and methods not only of sciences and methodologies but also of the arts. ED is certain today to make use of computers and computer sciences. Developing the methodology, we place humans at the center of the whole system. Humanistic sciences help us to understand ourselves. One of the tasks of engineering training is to assist a future engineer to create a special “interface” for activity in the structure shown in Fig. 1 in (8): taking a broader view of things, to help to find a place in this structure. Not everyone can succeed equally in fundamental sciences, in creativity, in human sciences and dialogues. But we believe that every engineer must have some knowledge of all these sciences, be aware of their methods and means, integrate them into the ideology of a personality, and finally transform the latter into an individual creative style of productive activity.

Speaking about the global aims of higher engineering education, we could formulate them as follows: to satisfy cognitive needs, to disclose and/or preserve and develop creative needs of a personality, and to help the student to find the meaning of life (9) and to choose the modus vivendi (10).

4. INTEGRATION OF THE SYSTEM OF KNOWLEDGE, SYSTEM OF SKILLS AND CREATIVE ABILITIES INTO ENGINEERING INNOVATIVE ACTIVITY

The described in (11) ED structure offers an engineer freedom of creative self-realization in practically every procedure. Thus, even the multicriterion statement of the design problem naturally integrates an inventive (and innovative) activity with a systematic approach, with physical analysis and mathematical optimization. The result of this integration can be rightly called creative technology, which is necessary for innovative activity and for IE.

Even before procedure P1 - see Fig. 1 in (11) - creativity in the choice or generation of new needs and functions of an object is necessary. In P1 it is very important to divide all requirements into quotients of quality (QQs), conditions and restrictions (11). This division is always ambiguous; here the system of values and preferences of the designer matters. Nobody can deny the creative character of P2. But if the generation of a new pattern occurs after P8, creativity acquires a new quality - it embraces all that can be physically described and mathematically calculated. In P3 the creativity of an engineer consists in the choice of a model, rather in the choice of a set of models. Experience shows that three types of models are required:

- the quickest for the preliminary survey of the QQs space in P4;
- more accurate for investigating the region of non-worst solutions (NWSs) in P5;
- the most accurate for the verification in P6 of solutions chosen in P5.

Procedure P4 in simple situations can be completely formalized and even automated. But if the number of QQs chosen in PI was rather large and models in the P3 are sufficiently complicated, one cannot proceed P4 without the designer's experience and intuition. In the P7, the engineer's creativity takes quite a peculiar form. A certain shrewdness and a developed associative way of thinking are required here to choose the direction of the search of a new pattern in accordance with the form of multidimensional region of NWSs and to realize it in the P8.

Let us list some properties of the structure described above. 1) It is very important that "quality images" of various patterns, satisfying the accepted DS, could be compared, because solutions with various patterns are not only described by various equations but also in various sets of parameters. 2) An engineer obtains an quantitative criterion of usefulness of the inventor's idea: a new pattern is useful if it possesses the best quality image. The inventor turns out to be more interested in carrying out the procedure of multicriterion truncation in this case. A comparison of images for various patterns and a choice of the directions of the search of new patterns is natural for a human. 3) One more sphere of the manifestation of creativity and individuality of an engineer's personality is due to the fact that after P4 an engineer obtains, as a rule, not a single solution but a whole set of them. It gives three new qualities:

- ➢ Unconventional choice implies creativity at a qualitatively new level.
- ➢ An engineer in P5 has to examine ecological and other aspects besides formalized in DS.
- ➢ If the decision object satisfies the needs of a human not alone but in the system with other objects, an opportunity is offered for making easier the choice from them.

5. CONCLUSION - ED, IE AND INNOVATIVE ACTIVITY

It was presented to us, that the described conceptual properties of the ED methodology can be applied as base for development of both methodologies for the IE concept, and methodologies for the innovative activity in general. However, in the structure Fig.1 of our work (11) many aspects, important for innovative activity are not taken into account. In particular:

- Specific social and cultural aspects of activity, characteristic both for sphere of education, and for sphere of educational management,
- The fact that innovative management always is a joint activity, as many people participate in it, and between them there is a continuous interaction,
- Opportunity and necessity both to operate process during its development and to carry out the control of its course (monitoring).

Developed in view of named and other specific requirements is the structure of innovative activity borrowed from monograph (7) written in 1999 by one of the authors (LR), given in fig. 2 at the end of the report. We ask the reader himself to establish a degree of connection between this structure and ED structure from our paper (11).

REFERENCES
(1) Shukshunov, V.(VS), Vzyatyshev, V.(VV), Romankova, L.(LR): "Through development of education - to new Russia", Moscow, IHEAS, 1993, 44pp.
(2) VS, VV, LR: "Innovative education: ideas, principles, models", IHEAS, 1996, 65pp.
(3) VS., VV., LR, Sergievskiy, V.: "From comprehension of paradigm to educational practice", *Higher education in Russia*, 1995 (**3**), pp. 35-44.
(4)VV, RL "Societal technologies in education", *Higher education in Russia*, 1998 (**1**), p28-37
(5) Korolkov V., VV., LR: "Personnel selection in a higher school: problems and tendencies", *Higher education in Russia*, 1999 (**2**), pp. 7-17.
(6) IHEAS: "Education and science on the eve of the XXI century", Academic Readings: 1994 – St. Petersburg, 1995 – Kiev, 1996 – Minsk, 1998 – Kishinev, 1999 – Kazan.
(7) Romankova, L.: "Higher school: societal technologies of activity", Moscow, 1999, 256pp.
(8) Vzyatyshev, V., Kandyrin, Yu., Pokrovskiy, F.: - in this Conf. Proc.
(9) Frankl, V.: "Man's Search for Meaning" (Simon and Shuster, 3rd edition, 1984).
(10) Fromm, E.: "To have or to be? Psychoanalysis and Religion", (New York, 1976).
(11) Vzyatyshev, V.: - in this Conf. Proc.
(12) Morgunov, G., Zuev, Yu. - in this Conf. Proc.

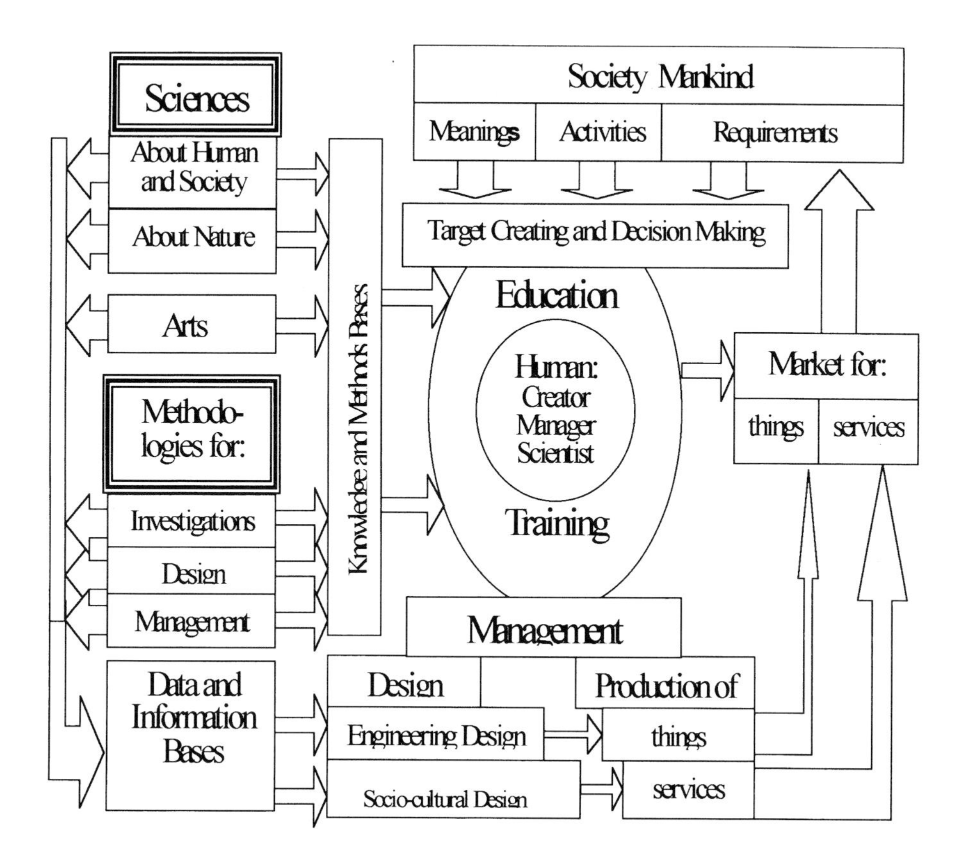

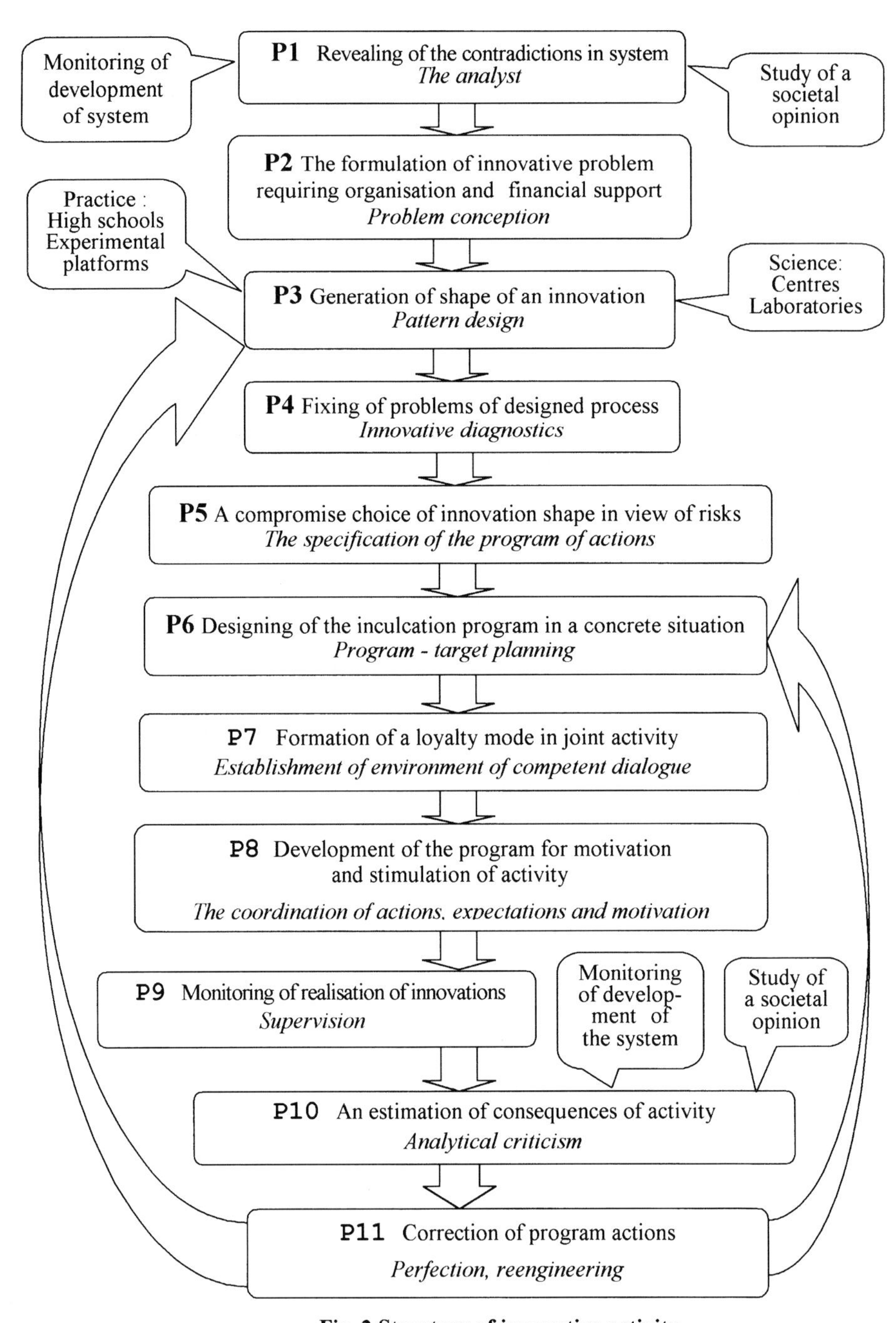

Fig. 2 Structure of innovative activity

Engineering design and 'activity concept' of professional pedagogic in engineering education

V VZYATYSHEV
Engineering Design Centre, Moscow Power Engineering Institute, Russia

SYNOPSIS

The key principles for a group of reports (1-4) from the Moscow Power Engineering Institute (MPEI) Engineering Design Centre (EDC) are stated. The main idea is: Engineering Design (ED) is a specific and extremely important form of intellectual activity. The principal concepts of ED are: a combination of parameter and pattern in object descriptions, a system approach, new patterns as a means of resolving contradictions. ED structure includes procedures such as: multicriterion statement of the problem, pattern building, modeling, optimization, search and resolving contradictions etc. ED is natural base for innovative engineering education (3).

1. INTRODUCTION

The formation of the EDC MPEI methodological school is connected to the issues in the paper collection "Activity concept of professional pedagogic in engineering education" (January, 1989). This includes approximately one hundred papers on problems of Engineering Design, published during four years in the magazine "Bulletin of a higher school", in the bulletin "SDS: problems and decisions" and in the newspaper "Energetic".

It is difficult to find words to describe this better than the ones chosen for the Foreword by the corresponding member of the Russian Academy of Science P. Krasnoschekov: "The authors of this collection develop the Activity Concept of the engineers professional training and are capable of creating technology for new generations. All of us know that the practice of using engineers in a national (Russian) economy today is varied. But the scientific approach demands the choice of typical ones. Such types of the engineer's activity determine their social role, and the creation of the new techno-sphere elements. From here key words of the collection are "Engineering Design (ED)". From here the key concept of professional engineering pedagogic is "The ED Methodology". The engineers engaged in research, production and operation of techno-sphere elements, also should own ED procedures and ED concepts, focusing their professional activity on satisfaction of needs of the Human and Society. Features of ED as subject of scientific-pedagogical researches are:

- ED as a specific kind of activity connecting science with production;
- the psychological inertia, caused by it, of public consciousness;
- urgency and movement of interdisciplinary problems, hopefully involving in the discussion the maximum number of the teachers, engineers and scientists.

In these conditions wide use of press is a successful tool of the MPEI Engineering Design Center (EDC). Long term (8-12 days) and published in large edition (3000 copies) newspaper publications, these prove an excellent supplement to scientific and methodological seminars of the EDC and to systems of improvement of professional skill of the teachers. The main value of the collection uniting works of MPEI scientific-pedagogical collective is that it allows the ED problem to be seen as the whole, in all its variety and discrepancy. Perhaps any of today's books, including translated ones, do not give such opportunity".

The named and discussed collection is possible to be considered as the original presentation of the MPEI Engineering Design Center, established by the order of the MPEI rector Prof. I.N. Orlov in March, 1988 (1). The information about the interaction of the MPEI EDC in the area "Activity concept" with British ED methodologists can be read in the "Energetic" N 26, 1988 in Prof. F. Pokrovskiy's article and in the shorthand report of discussion "An Engineering Design at British Universities" (5).

2. ENGINEERING DESIGN: DUALISM OF DESCRIPTION, MULTICRITERION APPROACH TO THE STATEMENT, RESOLVING CONTRADICTIONS

The ancients believed the world to rest on three whales. We want to make use of this model when building the picture of engineering activity. We are to select three key elements of the methodology. The search for a technical solution (TS) is realized by means of TS generation: by first broadening the search region and then narrowing it by discarding the worst TSs - those which are less efficient than the others. We believe that the key elements of this strategy are:

- the way of organizing the search region —a set of solutions;
- the criteria for narrowing the region — truncation of the set of solutions;
- the principle of organizing the search cycles.

The formation of these techniques is an important task of engineering training.

2.1 The first professional technique is a "dual" description of technical solutions

This is a technique of organizing the search region of TS. It is obvious that it is necessary to specify the space of their signs. We consider it methodologically necessary to divide those signs into qualitative and quantitative ones (see Table 1). It is natural to name qualitative signs TS parameters. These include geometrical dimensions of the elements of the construction, the nominal values of the electronic circuit elements, etc. A set of TSs differing only in the values of their parameters is characterized by a certain qualitative generality. We name it the TS pattern. No mathematical operation describing the familiar pattern of TS can result in the formation of a new pattern. Hence, the influence of optimization on technical progress is strictly restricted by the known set of patterns. The principal mechanism of progress is the perfection of patterns.

Table 1 "Dualism" in engineering activity

Type of description: Signs	Parametric	Pattern
Character of description	Quantitative	Qualitative
Character of processing	Analytical	Pattern
Character of innovation	Continuous	Stepwise
Mechanism of innovation	Optimization	Heuristic search
Innovation result	Optimal	Invention
Effect estimation	20 7 2 %	2 10 100 times

2.2 The second professional procedure - multicriterion statement and truncation

The conceptual aim of the truncation is to provide the highest possible TS quality, i.e. the best functions, the maximum resources expenditure and the least number of harmful effects. We believe: to accomplish such a description of qualities a set of parameters is needed; let us call them quotients of quality (QQs). QQ is a quantitative parameter of the designed object, which is monotonically related (with the condition of other QQs constancy) with its quality (6). QQs include mass, dimensions, strength, productivity, reliability, cost, expenditure of materials, etc.

QQ has been introduced for the formalization of the choice and truncation. One should retain those variants of TSs with higher QQs, and discard those which are worse. It should be stressed that if the number of QQs is two or more, the notion "worse/better" becomes less obvious than in the case of one QQ (6,7). The principal factor specifying an engineer's creative role is the number of QQs. If one QQ has been chosen or several QQs are reduced to one criterion, the problem loses its creative character to a considerable degree (see Table 2).

Table 2 Comparison of quality concepts

Number of QQs: Quotients	One	Several
Principle of processing	Optimization	Truncation
Processing result	One - the best	"Quality Image" of the pattern
Pattern comparison	Trivial - in numbers	Visual
Contradictions	Hidden	Disclosed
Taking a solution	Not necessary	Necessary
Information for further search	Almost lacking	Completely available

The TS pattern can be characterized by a so-called "quality image" (QI), a multidimensional surface in the space of QQs consisting of non-worst solutions. Comparison of various TS patterns is reduced to the comparison of their QI. Operations with images are more natural for a design of new patterns. It allows the designer to make use of the third technique.

2.3 The third professional technique - analysis of technical contradictions

Technical contradiction (TC) between two or more QQs is considered in such a situation when any change of the parameters aimed at the change of one QQ within accepted restrictions unavoidably leads to the worsening of a minimum of one of the other QQs.

The above named QI discloses and quantitatively describes the system of technical contradictions. An engineer observes which pairs of QQs are in contradiction and where the contradiction lies - between QQs or between QQ and the restrictions assumed in the statement of the problem. Contradictions, which are quantitatively described by QI, are likely to be those TCs which Altshuller (8) qualitatively describes. Altshuller rightly believes their solving to be the main aim of inventions. The system of TCs is immovable within the frames of a fixed TS pattern. Only by changing the TS pattern you can change or eliminate TCs. But pattern change is an invention. The advantage of the third professional technique is that an engineer processing the result of the multicriterion truncation - QI - is given an opportunity to create at a principally new level – "standing on the shoulders of a giant", i.e. in our case – physical and mathematical analysis.

Having singled out and discussed the main professional techniques of engineering creativity, we can now build the general structure of the professional engineering activity.

3. STRUCTURE OF ENGINEERING DESIGN: SUBJECT, OBJECT, STATES, PROCEDURES AND SYSTEM OF CONNECTIONS

Fig. 1 shows the Engineering Design structure (9) developed in the EDC in accordance with the views described above. An engineer is the subject of activities (not shown in the figure).

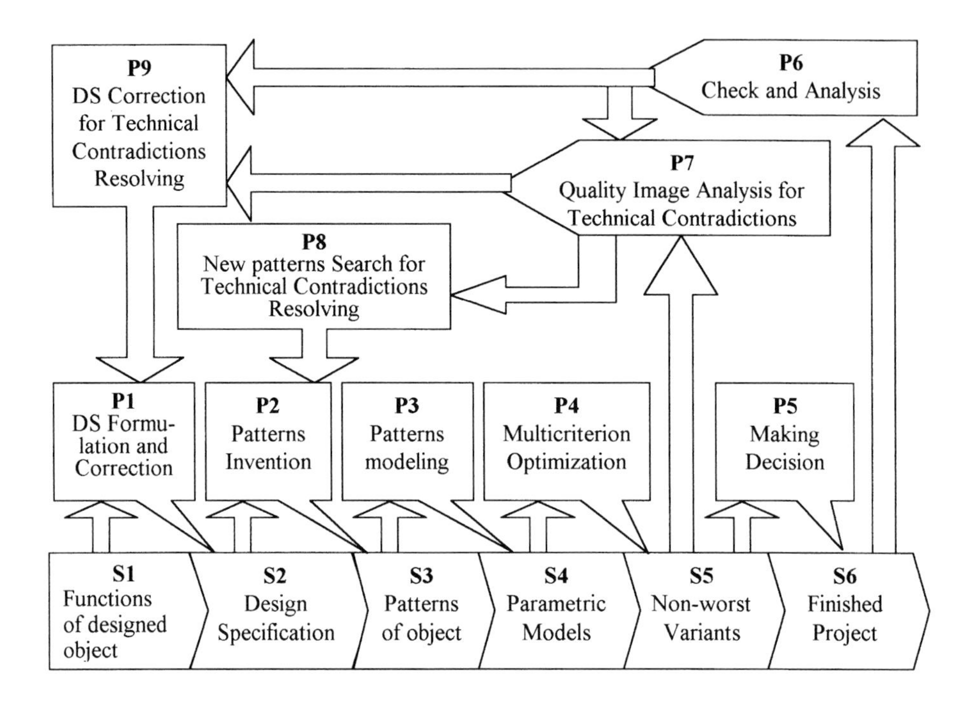

Fig. 1 The Engineering Design Structure

We understand the ED process as a set of purposeful procedures (P1-P8), transferring the object of ED from one state into another (S1-S6). A state is a combination of the designer's systematic notions conditioned by the aims of the given ED procedure. Before ED begins (state S1) the "tree of needs" must be conceptually understood.

3.1. States of ED object and ED procedures

To obtain a quantitative comparison of the TS variants it is necessary to convert conceptual requirements of the object into a quantitative form - to formulate a Design Specification (DS): state S2. Experienced designers state that a good DS predetermines 60% of success. An engineer's intention to create a competitive ED object is to be reflected first of all in the DS. One must make use of QQs because they play the primary role. It is QQs that make up the core of the ED problem: to create an object fulfilling the functions in the best possible way with the minimum negative consequences and the least possible expenditures. Of particular importance is the fact that every ED object requires many QQs to be taken into consideration. Reduction of the QQs total number and the conversion of some of them into restrictions is a forced measure - for the simplification of the models and facilitation of the optimization procedure. Therefore, the choice of the set of QQs is one of the most important conceptual tasks.

The next particular aim of ED is creative in its essence. This is the invention (generation) of the maximum number of patterns: ideas, schemes, and kinds of construction (S3). To make powerful mathematical apparatus work, one has to describe all patterns by models (S4). The designation of the models is to connect constructive (inner) parameters of an object with QQs and other parameters of DS. States S3 and S4 are the most capacious object states. In the choice of their capacities lies one of the ED contradictions: between a probability of finding the solution of the desired level of quality and the materials necessary for the search, labor and time resources.

The aim at the next stage is to significantly reduce the search region. This can be simply achieved by eliminating all superfluous things; the so-called "worst solutions" - both the worst patterns and the worst combinations of parameters. In state S5 some set of "incomparable" patterns is retained, and each of them is represented by a set of non-worst combinations of parameters.

3.2 ED procedures and ED strategies

The achievement of particular design aims, that is the conversion of an object from one state into another, can be realized by purposeful "technological" actions, which we call ED procedures. ED strategies make up the consequence of procedures. So-called "direct" procedures (P1-P5) form a linear strategy of the direct solution of the ED problem. But this is, as a rule, the worst strategy. The best and least time-consuming results can be obtained by means of cyclic iterative and branched strategies including "reverse" procedures (P6- P9).

Formulation of DS (P1) is a responsible stage of the design. The principal "tool" at the disposal of an engineer in this procedure is his or her conceptual thinking. Procedure P2 is heuristic and inventive. The principal "tools" in it are engineer's ability to invent, vision and intuition. Procedure P3 - the description of patterns in terms of models - is the most formalized one. Both analytical and conceptual thinking is necessary at this stage: for selecting the style of description, important factors and processes, for the substantiation of the possibilities of the simplification of a model. The aim of procedures P4 is a significant reduction of a set of variants – ideally to one. But, unfortunately, the world is not as simple as that! If the number of QQs in S2 is more than one, one can choose the non-worst set S5 only by means of formal methods of multicriterion optimization. In complicated problems the search for non-worst solutions demands the heuristic organization of the process and the choice of the directions of the search.

3.3 Decision-making

The final procedure P5 consists of choosing one or several solutions S6 from a collection of non-worst solutions S5. The peculiarity of the procedure lies in the impossibility of using formally the DS requirements (all of them have already been made use of in P4!). Here the main "tool" is the human ability to take decisions. As a person taking a decision in P5, a designer takes into account aesthetic aspects, requirements of ecology and security, the values of the user, and many other factors. In short, the engineer's technical and general culture, understanding of the meaning of life and modus of existence, are the factors, which count at S6.

It should be noted that the form of the region of non-worst solutions gives unbiased information about the expediency of making one type of article or a number of types of article (7). But it is a human who must take a decision, and be responsible for it, even in the case when a computer is being used. Iterative strategies with a return by way of "reverse" procedures serve for the verification and improvement of the results for the change of direction of the search and making it more accurate. They are more efficient than direct ones, but more complicated.

If the result could not stand the test in P6, the process comes back to P4 (not shown at the Fig.1), when there is some hope for improving the result with the help of one pattern out of the set S3, or to P2 (when new patterns are desired) or even to P9 (with the aim of DS correction). The improvement of the quality of TS solutions in ED is almost always accompanied by the situation in which attempts to improve this or that QQ of TS cause the worsening of other QQs or are blocked by the necessity to fulfil DS conditions and DS restrictions of the assignment. Namely such a situation is a technical contradiction (TC). Quantitatively, the set of TCs of an object is characterized by the form of the region of non-worst solutions in QO space.

A TC can be weak or strong, removable (by accepting a compromise) and unremovable. The character of the TCs is a "portrait" of the design situation. It is most important for an engineer that TCs, as a rule, are specified by the pattern of the object. By changing the patterns, one can purposefully affect the character of a TC and in this way find ways of advancing. This is one of the key laws of the developments of ED objects. We completely agree with G. Altshuller (8), who was one of the first to show that inventions are designated for resolving contradictions. Procedures P7 and P8 are the key ones in the technology of improving the object to be designed; they are of primary importance at the stage of pre-designing research.

4. CONCLUSION ENGINEERING DESIGN AND INNOVATIVE HIGHER EDUCATION

The presented ED structure, based on the described professional techniques, offers an engineer freedom of creative self-realization in practically every procedure. Thus, the multicriterion statement of the design problem naturally integrates invention with a systematic approach, with physical analysis and mathematical optimization. The result of this integration can be rightly called engineering creativity. It is represented to us, that the conceptual properties of the ED methodology can be used in a basis of methodology of innovative education. We discuss this opportunity together with L. Romankova in more details in the paper (3).

REFERENCES

(1) Vzyatyshev, V., Kandyrin, Yu., Pokrovskiy, F.: - in this Conf. Proc.
(2) Morgunov, G., Zuev, Yu.: - in this Conf. Proc.
(3) Romankova, L., Vzyatyshev, V.: - in this Conf. Proc.
(4) Korj, D., Khoroshev, A.: - in this Conf. Proc.
(5) "Energetic", MPEI newspaper. - Moscow, 1988, (26).
(6) Gutkin L.: "Proektirovanie radioelektronnykh ustroistv po sovokupnosti pokazateley kachestva", Moscow, (Sov. Radio, 1975) (in Russian).
(7) Krasnoschekov, P., Petrov, A., Fedorov, V.: "Novoe v zhizni. nauke i tekhnike",

Matematika, Kibernetika, 1986, (10). (Informatika i projektirovanie) (in Russian).
(8) Altshuller, G.: "Algoritm izobretenij", Moscow, (Moskovskiy rabochiy, 1973) (in Russian).
(9) Vzyatyshev, V.: "Struktura protsessa inzhenernogo proektirovaniya i problemy podgotovki radioinzhenerov", *Radiotekhnika,* 1989, (3), pp. 91-95 (in Russian).

Authors' Index